Electroluminescence II

SEMICONDUCTORS
AND SEMIMETALS
Volume 65

Semiconductors and Semimetals

A Treatise

Edited by R. K. Willardson
CONSULTING PHYSICIST
SPOKANE, WASHINGTON

Eicke R. Weber
DEPARTMENT OF MATERIALS SCIENCE
AND MINERAL ENGINEERING
UNIVERSITY OF CALIFORNIA
AT BERKELEY

Electroluminescence II

SEMICONDUCTORS
AND SEMIMETALS

Volume 65

Volume Editor

GERD MUELLER

COMMUNICATIONS AND OPTICS RESEARCH LABORATORY
HEWLETT-PACKARD LABORATORIES

ACADEMIC PRESS
San Diego   San Francisco   New York   Boston
London   Sydney   Tokyo

This book is printed on acid-free paper. ∞

COPYRIGHT © 2000 BY ACADEMIC PRESS

ALL RIGHTS RESERVED.

NO PART OF THIS PUBLICATION MAY BE REPRODUCED OR TRANSMITTED IN ANY FORM OR BY ANY MEANS, ELECTRONIC OR MECHANICAL, INCLUDING PHOTOCOPY, RECORDING, OR ANY INFORMATION STORAGE AND RETRIEVAL SYSTEM, WITHOUT PERMISSION IN WRITING FROM THE PUBLISHER.

Requests for permission to make copies of any part of the work should be mailed to: Permissions Department, Harcourt, Inc., 6277 Sea Harbor Drive, Orlando, Florida, 32887-6777

The appearance of the code at the bottom of the first page of a chapter in this book indicates the Publisher's consent that copies of the chapter may be made for personal or internal use of specific clients. This consent is given on the condition, however, that the copier pay the stated per-copy fee through the Copyright Clearance Center, Inc. (222 Rosewood Drive, Danvers, Massachusetts 01923), for copying beyond that permitted by Sections 107 or 108 of the U.S. Copyright Law. This consent does not extend to other kinds of copying, such as copying for general distribution, for advertising or promotional purposes, for creating new collective works, or for resale. Copy fees for pre-1999 chapters are as shown on the title pages; if no fee code appears on the title page, the copy fee is the same as for current chapters. 0080-8784/00 $30.00

ACADEMIC PRESS
A Harcourt Science & Technology Company
525 B Street, Suite 1900, San Diego, CA 92101-4495, USA
http://www.academicpress.com

ACADEMIC PRESS
24–28 Oval Road, London NW1 7DX, UK
http://www.hbuk.co.uk/ap/

International Standard Book Number: 0-12-752174-7
International Standard Serial Number: 0080-8784

PRINTED IN THE UNITED STATES OF AMERICA
99 00 01 02 03 04 EB 9 8 7 6 5 4 3 2 1

Contents

PREFACE ix
LIST OF CONTRIBUTORS xvii

Chapter 1 Polymeric and Molecular Organic Light Emitting Devices: A Comparison 1
V. Bulović and S. R. Forrest

    I. INTRODUCTION 1
   II. POLYMERS VS MOLECULAR ORGANIC MATERIALS 2
      1. *Structural and Electrooptic Properties of Conjugated Polymers* 2
      2. *Purification of Organic Materials* 4
      3. *LED Structures* 6
      4. *Thin Film Patterning* 8
  III. POLYMER LED STRUCTURES 10
  IV. CURRENT–VOLTAGE–LUMINANCE CHARACTERISTICS 11
      1. *Polymer–Metal Interfaces* 12
      2. *Conduction in Polymer Thin Films* 13
      3. *PLED Performance* 16
   V. THE GOOD AND THE BAD OF ORGANIC MATERIALS 17
      1. *Advantages of Organic Materials* 17
      2. *Stability of Polymeric and Molecular Organic Materials* 19
      3. *Mechanical and Electronic Concerns* 19
  VI. A LIGHT AT THE END OF THE TUNNEL? 22
     REFERENCES 24

Chapter 2 Thin Film Electroluminescence 27
Regina Mueller-Mach and Gerd O. Mueller

    I. INTRODUCTION 27
   II. PHENOMENOLOGY 31
  III. BASIC MECHANISMS IN THIN FILM ELECTROLUMINESCENCE 41
      1. *The Simple Model* 44
      2. *Electrical Properties* 45

	3. Optical Properties	55
	4. Luminescence Properties	60
	5. Efficiency	68
	6. High-Field Electronic Transport	73
	7. Design Rules	77
IV.	THIN FILM EL DEVICES	80
	1. Direct View Displays	81
	2. AMEL	89
	3. Nondisplay Applications	91
V.	MEASURING TFEL	93
	1. Measurements on Test Areas	93
	REFERENCES	101

Chapter 3 Materials in Thin Film Electroluminescent Devices . . . 107

Markku Leskelä, Wei-Min Li, and Mikka Ritala

I.	INTRODUCTION	107
II.	DEVICE STRUCTURES AND CHARACTERISTICS	108
III.	FILM DEPOSITION TECHNIQUES	111
IV.	TFEL PHOSPHORS	112
	1. Requirements for the Phosphors	112
	2. Dominant Materials	117
	3. Emerging Materials	136
	4. Materials of Minor Importance	139
	5. Color by White	152
V.	OTHER MATERIALS	154
	1. Substrates	154
	2. Transparent Electrodes	156
	3. Metal Electrodes	157
	4. Insulators	158
VI.	CONCLUSIONS	165
	REFERENCES	166

Chapter 4 Microcavities for Electroluminescent Devices 183

Kristiaan Neyts

I.	INTRODUCTION	183
II.	THEORETICAL DESCRIPTION	185
	1. Light Waves in a Microcavity	186
	2. Dipole Antenna in a Microcavity	188
	3. Dipole Transitions in a Microcavity	198
III.	BASIC EXAMPLES	200
	1. Mirror Characteristics	200
	2. Radiation from Dipole Antennas	202
	3. Dipole Antennas Near a Mirror	204
	4. Semitransparent Mirror Reflectivity	207
	5. Double Mirror Microcavities	210
IV.	APPLICATIONS	214
	1. Inorganic EL Devices	215

2. *Light-Emitting Diodes*	220
3. *Organic Electroluminescent Devices*	226
V. CONCLUSION	230
REFERENCES	232
INDEX	235
CONTENTS OF VOLUMES IN THIS SERIES	239

Preface

Electroluminescence is the direct conversion of electrical energy into radiation. This broadest definition of the term makes a sharp distinction from thermal radiation, but it leaves open the question of the material to be used. What we are discussing here is *electroluminescence of solids* as distinct from gases, plasmas, and, if possible at all, fluids. We will avoid the narrowing "of solids," but take it as implicit for the purpose of this volume.

The idea for these volumes was to have experts of the various subfields or technologies of electroluminescence present an account of the present status, covering the most important materials for electroluminescence — inorganic and organic. The expression "most important" always brings up the question of "for whom," "by what," or "when." Admitting from the beginning that the selection is highly subjective, nevertheless, some discussion is appropriate before leaving the selection as well as the whole book to the judgment of the reader.

Electroluminescence (of solids) was first described in the scientific literature by G. Destriau in 1936 (*J. Chim. Physique* **33**, 587). In fact, the original paper concerns photoelectroluminescence from ZnS powder embedded in a dielectric matrix under the action of a high electric field, and for many years thereafter electroluminescence (EL) was mainly attempted and executed on (inorganic) powders. In the 1960s and 1970s, a rather complicated mechanism of electric excitation of suitable dopants (centers) gained some industrial use in EL powder lamps, and in the late 1980s in displays. Even today inorganic powder EL is used in emergency guide lights in airplanes and a variety of niche applications. Thus, it was initially on the list of possible topics for this volume. However, because it has not been documented as "in strong progress" by papers at recent international conferences or product release news, it was considered as "covered" in the prior literature. As far as I understand its status, powder EL — DC as well as AC — suffers from short-term, drive-dependent degradation, and its mechanism as the processes in much more homogeneous systems such as thin films or even single crystals is not well understood.*

Implicitly, the preceding paragraph identifies some of the selection criteria:

- Applications or new potential applications
- New physical concepts or improved understanding
- Examples of a variety of different mechanisms and materials

Of course, the five different examples covered in these volumes are in different stages of development, but all are of present interest from an applications point of view. High sophistication in understanding and manufacture of inorganic light-emitting diodes (LEDs) has been reached. This is true in slightly different degrees for the classic GaHP-based red- and amber-emitting phosphide devices and the relatively new GaN-based blue and green ones. One chapter is devoted to each of them, and one chapter points out the system aspects and the differences and similarities. For organic LEDs, three chapters address the two groups of materials — small molecules and polymers — implementing much the same mechanism, and one chapter describes system aspects and applications. Thin film electroluminescence has received much less attention in the literature. What has been written primarily aims at high-information-content displays, and with some $100 million/year total revenue is dwarfed compared to LEDs. However, from the point of view that a rather limited selection of materials (treated in one chapter) is able to realize the special mechanism (a second chapter), it appears interesting enough to be included.

All EL devices suffer substantial difficulty in bringing the light generated in a high-refractive-index material into the outer world. Optical interference always accompanies this process, and microcavities can improve or at least influence it, as outlined in Volume 65.

The extreme case of a microcavity is a laser resonator. Semiconductor lasers are applications of the electroluminescence processes, usually very much the same as in corresponding LEDs, however strongly coupled to the radiation field in the resonator. Recently, blue H and UV lasers were introduced based on AlInGaN, which by far outdated the performance of blue II–VI compound lasers. Infrared lasers, GaAs- and InP-based ternary or quaternary compounds of the (Ga,In)(As,P) system, are now and will be even more in the near future, the backbones of digital communication. Data storage (CD, DVD) and laser printing are big markets for electroluminescent devices. However, as every volume of *Semiconductors and Semimetals* is

---

*Shortly after writing this preface, I was confronted at the exhibition of the Meeting of the Society of Information Displays in San Jose with new powder EL products, which were developed without much publicity outside the patent literature.

limited to a maximum number of pages, lasers have been excluded from the scope of these volumes as an explicit subject.

The nine chapters in Volumes 64 and 65 review a substantial fraction of the present research and development activities undertaken worldwide and aimed at signaling, imaging, and display, but recently in an increasing proportion also on solid state lighting. This part of the palette of possible applications seemed out of reach for quite some time after the dream of "cold light" from powder electroluminescence failed in the 1970s. The advent of fairly efficient blue and green LEDs, and the dramatic increase in the efficiency of the red LEDs has opened up a completely new avenue to illumination by mixing white from the three primary colors. Even blue diodes alone can be used to excite a broadband-emitting phosphor to produce white light by complementing its own only partially absorbed emission.

Readers who are especially interested in one technology or topic should stop reading here and go to the appropriate chapter. But then perhaps one should come back to this point and become interested in other topics by the short, subjective outline of the highlights provided next. These topics are certainly contained in the respective chapters, but are not necessarily immediately obvious. And again, of course, they are subjectively chosen.

Light-emitting diodes have been industrially produced over the last three decades. The overall consumption since their inception has been about 50 LEDs per human. Thus, every reader can make use of about 100 LEDs, as there are certainly others using fewer. In addition, many LEDs have probably been scrapped and replaced by new and better ones. The first chapter provides some information about the tremendous improvements in LEDs achieved over the years, but one should keep in mind that a substantial fraction of today's production of amber and red diodes is not highly efficient TS–AlGaInP. The reason is twofold—lower price and no need for high output or efficiency. Many signaling applications can function using the cheapest diode available.

Efficiency in the red–amber LED is a matter of avoiding radiationless transitions, and this is essentially a matter of crystallinity. The new, quaternary III–V compound became interesting after the discovery of the almost perfect lattice match to (available) good GaAs substrates of $(Al,Ga)_{0.5}In_{0.5}P$ and the possibility to tune its bandgap–emission by the Al:In ratio. The second achievement contributes to the high efficiency and comes from a completely different direction. While the first one ramps up the internal efficiency, the transparent substrate increases the light extraction dramatically. And this transparent substrate (TS) is a replacement of the original GaAs substrate by a GaP "carrier," which no longer must be lattice-matched, as now nothing is "growing." Thus, the newer history of the red–amber LED is a fascinating example of improvement of *both* factors governing the overall efficiency—internal quantum efficiency *and* light extraction. To appreciate the achievement represented by an external power efficiency of 28% [W/W], one should keep in mind that the light extraction from a material with a refractive index >3.0 is $<5\%$. The story repeats to some extent or at least in principle in the (Al,Ga,In)N case. The outcoupling is not as low and the substrate is not absorbing from the beginning, as no lattice-matching substrate is available. But there is a gift of nature: (Ga,In)N grows on (transparent) sapphire (Al_2O_3) substrates, in spite of a 13% mismatch, and consequently with a large dislocation density, but with a reasonable internal quantum efficiency. In addition, at least some papers claim that a drastic reduction of imperfections does not increase EL light output—one of the points that justifies hopes for tangible increases of efficiency.

Thus, comparing AlGaInP and AlGaInN, the phosphide exhibits extremely high internal quantum efficiencies—up to 80%—combined with very low outcoupling of the light (if no special tricks are played), but the opposite is true in the nitride devices. Another difference goes relatively unnoticed: in the phosphides the dependence of efficiency upon drive is "normal," nonradiative channels must saturate before reaching the ultimate values. In the nitrides, the efficiency drops with increasing drive current, possibly because of overflow of the quantum wells or "dots."

Rather unobserved by a large fraction of even the engineering community, the red LEDs have matured to an extent, which has enabled them to enter the traffic light market. In the San Francisco Bay area, a large-scale replacement of the maintenance-intensive incandescent bulbs, which also consume much more energy, has begun. Brake lights on automobiles and many other items will also fall to LEDs quite soon.

Summing up on inorganic LEDs, the physics is "simple": forward current driven over a $p-n$ junction and electron–hole recombination through shallow states or in quantum wells. Suitable materials have been found that have the bandgaps in the right place for visible emission (or infrared emission in the communications area). Only single-crystal material is capable of exercising this kind of recombination radiation efficiently enough for practical use. To be manufacturable at reasonable costs, epitaxy on affordable single-crystal substrates had to be developed. But even then, only since 1996 have LEDs been considered for applications beyond the traditional indicator functions and assembled displays with a limited number of pixels.

Because of this limitation of LED use, researchers looked for other forms of electroluminescence and materials that could work in amorphous or polycrystalline films. Large-area displays were the primary target of those activities. Thin film electroluminescence was a consequent extension of the research done on powders. Direct current and alternating current versions were tried. What survived and is a stagnating niche market of some $100 million/year over the last years is AC. Thin film EL (TFEL) is addressed in Chapters 2 (mechanisms) and 3 (materials) of Volume 65. It differs in almost all of its essential features from LEDs, inorganic as well as organic. The only similarity is the difficulty of light extraction, a negative aspect. While (radiative) recombination of free electrons with free holes is the luminescence process in an LED, and presence and transport of electrons and holes are equally important, TFEL relies on majority carriers (electrons) only, and the radiative transition is an innershell transition in a metal or a rare-earth ion. It works only in materials that allow electrons to achieve kinetic energies of some 3 eV in an electric field of about 2 MV/cm. The energetic electrons impact–excite the innershell electrons of the dopant, without removing them from the individual ion. The removal of the excited electron (exciton) from its ion is usually a loss process, as is the generation of electron–hole pairs by impact–ionization of valence band electrons. Internal quantum efficiencies up to 5% have been reported, which convert into external power efficiency values around 1%. This does not appear very attractive from a light generation point of view, and is not a potential method for solid state illumination. However, there are two properties that could make the mechanism a winner in the display world. One is multiplexability, which is so poor in LCDs, is no problem even at 1000 lines or HDTV. The other is the simplicity of design and manufacture of five thin films on glass, of which only the electrode stripes must be patterned. This is beatable only (eventually) by polymer organic LEDs (OLEDs). Monochrome, 19-in. diagonal XGA displays have been manufactured on a large scale. However, two problems persisted for many years: blue was difficult to generate with adequate efficiency, and the driver cost is high, as about 200 V must be used. The problem of blue deficiency is about to be solved, and by the time these volumes are published the problem may have disappeared. The driver costs might be driven down by the plasma displays' consumption of increasing numbers of (almost) the same chips. From a scientific point of view, some unsolved questions about the special high-field transport mechanism remain, as well as interesting problems with interface-related degradation. The latter could turn out to be a model case for other EL phenomena.

As easy to manufacture, but possibly more difficult to control, the advances in the EL of polymeric light emitting devices (PLEDs) have drawn even more attention to OLEDs in general. In Chapter 4 of Volume 64 the system aspects and the applications issues of small molecule and polymeric organic LEDs are addressed. Questions such as driving schemes of large arrays of organic LEDs, and their reliability, as well as the various possible schemes for multi- or full-color displays are discussed. Chapter 5 of Volume

64 provides a rather detailed account of the status of OLEDs based on small molecules, and Chapter 1 of Volume 65 compares polymeric and molecular LEDs.

The big run for organic electroluminescence started after almost two decades of stagnation, when a Kodak team published breakthrough results in 1987, based on a novel thin film structure. It was only 3 years later that a group from the University of Cambridge reported the first EL results on polymer films. After that, industrial interest centered for some time on small molecules, but now seems to lean toward polymers, and "best values" are being reported in short time sequence by both "parties" with changing top positions.

The term LED used for the device structure is well justified in that the basic mechanism is an injection of electrons and holes into a heterojunction, and a radiative recombination of excitons formed from the electrons and holes. There are, however, many properties differentiating the classic LED, which some people now like to call ILED (for inorganic LED), from the OLED. The main point is the extremely low mobilities of the carriers, about six orders of magnitude lower than in the III–V compounds. This is caused not only by the fact that the small-molecule OLEDs use amorphous films, and the polymer films cannot be regarded as crystals either; it is also influenced by the nature of the electronic transport — mainly hopping — that appreciable voltage drops over film thicknesses of some 100 nm occur. There are some distinctions between the two classes of materials, and polymers are preferable from this point of view. A big advantage of the organic materials is their low refractive index, smaller than 1.7, often as low as 1.5. Thus, light extraction is much easier. Quantum efficiencies are constant over many orders of magnitude, so that efficiency is often given in candelas per ampere, which tells only part of the story. The low mobility causes appreciable increases in voltage drop with increasing current and reduces the power efficiency with drive drastically. Under these circumstances multiplexing is accompanied by a substantial efficiency penalty. This is one of the reasons why a strong tendency toward active matrix addressing developed in the community.

Going to thinner films could in principle reduce the voltage drop. However, early experiments have shown that a certain distance of the recombining electron–hole pairs or excitons from the metal electrodes is necessary to maintain high radiative probability. This given, the internal quantum efficiency can approach 25%, the limit set by the fraction of singlet excitons, and triplet excitons annihilating radiationless. Attempts to increase this fraction by phosphorescence based on triplet excitons were not very successful yet.

With 25% internal quantum efficiency and 25% light extraction from a material with an index of 1.5, we can expect 6.2% external quantum efficiency, which is much more than in the TFEL case, but much less than in many over-the-counter LED lamps. This best value has been approached mainly in the green part of the visible spectrum.

Every technology seems to have a sweet spot, or better, a "sweet color": while red is the best for LEDs, yellow is best for TFELs, and OLEDs prefer green. Blue has been critical for all of them, but very recently power efficiency values of 22% have been reached in blue LEDs. PLEDs have had some very recent progress in blue and seem now to be slightly better than OLEDs there.

The differences in spectral widths are also interesting: LEDs are the narrowest emitters, with linewidths that range from 25 nm in the green–blue to 10 nm in the red; OLEDs and PLEDs have the widest emission bands, seldom narrower than 100 nm. TFEL devices can span the whole range from narrow—from 10 nm in the case of $4f$–$4f$ transitions of rare-earth dopants—to as wide as 70 nm in allowed $5d$–$4f$ transitions of Ce^{3+}. This question of linewidth is not only interesting from a physics point of view, but is of some practical importance. Color saturation in displays depends very much on it, and sometimes as in lighting one would prefer broader emission.

When talking about colors and emission bandwidth, one must keep in mind that all the structures of EL devices resemble an interference filter—films with different refractive indices and thicknesses of the order of one wavelength of visible light are stacked above each other. This is exactly the structure now termed microcavity. Chapter 4 of Volume 65 provides a systematic and very subject oriented treatise of light emission from microcavities. What should be stressed once more and over all is the angular dependence of radiation from microcavities, which very often has not found adequate attention in the measurement of brightness or efficiency of test structures and devices. It is standard practice to measure in or into integrating spheres in LED labs. This is not as much the case in TFEL and OLED labs. And if for some OLEDs external power efficiency values above 7% are reported, the packaging and/or measuring conditions should be included in the paper.

Microcavities can greatly enhance the usage of light generated inside a dielectric stack, if only a certain aperture accepts the light—*and* if the linewidth is not too large. Chapter 4 of Volume 65 gives some very useful examples for any of the three technologies, which can be extended easily on the basis provided.

Even some of the authors have asked which is the most promising technology. So one can expect some of the readers to also ask this. The answer is inevitably not a simple one and it mainly depends on the application. It appears very likely that LEDs will go for solid state lighting (SSL). Recently formed ventures of LED manufacturers with lighting companies—Emtron and General Electric, Siemens and Osram, Hewlett-

Packard and Philips — underline the seriousness of the issue. And a market size well above $10 billion/year does not play it down. All colors are now reasonably well developed to generate "white." Very soon the efficiency of white solid state lamps will surpass that of incandescent bulbs, but it will be quite some time — if ever — before retail prices can come close. The availability of all basic colors from LEDs has enabled the production of huge displays for sports stadiums and general advertisements. As soon as the cost of these mostly hand-assembled giants comes down, we will see them more, also outside Japan.

On the other hand, the use of LEDs in small displays such as, for instance, in cellular phones might be cut back by the advent of cheaper direct multiplexed PLED or OLED devices. However, any forecast of the development of electroluminescent devices in the display sector is extremely difficult as factors other than performance criteria will play an important role. There is little doubt that PLEDs and OLEDs will conquer certain niche markets, as TFEL has done and might continue to do. The costs of large-scale manufacture of the displays have not yet been established, and they might depend not so much on the specific material.

It appears certain that all technologies discussed in these volumes will obtain continued attention and funding, and especially in the arena of organic LEDs, both molecular and polymeric, the materials list is in its infancy.

Finally, thanks are expressed to all the contributors for their great efforts and all the wonderful accomplishments they have brought to these volumes. Thanks are also due to the series editor, Eicke Weber, to the mentor at Academic Press, Zvi Ruder, and his staff, and to my wife, Regina, who tolerated the long hours I have worked on this volume.

GERD MUELLER

List of Contributors

Numbers in parentheses indicate the pages on which the authors' contribution begins.

V. BULOVIĆ (1), *Center for Photonics and Optoelectronic Materials (POEM), Department of Electrical Engineering and the Princeton Materials Institute, Princeton University, Princeton, New Jersey*

S. R. FORREST (1), *Center for Photonics and Optoelectronic Materials (POEM), Department of Electrical Engineering and the Princeton Materials Institute, Princeton University, Princeton, New Jersey*

MARKKU LESKELÄ (107), *Department of Chemistry, University of Helsinki, Helsinki, Finland*

WEI-MIN LI (107), *Department of Chemistry, University of Helsinki, Helsinki, Finland*

GERD O. MUELLER (27), *Communications and Optics Research Laboratory, Hewlett-Packard Laboratories, Palo Alto, California*

REGINA MUELLER-MACH (27), *Communications and Optics Research Laboratory, Hewlett-Packard Laboratories, Palo Alto, California*

KRISTIAAN NEYTS (183), *Electronics and Information Systems Department, University of Ghent, Ghent, Belgium*

MIKKO RITALA (107), *Department of Chemistry, University of Helsinki, Helsinki, Finland*

CHAPTER 1

Polymeric and Molecular Organic Light Emitting Devices: A Comparison

V. Bulović and S. R. Forrest

CENTER FOR PHOTONICS AND OPTOELECTRONIC MATERIALS (POEM)
DEPARTMENT OF ELECTRICAL ENGINEERING AND THE PRINCETON MATERIALS INSTITUTE
PRINCETON UNIVERSITY, PRINCETON, NEW JERSEY

I. INTRODUCTION	1
II. POLYMERS VS MOLECULAR ORGANIC MATERIALS	2
1. Structural and Electrooptic Properties of Conjugated Polymers	2
2. Purification of Organic Materials	4
3. LED Structures	6
4. Thin Film Patterning	8
III. POLYMER LED STRUCTURES	10
IV. CURRENT–VOLTAGE–LUMINANCE CHARACTERISTICS	11
1. Polymer–Metal Interfaces	12
2. Conduction in Polymer Thin Films	13
3. PLED Performance	16
V. THE GOOD AND THE BAD OF ORGANIC MATERIALS	17
1. Advantages of Organic Materials	17
2. Stability of Polymeric and Molecular Organic Materials	19
3. Mechanical and Electronic Concerns	19
VI. A LIGHT AT THE END OF THE TUNNEL?	22
REFERENCES	24

I. Introduction

The first light-emitting devices (LEDs) containing luminescent polymer thin films were demonstrated in 1990 (Burroughes *et al.*, 1990), just 3 years after reports of the demonstration of efficient molecular organic LEDs (OLEDs) (Tang *et al.*, 1987). Since then, the development of polymer LEDs (PLEDs) and small-molecule-based OLEDs proceeded in parallel. The two classes of devices have similar physical properties and operating characteristics, which is not surprising considering the close relationship between the optical and electronic properties of polymeric and molecular organic materials. To avoid repeating concepts described in the chapter on electroluminescence of molecular OLEDs, this chapter describes the operating properties of PLEDs by focusing on notable differences between PLED and OLED technologies. Therefore, in Section II we first introduce structural and electrooptic properties of polymers, describe methods of growing thin films, and emphasize differences with molecular organic thin films. Methods of patterning both classes of organic materials are also discussed. The structure of a typical PLED is described in Section III, and their operating characteristics are presented in Section IV. Section V outlines the advantages and disadvantages of both polymeric and molecular organic materials in comparison to more conventional inorganic semiconductors. The chapter concludes in Section VI by discussing the prospects for the use of polymer and molecular organic LED technology in full-color displays. For further discussion of structure, operating characteristics, and technology of PLEDs and OLEDs we direct the reader to a number of references (Forrest *et al.*, 1998; Friend *et al.*, 1999; Greenham and Friend, 1995; Sheats *et al.*, 1996; Vincken, 1998).

II. Polymers vs Molecular Organic Materials

Differences in the physical structure of polymeric and molecular organic materials are the reason for their different purification, deposition, and patterning methods. These determine the sets of advantages and limitations associated with both classes of materials. Recognizing trade-offs inherent in the choice of a particular material system is essential for practical utilization of both polymeric and molecular organic thin films. In this section, we first describe polymeric materials and then contrast their physical properties and processing methods with those of molecular organics.

1. STRUCTURAL AND ELECTROOPTIC PROPERTIES OF CONJUGATED POLYMERS

A polymer consists of a chain of covalently bonded molecular units, or monomers. Opposite ends of a monomer have an incomplete electron orbital that readily bonds to a next monomer unit. In the condensed phase, such extended chain structures can result in lightweight materials of high physical strength that are easily processable by molding and extrusion through a die into plastic sheets and rods. Over the years, attractive mechanical properties of plastics have displaced traditional materials such as metals, ceramics, glass, and wood in many applications. The demonstration of electroluminescence in conjugated polymers reinvigorated investigation of electrical properties of these materials. The term *conjugated* refers to the alternating sequence of single and double bonds that is responsible for both the metallic and semiconducting character of this class of solids. Overlap of the p_z orbitals of the double (or triple) bonds results in formation of a delocalized π-electron system along the polymer backbone. Bonding and antibonding π orbitals form the equivalent of conduction and valence bands, respectively, allowing conjugated polymers to support positive and negative charge carriers with high mobilities along the polymer chain. As in molecular organic solids, the motion of carriers strongly influences the neighboring electronic structure and is often described in terms of polarons rather than free carriers.

To the polymer backbone, luminescent or photoabsorptive functional units (also known as side chains) can be attached (Greenham and Friend, 1995) facilitating substantial versatility in their design. The energy bandgap and the redox potential of polymers can easily be tuned by attaching electron-withdrawing or electron-donating side chains (Cornil *et al.*, 1996; Greenham *et al.*, 1993). As an example, Fig. 1 shows photoluminescence spectra of a series of polymers with the same general formula (indicated in

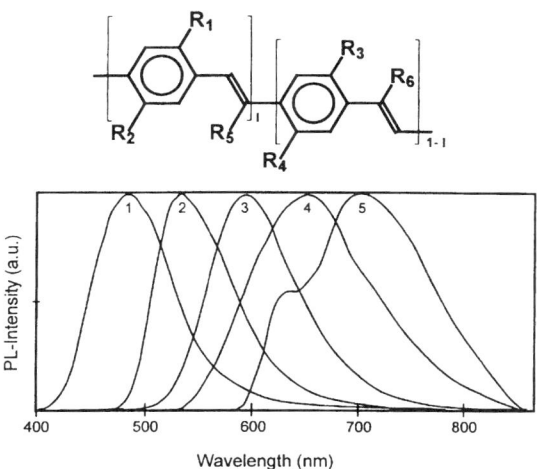

FIG. 1. Normalized photoluminescence spectra of side-chain-modified PPV with the general formula shown on the top: (1) R_1, R_2, R_3, R_4 = alkyl, R_5, R_6 = CN; (2) R_1, R_2 = alkoxy, R_3, R_4 = alkyl, R_5, R_6 = CN; (3) R_1, R_2 = alkoxy, R_3, R_4 = alkoxy, R_5, R_6 = H; (4) R_1, R_2 = alkoxy, R_3, R_4 = alkoxy, R_5, R_6 = H annealed; and (5) R_1, R_2 = alkoxy, R_3, R_4 = alkoxy; R_5, R_6 = CN (after Schoo and Demandt, © 1998, with permission from Elsevier Science).

the figure) (Schoo and Demandt, 1998). Indeed, the emission color of this and other polymeric systems can be tuned over the entire visible spectrum by choosing the appropriate side chains, where electron-donating groups shift the luminescence to lower energy, and electron-accepting groups give a blue shift. This property is essential in optimizing the PLED emission spectrum. We note, however, that similar to molecular OLEDs, due to strong exciton–phonon coupling, it is challenging to obtain saturated PLED colors due to the resulting considerable (~ 100 nm) spectral width. Indeed, PLED emission tends to be further broadened due to dispersion in conjugation lengths, providing a range of excited state energies.

Doping can alter the conduction properties of some polymers over many orders of magnitude by introducing energy states in the electron energy bandgap. For example, doping some semiconducting polymers can transform them from insulators to conductors (Heeger et al., 1988). Polyethylene, polystyrene, polytetrafluoroethylene (Teflon), and polyethylene terephthalate all have conductivities of $\sigma < 10^{-16}\,\Omega^{-1}\,\text{cm}^{-1}$ (Pope and Swenberg, 1982). In contrast, doped conjugated polymers such as polyaniline have achieved $\sigma > 10^2\,\Omega^{-1}\,\text{cm}^{-1}$ (Kohlman et al., 1997). Many polymers have been employed as photoconductors in xerographic copiers and printers (Borsenberger and Weiss, 1993). One example is polyvinylcarbazole (PVK), which was shown to have an electric-field-dependent photogeneration efficiency that can reach 20% (Pope and Swenberg, 1982).

Linear polymer chains can also be stretched or "poled" under the influence of an electric field when heated above their glass transition temperature. This can give rise to considerable anisotropy in their electrical and optical properties. The conductivity of stretched polyacetylene, for example, is ~ 10 times higher along the stretch direction (Park et al., 1979). The anisotropy of poled polymers has been utilized in nonlinear optic devices such as optical modulators (Hasegawa et al., 1992).

2. Purification of Organic Materials

Commercially available organic compounds are often cited as 99.9% pure. This reflects both the extent of purification of some materials and the uncertainty in the analytical techniques for detecting impurities in organic solids. Such low material purity implies that in a three-dimensional lattice, 1 molecule in 10 along a particular linear direction is likely to be an impurity. The need for further purification of commercially obtained materials is therefore evident if the electrooptical properties of the material are to be predictable and controllable from device to device.

For molecular solids, the most effective technique of purification is zone refining (Warta *et al.*, 1995). This process was originally introduced for purification of inorganic semiconductors and involves repeated passage of a narrow molten zone along a rod of the material to be purified. It requires that the host can melt without decomposition and that the distribution coefficient of the impurity is higher in the liquid than in the solid phase. As the molten zone is swept slowly along the length of the rod, the impurities are pulled into the melt and hence are segregated from the highly pure regions. Using zone refining, the impurity concentration can be substantially reduced.

If the molecular material sublimes on heating, purification by fractional sublimation is used (Gutmann and Lyon 1981). In this process, the unpurified source material in powder form is initially loaded into the hot end of a continually evacuated tube that is heated in a temperature gradient. Volatile impurities are released at a high temperature and removed by the pumping system. Less volatile materials redeposit in different, cooler zones of the tube according to their sublimation temperatures. At the end of the purification cycle, a section of the tube contains the purified material that can be extracted. The process is repeated several times to achieve the desired level of purity, where the temperature gradient and rate of sublimation is kept low to achieve the highest purity.

The purity of polymers is determined by the process of chemical synthesis. The concentration of impurities and defects incorporated in the final material is strictly dependent on the nature, sequence, and yield of individual chemical reactions comprising the synthetic route. When the polymer precursor is synthesized, impurities such as salts, monomer residues, and low-molecular- weight reaction products can be removed through chemical purification or dialysis. In contrast, defects in the final polymer structure are integral to the material and are very hard, if at all possible, to remove. To obtain high-quality material for electrooptic applications it is essential to carefully control reaction conditions, monomer purity, type of solvent, reaction temperature, and reactant concentration. It has been shown that the high luminescence efficiency of conjugated polymers is directly related to high material purity and is an important factor in long-term stability (Carter *et al.*, 1997). Ionic impurities and transition metals can initiate photochemical reactions, while the presence of oxygen in the films during and after conversion may cause the formation of carbonyl groups, which may quench photoluminescence (Harrison *et al.*, 1996; Papadimitrakopoulos *et al.*, 1994) and trigger train scission reactions (Cumpston and Jensen, 1995; Scott *et al.*, 1996; Scurlock *et al.*, 1995). Progress in the development of active polymeric materials and their purity resulted in astonishing improvements of the external luminescence efficiency of from 0.01 lm/W just a few years ago to >20 lm/W for the latest PLEDs (Friend *et al.*, 1999).

3. LED Structures

The principal difference between polymeric and molecular organic electroluminescent devices is the method of fabrication of these thin film structures. OLEDs are grown by evaporating a sequence of layers in a high-vacuum environment. For PLEDs, the layers are deposited by spin- or dip-coating a solution of a soluble polymer onto a substrate. While the application of the polymer layers occurs by such potentially low cost wet-chemistry processes, the cathode (typically calcium or other low work function metal) is deposited in vacuum, similar to the case of molecular OLEDs.

For the spin-on method, the substrate is mounted onto a rotating chuck. A drop of a solvent containing the dissolved polymer is placed at the center of the rotating substrate, with the centripetal force spreading the drop radialy. Adhesion to the substrate and the surface tension of the solvent act to form a thin uniform film. Subsequent baking of the substrate removes excess solvent, leaving behind a thin (~ 1000-Å) organic film.

Solubility of deposited polymers will, in part, determine the quality of the spin-coated films. For example, poly-(*para*-phenylenevinylene) (PPV) and its derivatives are most commonly used in PLEDs despite the fact that PPV is insoluble. To generate a PPV film, a soluble precursor is deposited and converted by thermal treatment, Fig. 2a (Burroughes *et al.*, 1990). However, this is not a preferred method of polymerization, since the quality of thermally treated material is not reproducibly attained. Alternatively, the introduction of side chains can lead to more soluble polymers such as MEH–PPV and OC_1C_{10}–PPV (Fig. 2b), which can be processed by spin-coating, and have the advantage of being fully conjugated prior to deposition (Schoo and Demandt, 1998). This later class of polymers typically yields better quality devices.

The spin-on method is performed in an inert atmosphere to avoid incorporating high levels of impurities that can significantly affect the electronic and optical properties. For example, even a low concentration of impurities introduced during the film casting process can change the conductivity or the quenching rate of excited states of the polymer. Since thermal evaporation of organic materials in vacuum can produce thin films of very high purity, an alternative although potentially more costly means for polymer film growth is to sublime monomers, which are subsequently polymerized by baking or exposure to ultraviolet light (Vaeth and Jensen, 1997a, 1997b).

Controlling film uniformity and thickness to the dimensions needed in single- and double-heterostructure OLEDs is straightforward using vapor deposition, while it is more challenging with spin-on techniques. For thermally evaporated–sublimed materials, film thickness can be precisely

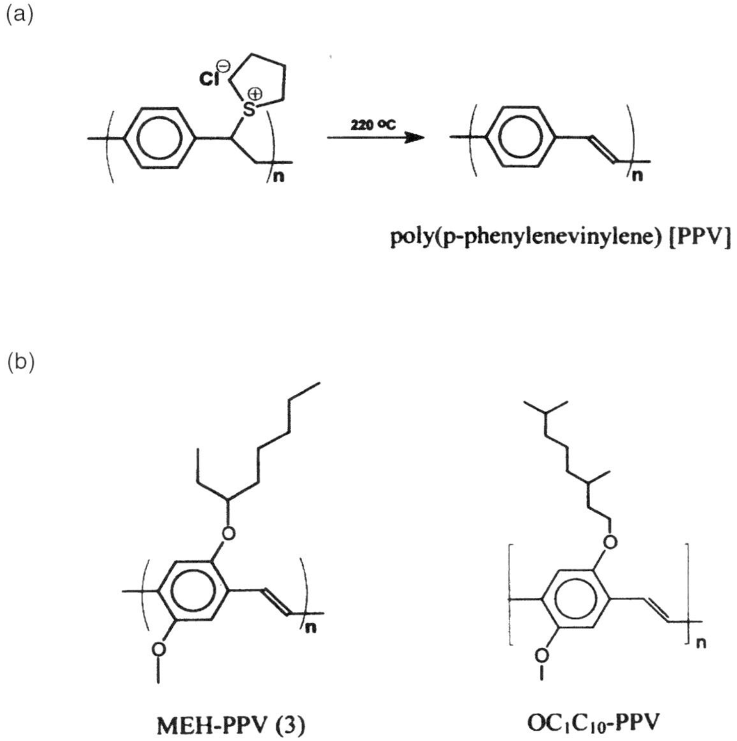

FIG. 2. (a) The soluble precursor is converted to poly(p-phenylenevinylene) by thermal treatment. (b) MEH–PPV and OC_1C_{10}–PPV are completely soluble due to the presence of the side chains (after Schoo and Demandt, © 1998, with permission from Elsevier Science).

controlled and monitored during deposition using a quartz microbalance with a relative resolution of 10 Å. The thickness of a spun-on film, however, depends on the speed of rotation, solvent viscosity, and the amount of dispensed liquid. Furthermore, since polymers are cast from solution, care must be taken in multilayer structures to ensure that the solvent used for the second layer does not dissolve the first polymer. Hence, chemical compatibility between successively applied polymers ultimately limits the complexity of the devices made using these materials. However, a number of novel deposition techniques such as ink-jet printing (Hebner et al., 1998) and masked dye diffusion (Pschenitzka and Sturm, 1999) of active layers is opening new possibilities in the high-resolution patterning of polymer thin films. The next section describes some of these advances.

4. Thin Film Patterning

In contrast to the well-developed procedures for handling and processing of inorganic semiconductors (such as silicon), standardized procedures for handling organic materials are considerably less well defined. This presents a formidable challenge to the commercialization of devices based on organic materials, and is therefore one of the research topics of greatest current interest.

Presently, patterning of inorganic semiconductor microelectronic circuitry relies on photolithographic techniques. Unfortunately, the same methods can not be used with organic thin films. Solvents used in standard photolithography such as acetone, methanol, and propanol interact with many organic materials, either by dissolving them or by causing swelling, thereby modifying their electronic and mechanical properties. As a result, new methods of patterning organic films have been proposed, including the use of prepatterned substrates (Taylor et al., 1995) with the directional deposition of sublimed films (Tang, 1994), or translational shadow masking (Tian et al., 1997) described in an earlier volume in this series (see Vol. 64, Ch. 5) in the context of patterning molecular organic thin films.

Proposed low-cost patterning techniques such as stamping (Kim et al., 1995; Xia et al., 1996), printing (Garnier et al., 1994), screen-printing (Bao et al., 1997), ink-jet printing (Bharathan and Yang, 1998; Hebner et al., 1998; Perçin et al., 1998), and masked dye diffusion (Pschenitzka and Sturm, 1999) are uniquely suited for fast, cheap, large-area patterning of organic thin films into active device areas. Among these, ink-jet printing and masked dye diffusion are currently the most promising methods for obtaining polymer-based full-color displays.

In the ink-jet-printing method, the ink cartridge of a standard printer is replaced with one loaded with polymer suspension in a solvent. In principle, different color PLEDs and OLEDs can be printed by using polymer solutions of different composition. In practice, some organic solvents tend to dissolve key components of standard printers, emphasizing the need for custom-made printer technology. In this method, issues of uniform thickness of printed layers and material purity must be addressed in order to develop high-efficiency, cheap, full-color displays with side-by-side patterned pixels. Alternatively, pixels of uniform thickness were demonstrated when droplets of fluorescent coumarin 6 dyes dissolved in acetone were printed onto previously fabricated PVK thin films (Hebner and Sturm, 1998). The electroluminescence of doped regions changed from blue PVK emission to green, demonstrating the capability of locally tuning the color. Figure 3 (see color insert) shows a photograph of the ink-jet-printed luminescent letters of nile-red- doped PVK excited by an ultraviolet lamp.

The masked dye diffusion process allows for uniform doping of large areas in a single diffusion step. For large emissive displays, this process should be faster than the serially applied ink-jet printing. The technique is sketched in Fig. 4. Upon heating the diffusion source, dopant molecules desorb from the source film and diffuse into the target film. The lateral pattern defined by the intermediate shadow mask can form pixels with linear dimensions on the order of 200 μm. Through this nonvacuum process, both red and green dopants were introduced in the blue-emitting PVK (71.5% by weight)–PBD (28.5%), demonstrating the capability of generating all three color pixels (Pschenitzka and Sturn, 1999).

At present, all polymer display technologies require that the metal contact be deposited in vacuum as the final processing step. This introduction into the vacuum environment ultimately reduces the potential cost savings of using spin-coating or other liquid phase deposition processes. Hence, it is important to find means for cathode deposition *ex vacuuo*. For example, it is of interest to demonstrate the compatibility of conductive inks with the ink-jet technology deposition of electrodes. If successful, the complete PLED structure could then be printed under inert atmospheric conditions. We note that the typical size of 100-μm-diameter subpixels corresponds to a density of 250 dots per inch (dpi), which is feasible with present-day ink-jet technology. It is, therefore, likely that with some development, a combination of the preceding methods will provide an inexpensive method of patterning full-color high-resolution PLED displays.

III. Polymer LED Structures

Energy level diagrams of typical single organic layer and heterojunction PLED structures are shown in Fig. 5. The first PLEDs (Fig. 5a) consisted

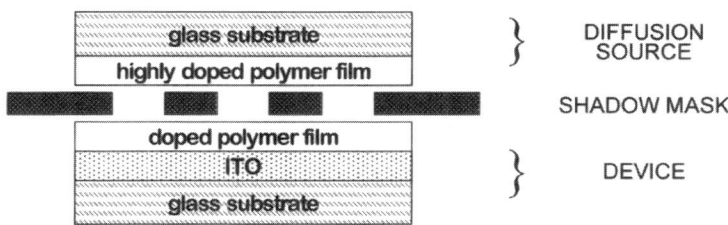

FIG. 4. Schematic diagram of the dye diffusion patterning method. The patterned masking layer is sandwiched between the diffusion source and the device substrate (after Pschenitzka and Sturn, 1999).

of a thin film of PPV spun on an ITO anode, with a low-work-function metal cathode such as Al, Mg, or Ca (Burroughes *et al.*, 1990). Under forward bias, electrons and holes injected into PPV from the cathode and the anode, respectively, move through the polymer layer until they combine to form excitons that subsequently radiatively recombine. Quenching sites in the PPV layer limit the luminescence efficiency η of these devices to ~0.01%. Insights in material properties and consequent improvements have

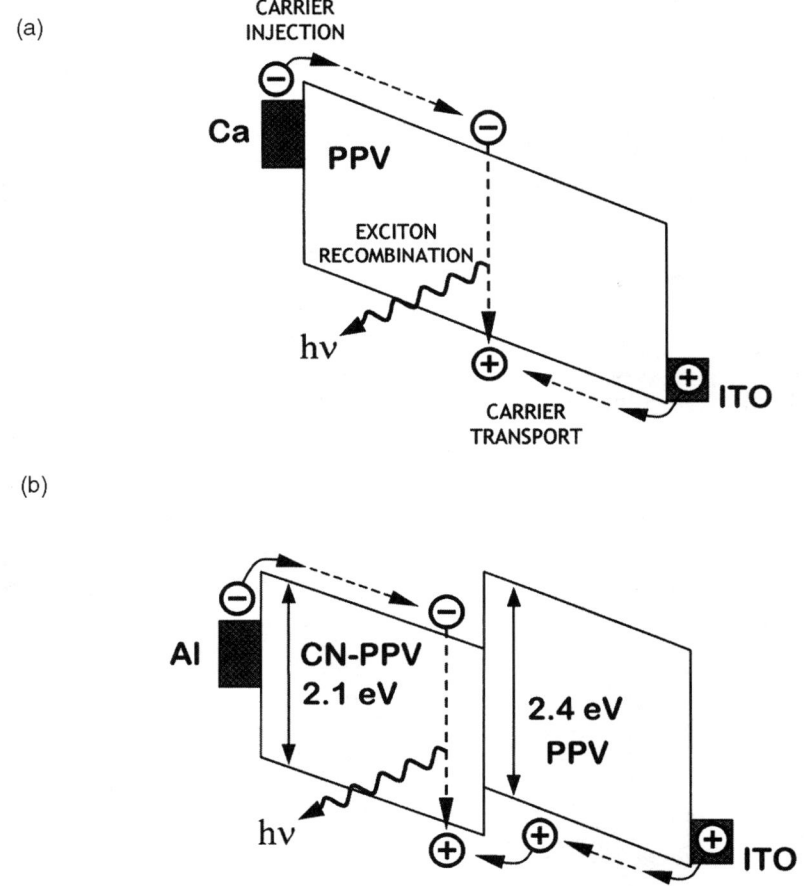

FIG. 5. Schematic energy level diagram of (a) single-polymer-layer and (b) heterostructure PLEDs with some typical polymer and electrode materials indicated in the figure.

recently led to $\eta > 1\%$ in related conjugated polymers (Friend et al., 1999). PPV PLEDs emit green light, but since these early demonstrations, researchers have made rapid progress in developing organic compounds and efficient devices that also emit in the red and blue regions of the spectrum (Cambridge Display Technology).

Another efficient polymer-based structure employs a bilayer (Baigent et al., 1994; Becker et al., 1997; Greenham et al., 1993) similar to molecular organic heterostructures, or even blends of electron- and hole-transporting materials (Wu et al., 1997). In heterostructure devices (Fig. 5b), the first polymer layer, which is in contact with ITO surface, can serve solely as a hole injecting–conducting layer, as in the case of PPV, while the second organic layer, such as cyano-PPV (CN-PPV), transports electrons to the polymer–polymer interface. The carriers are confined at the interface due to the energy level offset between the organic layers, which increases the likelihood of electron–hole capture to form an exciton that radiatively recombines. Excitons form in both organic layers but those in PPV can nonradiatively transfer to the smaller bandgap CN-PPV, where they radiatively recombine. The presence of the electron-withdrawing cyano groups in CN-PPV is responsible for the larger electron affinity and smaller bandgap of this layer as compared to PPV. Adjusting the electron-withdrawing character of individual materials tailors the polymer–polymer energy level offset, which in turn tunes the device performance as discussed in the following section.

Due to the inherent mechanical flexibility of thin polymer layers, flexible PLEDs on ITO-coated polyester substrates have been demonstrated (Gustaffson et al., 1993; May, 1996). Just as for flexible OLEDs (Gu et al., 1997), such devices create the potential for roll-up displays, or conformable displays on curved surfaces. Vacuum deposition on plastic sheets also allows for mass production of large-area PLEDs and OLEDs via roll-to-roll (web) processing. The successful development of such a process could result in extremely low-cost manufacture.

IV. Current–Voltage–Luminance Characteristics

The current–voltage (J–V) characteristics of PLEDs are limited by either carrier injection at the contacts and/or carrier transport through the polymer bulk. Discussion of contact-limited operation was presented in the context of molecular OLEDs and similarly applies to the case of PLEDs. Rather than repeat the concepts described earlier (see Vol. 64, Ch. 5) for molecular organic–metal contacts, in Subsection 1 of this section we discuss specific examples of conjugated polymer–metal interfaces. Theories of polymer bulk transport are described in Subsection 2 with a focus on transport in PPV thin films. In Subsection 3, electroluminescence processes are discussed, and the luminescence efficiencies of PLEDs and OLEDs are compared for the best devices reported to date.

1. POLYMER–METAL INTERFACES

To contribute to the formation of excitons in PLEDs, a carrier is injected over the energy barrier at the polymer–electrode interface. For large energy barriers, injection dominates the J–V characteristics of the device. Comparing the experimental J–V measurements to the injection dominated mechanisms such as Fowler–Nordheim tunneling through (Marks et al., 1993; Parker, 1994) and thermionic emission over (Gmeiner et al., 1993) the organic–metal interface energy barrier, agreement is found only over a limited range of operating currents and temperatures. Indeed, quantitative predictions of the Fowler–Nordheim theory exceed the experimentally observed currents by several orders of magnitude. Nevertheless, the tunneling model inspired the concept of reducing the energy barrier at the organic–metal contact in order to inject carriers more efficiently. Lowering the injection barrier leads to a larger current and higher light output at the same voltage. Hence, low work function cathodes such as Ca, Mg, In, or Al (with work function $\Phi = 2.9$, 3.7, 4.1, and 4.3 eV, respectively) in conjunction with a high-work-function anode such as Au or ITO ($\Phi = 5.1$, ~ 4.8 eV, respectively) are typically chosen for the highest performance (*Handbook of Chemistry and Physics*, 1999). Theoretical modeling and experimental analysis suggest that injection limited conduction is dominant for devices with injection barriers that exceed 0.3–0.4 eV (Campbell et al., 1997; Davids et al., 1997). In this case, charge backflow due to the low carrier mobilities has been suggested as a mechanism that strongly affects the J–V characteristics (Davids et al., 1997; Vestweber et al., 1995). For smaller injection barriers, the J–V characteristics are bulk transport limited, as described in the next subsection.

Experimental and theoretical analyses of the most common metal–polymer (Brédas, 1995; Salaneck and Brédas, 1996; Salaneck et al., 1996) and ITO–polymer (Johanssen et al., 1998) interfaces indicate that chemical processes between the polymer and the electrode largely determine the properties of the contact. For example, both Al (Logdlund and Brédas, 1994) and Ca (Dannetun et al., 1994) diffuse 20–30 Å into oxygen-free surfaces of PPV derivatives, forming covalent and ionic bonds, respectively.

For oxygen rich surfaces the Ca forms a 20- to 30-Å-thick calcium oxide interfacial insulating layer that can affect the PLED operating characteristics (Gao et al., 1993). For the anode, ITO is most commonly used, however, inherent instabilities in this metal oxide alloy have promoted the search for alternative transparent contact materials. Several p-doped conjugated polymers including polypyrrole, polythiophene derivatives, and polyaniline, have been considered as hole-injecting electrodes (Hayashi et al., 1996; Yang and Heeger, 1994) due to their high conductivities. With these materials, improvements in both PLED efficiencies and lifetimes have been reported (Carter et al., 1997; Karg et al., 1996; Liedenbaum et al., 1997).

2. CONDUCTION IN POLYMER THIN FILMS

The carrier mobilities in both conjugated polymers and molecular organic thin films are typically low (10^{-2} to 10^{-8} cm^2 V^{-1} s^{-1}) (Borsenberger and Weiss, 1993) due to both lack of extended transport states (i.e., energy bands) and carrier trapping in these amorphous materials. Organic materials also have low intrinsic carrier densities due to their large energy gap. As a result, in an intrinsic organic material (one with a negligible extrinsic dopant density) large injected carrier densities are required to carry current typical for PLED operation. In this case, the current–voltage characteristics are often space charge limited and the internal electric field due to the large density of the injected charge is significant.

Different space–charge-limited conduction (SCLC) mechanisms have been proposed to describe charge transport through a polymer thin film. They fall in two categories: (1) trap-free SCLC with a field-dependent carrier mobility (Blom et al., 1996, 1997), or (2) SCLC in the presence of traps (Campbell et al., 1998). Carrier transport in molecular organic thin films was described in Ch. 5 of Vol. 64 in terms of SCLC in the presence of traps. Therefore, in this chapter we discuss trap-free SCLC with a field-dependent carrier mobility by describing hole transport in single-layer PPV PLEDs. The electron (minority carrier) transport in these PPV films has been found to be trap-limited. Our treatment is largely drawn from the analysis of Blom and de Jong (1998). For a detailed comparison of different bulk limited transport models in polymer thin films, see Campbell et al., (1998).

a. *Hole-Only PPV Devices*

Carrier transport in ITO–PPV–Au devices is dominated by holes due to the large barrier to electron injection at both interfaces (inset Fig. 6a). The

J–V characteristics of such devices (Fig. 6a) indicate that, for low electric fields the current density changes as the square of the voltage, which is a signature of trap-free SCLC. Assuming the carrier mobility μ is field-independent and that there are no traps in a polymer thin film of thickness d and relative dielectric constant ε_r, then SCLC current (Lampert and Mark, 1970) through this film is given by

$$J = \frac{9}{8}\varepsilon_0\varepsilon_r\mu\frac{V^2}{d^3} \tag{1}$$

where ε_0 is the permitivity in free space and d is the film thickness. Using $\varepsilon_r = 3$ and a hole mobility $\mu_p = 5 \times 10^{-7}\,\text{cm}^2\,\text{V}^{-1}\,\text{s}^{-1}$ (Blom and de Jong, 1998), good agreement between the data in Fig. 6a and the SCLC theory (solid line) is obtained. At fields higher than 3×10^5 V/cm, however, the current gradually deviates from the theory becoming larger than expected, as seen in Fig. 6b (showing theoretical predictions as a dotted line). This discrepancy has been explained in terms of field-dependent mobility (Gill, 1972; Pai, 1970) namely,

$$\mu_p = \mu_0 \exp\left(-\frac{\Delta}{kT}\right)\exp(\gamma\sqrt{E}) \tag{2}$$

where μ_0 is a temperature independent prefactor, Δ is the zero-field activation energy, k is Boltzmann's constant, T is temperature, and γ is given by the empirical expression

$$\gamma = G\left(\frac{1}{kT} - \frac{1}{kT_0}\right) \tag{3}$$

From the temperature dependence of the low field J–V characteristics (Fig. 6b) one can determine the zero-field mobility $\mu_0\exp(-\Delta/kT)$, finding $\Delta = 0.48$ eV and $\mu_0 = 3.5 \times 10^{-3}$ m^2/V s (Blom and de Jong, 1998). The empirical constants G and T_0 are similarly determined from the fits to the data to be 2.9×10^{-5} eV(V/m)$^{-1/2}$ and 600 K, respectively, which is in good agreement with earlier measurements in PVK (Gill, 1972). Using these parametrized values for $\mu(E, T)$, improved fits to the data are obtained, as shown by the solid lines in Fig. 6b.

We note that while these fits to the data are compelling, a more rigorous test of the model can only be accomplished through direct measurements of the field and temperature dependence of the mobility using time-of-flight techniques. To our knowledge, such a test has not yet been done in the case either of trap-free SCLC or trap-limited conduction models discussed in the context of PLEDs and molecular OLEDs.

Chapter 1

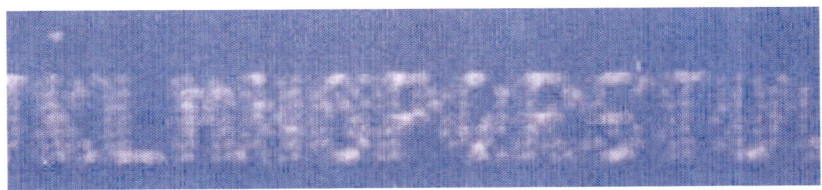

FIG. 3. Photograph of the ink-jet printed luminescent polymer letters of nile red doped PVK excited by an ultraviolet lamp.

Chapter 2

FIG. 1. Series of monochrome yellow TFEL displays (5.6 to 10.4 in.) of Planar Systems, USA, using a proprietary Integral Contrast and Brightness Enhancement technology (ICEBrite), leading to very high contrast without the use of filters.

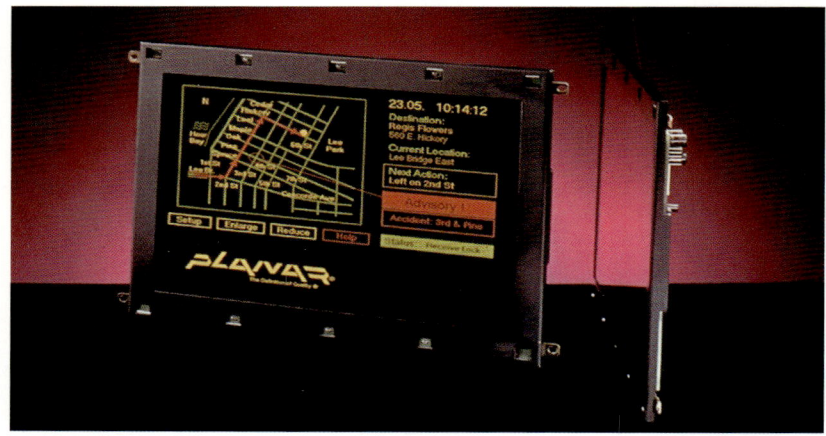

FIG. 2. 10.4 in. VGA multicolor, red, and green, TFEL display of Planar Systems, USA, with a typical luminance of 25 cd/m² and a contrast of 10:1 at 500 1x.

FIG. 3. A 4.8 in. quarter-VGA 320 (×3) × 240 full-color TFEL display of Planar Systems, which shows 10, 33, and 4 cd/m² for RGB and 45 cd/m² for white and a contrast ratio of 2:1 at 40,000 1x.

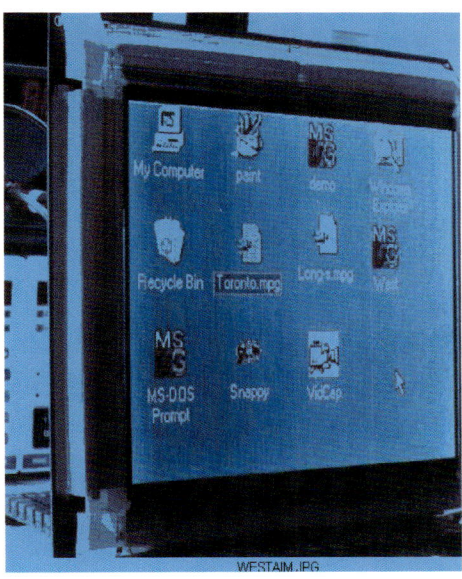

FIG. 61. Latest Westaim full-color display demonstrator, featuring 37, 77, and 21 cd/m^2 for R, G, and B at a 160-Hz frame refresh rate.

FIG. 64. Monochrome AMEL display of Planar Systems, Inc., featuring 340 cd/m^2 (yellow), a contrast ratio of 100:1, and 6 bits of gray scale. The active area is 0.61 × 0.45 in., resolution 640 × 480, 1000 lines/in. resolution (after King, 1996).

FIG. 65. Planar full-color AMEL display with 29, 78, and 29 cd/m² for R, G, and B; contrast 100:1, power input 1.3 W, typical (King, 1996).

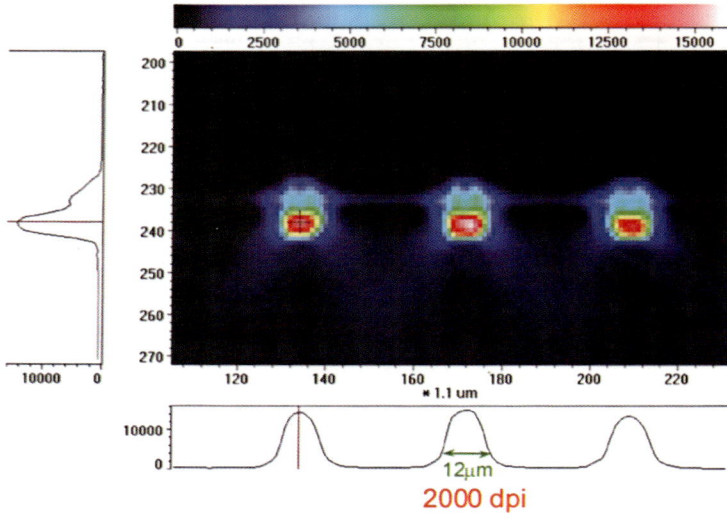

FIG. 67. Radiance distribution of an experimental 1200-dpi edge emitter array; every other pixel domed; false color scale at the top.

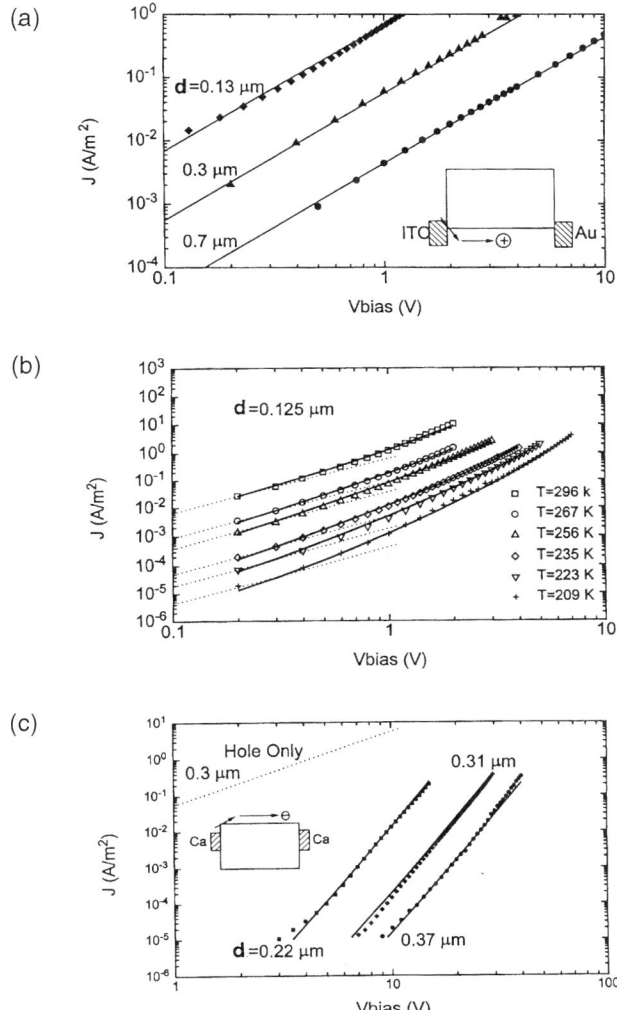

FIG. 6. (a) Room temperature J–V characteristics of a series of ITO–PPV–Au hole-only devices of various thicknesses. The hole transport for all samples is described in terms of SCLC (solid lines) of Eq. (1) with the hole mobility $\mu_p = 0.5 \times 10^{-6}$ cm^2 V^{-1} s^{-1}, and the dielectric constant $\varepsilon_r = 3$. (b) Experimental J–V characteristics of ITO–PPV–Au device with polymer layer thickness of $d = 0.125$ μm as a function of temperature. The data are fitted to the SCLC model with field-dependent mobility (solid lines) defined by Eqs. (2) and (3) using $\Delta = 0.48$ eV, $G = 2.9 \times 10^{-5}$ eV(V/m)$^{-1/2}$, and $T_0 = 600$ K. The low-field part of the data is fitted to the conventional SCLC model of Eq. (1) (dotted lines). (c) Room-temperature J–V characteristics of a series of Ca–PPV–Ca electron-only devices of various thicknesses. The data are fitted to the trap-limited SCLC (solid lines), with trap density $N_t = 1 \times 10^{18}$ cm^{-3}. For comparison, J–V characteristic of a hole-only device with $d = 0.3$ μm is also shown (dotted line) (after Blom and de Jong, © 1998, with permission of Elsevier Science).

b. Electron-Only PPV Devices

The electron-only devices consist of Ca–PPV–Ca, where the low work function of Ca facilitates electron injection into PPV. In Fig. 6c, the J–V behavior for a series of these devices with different thicknesses has been compared to the hole-only device with $d = 0.3$ μm (Blom and de Jong, 1998). Note that the current through the hole-only device is at least three orders of magnitude larger than that through an electron-only device. Furthermore, the electron current rapidly rises with the applied voltage, following a power law dependence, namely $J = V^7$, characteristic of trap-limited conduction described in Vol. 64, Ch. 5. Indeed, it was shown (Blom and de Jong, 1998) that the trap-limited conduction model agrees well with the data, from which carrier trap density is determined to be $N_t = 10^{18}$ cm^{-3} and trap temperature $T_t = 1500$ K. These values are similar to those obtained for molecular OLEDs, indicating a potentially close relationship between the electrical properties of polymeric and molecular organic thin films.

3. PLED PERFORMANCE

The electroluminescent (EL) efficiencies of PLEDs and OLEDs are proportional to the photoluminescence (PL) efficiency of the luminescent organic thin films. In PPV, for example, no side groups separate the polymer chains, and the excitation rapidly decays to the nonradiative interchain excitons, reducing the overall luminescence efficiency (Yan et al., 1994). Introduction of side groups to the PPV backbone reduces nonradiative processes, so that in ITO–MEH-PPV–Ca PLEDs an external efficiency of ~2 lm/W has been reported (Liedenbaum et al., 1997).

Emission of fluorescent devices originates from the radiative recombination of singlet states that are expected to comprise ~25% of all the electrically generated excitons (Greenham and Friend, 1995). The remaining 75% of the exciton population are triplet states. The triplet lifetime is typically long (microseconds to seconds) as their transition to the singlet ground state is spin symmetry forbidden, and their deexcitation is, in general, nonradiative. In some cases, nonradiative processes are not dominant, and phosphorescent emission can be observed. The advantage of phosphorescent devices is that both singlet and triplet excitons are utilized in EL, resulting in high quantum and power efficiencies. For green phosphorescent molecules, peak quantum efficiencies of 8% and power efficiency of 30 lm/W were obtained (Baldo et al., 1999).

The EL efficiency also depends on the balanced injection of electrons and holes into the recombination region. For single-layer PPV PLEDs, for example, the lower mobility of electrons as compared to holes results in an unbalanced carrier transport and formation of excitons in the vicinity of the cathode. The reduced intensity of the optical field near the metal cathode reduces the rate of radiative exciton recombination, while the defect states and nonradiative energy transfer to plasmon modes of the metal further quench the EL quantum efficiency (Bulović *et al.*, 1998). Higher EL efficiencies are often obtained for heterojunction devices where the proper tailoring of the organic–organic energy band offsets can confine the carriers at the organic heterojunction interface, far away from the quenching sites on the electrodes (Friend *et al.*, 1999). In these latter devices, space charge accumulates at the heterojunction, which can reduce the current while establishing a balance in the charge injection. Further refinement in carrier balancing is achieved by independently optimizing metal–organic interfaces of both electrodes.

Currently, the best reported luminance efficiencies of both PLEDs and OLEDs exceed that of incandescent light bulbs. For example, efficiencies of >20 lm/W have been reported for yellow-green emitting PLEDs (Fig. 7a) based on polyfluorine, (Friend *et al.*, 1999; Lacey, 1998; Liedenbaum, 1998) with somewhat higher (30 lm/W) (Baldo *et al.*, 1999) values attained for phosphorescent green molecular OLEDs (Fig. 7b). Table I enumerates the efficiencies of the best reported PLEDs and OLEDs of various colors. We note that device structures and material compositions of most of these LEDs are proprietary and hence are unknown to us. Nevertheless, this information demonstrates the quality of performance achievable with these organic light emitter technologies.

V. The Good and the Bad of Organic Materials

1. Advantages of Organic Materials

The relatively low cost growth and processing of organic materials is most often cited as their greatest advantage over conventional inorganic semiconductors. The organic source materials are typically assumed to be inexpensive, although the cost of some can be substantial due to chemical synthesis routes requiring multiple steps with low yield, or involved chemical purification. In general, however, large-area, bulk processing of organic materials is feasible, which differs from what can be achieved using crystalline inorganic semiconductor technologies.

High mechanical flexibility of organic thin films allows for compatibility with a large number of substrates. This is the case for both amorphous

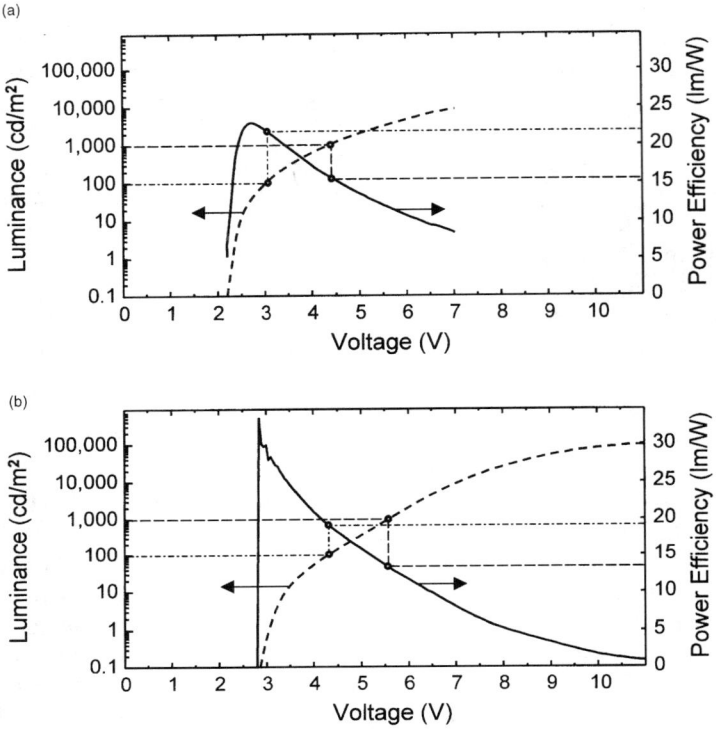

FIG. 7. Luminance and luminescence efficiency of presently most efficient (a) PLEDs (after Friend *et al.*, 1999; Cambridge Display Technology) and (b) OLEDs (after Baldo *et al.*, 1999). Both devices emit green light.

organic thin films, which typically conform to the underlying patterns, and crystalline organic thin films (Forrest, 1997). Growth of crystalline inorganic thin films requires a close lattice match to the underlying substrate due to the strong covalent or ionic bonds between atoms. This limits the combinations of inorganic materials that can be grown without inducing a high density of interface defects. In contrast, most crystalline organic solids can be grown with a substantial lattice mismatch to the underlying substrate (exceeding several percent) (Forrest, 1997). This is afforded by the considerable flexibility in intermolecular van der Waals interactions. As a result, a vast number of organic heterointerfaces is accessible, which is a considerable benefit in the design, refinement, and tuning of organic optoelectronic devices. Furthermore, organic films can be grown on top of inorganic substrates, facilitating integration with conventional optoelectronics (Taylor *et al.*, 1995).

Finally, a number of optical and electronic properties of organic materials are unique and do not have counterparts in inorganic semiconductors. For example, due to the large Franck–Condon shift in organic materials, they are transparent to their own luminescence. This property was utilized to develop transparent organic LEDs (Bulović *et al.*, 1996; Gu *et al.*, 1996; Parthasarathy *et al.*, 1998) for use in head-up displays and stacked, full-color pixels (Shen *et al.*, 1997; Parthasarathy *et al.*, 1999), as described in Chapter 5 (Vol. 64).

2. Stability of Polymeric and Molecular Organic Materials

In general, polymeric and molecular organic materials are known to readily react with water and oxygen. Exposure to atmosphere and other gases can significantly alter their electronic properties. As such, organics are ideal materials for gas and chemical sensing. However, this same property deleteriously affects the operation of PLEDs and OLEDs. Studies show that the stability of the OLED materials can be increased by sealed packaging (Burrows *et al.*, 1994), increasing the device lifetime to half its initial luminance of $100 \, cd/m^2$ of up to 50,000 h (Tang, 1996). Similar improvement is observed in PLEDs, with the best reported lifetimes exceeding 30,000 h (Liedenbaum, 1998).

Chapter 5 (Vol. 64) emphasized that the operating conditions together with the high reactivity of the cathode (needed to ensure efficient injection of electrons into the organic thin film) cause the generation and subsequent growth of "dark spots" within both PLEDs and OLEDs (Aziz *et al.*, 1998; Burrows *et al.*, 1994; McElvain *et al.*, 1996). Clarification of the nature and origin of dark spot defects is a major priority for future research.

The metallurgy of PLED and OLED contacts, such as wetting and annealing properties, needs to be investigated in parallel with their electrical characteristics. The sticking of the contact metal to organics is critical to its ability to inject charge. Encapsulation of devices is required to characterize these properties properly.

3. Mechanical and Electronic Concerns

Due to the weak van der Waals bonding, molecular organic materials are typically soft and susceptible to physical damage. This is less of a problem in polymers where strong intrachain covalent bonds provide mechanical stability. Indeed, many polymers can be cast, extruded, or rolled, and in

TABLE I
EFFICIENCIES OF THE BEST OF REPORTED PLEDs AND OLEDs

PLEDs

Reference	Parameter	Blue	Green	Orange	Red
Shimoda et al. (1999)					
	CIE x	0.187	0.394		0.696
	CIE y	0.235	0.57		0.303
At 100 cd/m^2	η_p	3.8	20		0.65
	Lifetime	>1000 h	>10,000 h		>8,000
Liedenbaum (1998)					
	V at 100 cd/m^2 (V)			3.5 V	
	η_p at 100 cd/m^2 (lm/W)			3.3 lm/W	
	Lifetime to 50% of initial			30,000 h	
Lacey (1998)					
	CIE x	0.16	0.39		0.7
	CIE y	0.1	0.57		0.3
At 3.5 V	Brightness (cd/m^2)		100		
	η_P (lm/W)		15		
At 100 cd/m^2	η_P (lm/W)	3.8			0.8
	Lifetime	>1000 h	6000 h at 140 cd/m^2		

Molecular OLEDs

Reference	Parameter	Blue	Green	Orange	Red
Miyaguchi (1998)	CIE x	0.14	0.28		0.62
	CIE y	0.12	0.67		0.38
	η at 300 cd/m^2 (cd/A)	3.9	16.3		3.8
	η_{QE} at 300 cd/m^2 (%)	4.5	4.4	2.6	2.6
	η_P at 300 cd/m^2 (lm/W)	2.2	11.4	1.4	1.4
Baldo et al. (1998, 1999)	CIE x		0.27		0.72
and O'Brien et al. (1999)	CIE y		0.63		0.28
	η at 100 cd/m^2 (cd/A)		26		0.4
	η_{QE} at 100 cd/m^2 (%)		7.5		2.2
	η_P at 100 cd/m^2 (lm/W)		19		0.6
Tang (1996)	CIE x	0.137	0.269	0.624	
	CIE y	0.084	0.628	0.374	
	η (cd/A)	3.47	18.8	9.1	
At 5 V	η_P (lm/W)	2.2	12.0	5.7	

some cases exhibit high tensile strength (Cao *et al.*, 1991). Van der Waals bonding is also responsible for the low stability of molecular organics at elevated temperatures, which is manifested in a low glass transition temperature (T_g).

The low mechanical and temperature stability limits the nature of deposition methods and range of processing techniques available to organic materials. For example, sputter deposition of inorganic materials on top of an organic thin film results in damage to the organic. In some cases, however, damage can be beneficial, as for the thermally evaporated metal cathode on top of an electron transporting organic layer of an OLED, where the resulting defect states facilitate electron injection into the organic film (Bulović *et al.*, 1997). In general, the low T_g limits the range of temperatures over which the organic materials can be processed.

Finally, the biggest challenge to use of organic thin films in electronic applications is the low carrier mobility intrinsic to van der Waals bonded solids. In molecular materials, the poor overlap of electron orbitals on neighboring molecules limits the carrier mobility to $<1\,\mathrm{cm^2/V\,s}$ for molecular crystals, and from 10^{-3} to $10^{-7}\,\mathrm{cm^2/V\,s}$ for amorphous molecular organic solids. Mobility in polymers is similar in magnitude, however, free carrier density is typically larger, resulting in consistently lower operating voltages for PLEDs as compared to OLEDs. Nevertheless, designing organic devices that operate below 10 V necessitates use of thin organic active layers, on the order of 10 to 100 nm. Such thin structures are susceptible to electrical shorts due to micron-size impurities introduced during substrate processing or to pinholes and other film discontinuities generated during layer deposition. Thin film defects can cause dark spots in the electroluminescent devices and can reduce yield (Burrows *et al.*, 1994).

VI. A Light at the End of the Tunnel?

Given the very strong economic factors driving flat panel displays (Forrest *et al.*, 1998), this field is currently experiencing a period of rapid research and development centered on several new and promising pixel technologies. Perhaps the technology with the greatest potential is based on organic thin film electroluminescent devices. Research has shown that both polymer and molecular LEDs can be efficient, bright, and reliable. The most attractive aspects of organic emissive displays are their potential for use in very large size displays and their suitability for low-cost manufacturing processes.

In the past few years, work published by over 80 laboratories worldwide (Burrows *et al.*, 1997; Gymer, 1996) has shown that arrays of vacuum-deposited, green OLEDs are sufficiently bright and long-lived to be a realistic alternative to conventional technologies such as backlit liquid crystal displays. OLEDs based on vacuum deposited organic thin films have the unique properties of transparency, flexibility, and simplicity of manufacture that are unobtainable using conventional, inorganic semiconductors. In addition, rapid progress in the development of polymer light-emitters indicates that these devices provide high brightness and a simple structure that will find application as low-cost monochrome and, eventually, full-color displays.

Today, numerous companies are rushing to commercialize displays based on OLEDs (Nakada and Tohma, 1996; Wakimoto *et al.*, 1996). These companies include such multinational corporations as Philips, Pioneer, and Kodak. However, as is the case for any emerging technology, this field has also provided significant opportunities for small, entrepreneurial firms. Numerous companies such as Universal Display Corp., Cambridge Display Technology, and Uniax have recently been founded to exploit the commercial possibilities of OLEDs. This intense worldwide effort to exploit PLED and OLED technologies appears poised to succeed in producing an entirely new generation of low-cost, flat panel displays with novel properties inaccessible to existing display concepts.

However, there remains much to be done if organics, whether they are based on polymer or vacuum-deposited molecular thin films, are to establish a foothold in the explosively growing display market. Clearly, achieving even higher efficiencies, lower operating voltages, and longer device lifetimes than have been demonstrated to date are all challenges that must be met before this can be realized. Given the aggressive efforts that are now developing worldwide, emissive organic thin films have an excellent chance of being the technology of choice for the next generation of high-resolution, high-efficiency flat panel displays.

Acknowledgments

It is a challenge to impartially compare the field of science in which you work to another one you observe. We thank Gerd Mueller for giving us the opportunity to test our impartiality. Although not emphasized in this chapter, our interest in molecular organic technology originates from innumerable hours of investigating this field with our colleagues at Princeton University and the University of Southern California. We thank these many scientists for the insights gained and shared. We are also grateful to AFOSR, DARPA, and Universal Display Corporation for their financial support of this research.

References

Aziz, H., Z. Popović, C. P. Tripp, N. X. Hu, A. M. Hor, and G. Xu. (1998). *Appl. Phys. Lett.* **72**, 2642.
Baigent, D. R., N. C. Greenham, J. Gruner, R. N. Marks, R. H. Friend, S. C. Moratti, and A. B. Holmes. (1994). *Synth. Met.* **67**, 3.
Baldo, M. A., S. Lamansky, P. E. Burrows, M. E. Thompson, and S. R. Forrest. (1999). *Appl. Phys. Lett.*
Baldo, M. A., D. F. O'Brien, Y. You, A. Shoustikov, M. E. Thompson, and S. R. Forrest. (1998). *Nature*, **395**, 151.
Bao, Z., Y. Feng, A. Dodabalapur, V. R. Raju, and A. J. Lovinger. (1997). *Chem. Mater.* **9**, 1299.
Becker, H., S. E. Burns, and R. H. Friend. (1997). *Phys. Rev. B* **56**, 1893.
Bharathan, J., and Y. Yang. (1998). *Appl. Phys. Lett.* **72**, 2660.
Blom, P. W. M., and M. J. M. de Jong. (1998). *Phillips J. Res.* **51**, 479.
Blom, P. W. M., M. J. M. de Jong, and J. J. M. Vleggaar. (1996). *Appl. Phys. Lett.* **68**, 3308.
Blom, P. W. M., M. J. M. de Jong, and M. G. van Munster. (1997). *Phys. Rev. B* **55**, R656.
Borsenberger, P. M., and D. S. Weiss. (1993). *Organic Photoreceptors for Imaging Systems.* (Marcel Dekker, New York).
Brédas, J. L. (1995). *Adv. Mat.* **7**, 263.
Bulović, V., G. Gu, P. E. Burrows, M. E. Thompson, and S. R. Forrest. (1996). *Nature (London)* **380**, 29.
Bulović, V., P. Tian, P. E. Burrows, M. Ghokale, S. R. Forrest, and M. E. Thompson. (1997). *Appl. Phys. Lett.* **70**, 2954.
Bulović, V., V. B. Khalfin, G. Gu, P. E. Burrows, D. Z. Garbuzov, and S. R. Forrest. (1998). *Phys. Rev. B* **58**, 3730.
Burroughes, J. H., D. D. C. Bradley, A. R. Brown, R. N. Marks, K. MacKay, R. H. Friend, P. L. Burn, and A. B. Holmes. (1990). *Nature* **347**, 539.
Burroughes, J., D. J. Lacey, I. Millard, C. Murphy, C. Towns. (1999). *Spr. Mtg. Mat. Res. Soc.*, B17.6, San Francisco.
Burrows, P. E., V. Bulović, S. R. Forrest, L. S. Sapochak, D. M. McCarty, and M. E. Thompson. (1994). *Appl. Phys. Lett.* **65**, 2922.
Burrows, P. E., S. R. Forrest, and M. E. Thompson. (1997). *Curr. Opin. Solid State Mat. Sci.* **2**, 236.
Cambridge Display Technology web site, http://www.cdtltd.co.uk.
Campbell, A. J., D. D. C. Bradley, and D. G. Litzey. (1997). *J. Appl. Phys.* **82**, 6326.
Campbell, A. J., M. S. Weaver, D. G. Lidzey, and D. D. C. Bradley. (1998). *J. Appl. Phys.* **84**, 6737.
Cao, Y., P. Smith, and A. J. Heeger. (1991). *Polymer* **32**, 1210.
Carter, J. C., I. Grizzi, S. K. Heeks, D. J. Lacey, S. G. Latham, P. G. May, O. Ruiz de los Paños, K. Pichler, C. R. Townes, and H. F. Wittman. (1997). *Appl. Phys. Lett.* **71**, 34.
Cornil, J., D. Beljonne, D. A. dos Santos, and J. L. Brédas. (1996). *Synth. Met.* **76**, 101.
Cumpston, B. H., and K. F. Jensen. (1995). *Synth. Met.* **73**, 195.
Dannetum, P., M. Fahlman, C. Fauquet, K. Kaerijama, Y. Sonoda, R. Lazzaroni, J. L. Brédas, and W. R. Salaneck. (1994). *Synth. Met.* **67**, 133.
Davids, P. S., I. H. Campbell, and D. L. Smith. (1997). *J. Appl. Phys.* **82**, 6319.
Forrest, S. R. (1997). *Chem. Rev.* **97**, 1793.
Forrest, S. R., P. E. Burrows, and M. E. Thompson. (1998). *Chem. Indus.* 1022.
Friend, R. H., R. W. Gymer, A. B. Holmes, J. H. Burroughes, R. N. Marks, C. Taliani, D. D. C. Bradley, D. A. Dos Santos, J. L. Brédas, M. Lögdlund, and W. R. Salaneck. (1999). *Nature (London)* **397**, 121.
Gao, Y., K. T. Park, and B. R. Hsieh. (1993). *J. Appl. Phys.* **73**, 7894.

Garnier, F., R. Hajlaoui, A. Yassar, and P. Srivastava. (1994). *Science* **265**, 1685.
Gill, W. G. (1972). *J. Appl. Phys.* **43**, 5033.
Gmeiner, J., S. Karg, M. Meier, W. Riess, P. Strohriegl, and M. Schwoerer. (1993). *Acta Polymer.* **44**, 201.
Greenham, N. C., and R. H. Friend. (1995). *Solid State Phys.* **49**, 1.
Greenham, N. C., S. C. Morati, D. D. C. Bradley, R. H. Friend, and A. B. Holmes. (1993). *Nature (London)* **365**, 628.
Gu, G., V. Bulović, P. E. Burrows, S. R. Forrest, and M. E. Thompson. (1996) *Appl. Phys. Lett.* **68**, 2606.
Gu, G., P. E. Burrows, S. Venkatesh, and S. R. Forrest. (1997). *Opt. Lett.* **22**, 172.
Gustaffson, G., G. M. Treacy, Y. Cao, F. Klavertter, N. Colaneri, and A. J. Heeger. (1993). *Synth. Met.* **57**, 4123.
Gutmann, F., and L. E. Lyon. (1981). *Organic Semiconductors*. (R. E. Krieger Publishing, Malabar, FL).
Gymer, R. W. (1996). *Endeavor* **20**, 115.
Handbook of Chemistry and Physics. (1999) (CRC Press, Boca Raton, FL).
Harrison, N. T., G. R. Hayes, R. T. Phillips, and R. H. Friend. (1996). *Phys. Rev. Lett.* **77**, 1881.
Hasegawa, T., Y. Iwasa, H. Sunamura, T. Koda, Y. Tokura, H. Tachibana, M. Matsumoto, and S. Abe. (1992). *Phys. Rev. Lett.* **68**, 668.
Hayashi, S., H. Etoh, and S. Saito. (1996). *Jpn. J. Appl. Phys.* **25**, 773.
Hebner, T. R., and J. C. Sturm. (1998). *Appl. Phys. Lett.* **73**, 1775.
Hebner, T. R., C. C. Wu, D. Marcy, M. H. Lu, and J. C. Sturm. (1998). *Appl. Phys. Lett.* **72**, 519.
Heeger, A. J., S. Kivelson, J. R. Schrieffer, and W.-P. Su. (1988). *Rev. Mod. Phys.* **60**, 781.
Johansson, N., F. Cacialli, K. Z. Xing, G. Beamson, D. T. Clark, R. H. Friend, W. R. Salaneck. (1998). *Synth. Met.* **92**, 207.
Karg, S., J. C. Scott, J. R. Salem, and M. Angelopoulos. (1996). *Synth. Met.* **80**, 111.
Kim, E., Y. Xia, and G. M. Whitesides. (1995). *Nature (London)* **376**, 581.
Kohlman, R. S., A. Zibold, D. B. Tanner, G. G. Ihas, T. Ishiguro, Y. G. Min, A. G. MacDiarmid, and A. J. Epstein. (1997). *Phys. Rev. Lett.* **78**, 3915.
Lacey, D. (1998). "High Efficiency Polymer Light-Emitting Diodes." Paper Presented at 9th International Workshop on Inorganic and Organic Electroluminescence, Bend, Oregon.
Lampert, M. A., and P. Mark. (1970). *Current Injection in Solids*. (Academic Press, New York).
Liedenbaum, C., Y. Croonen, P. van de Weijer, J. Vleggaar, and H. Schoo. (1997). *Synth. Met.* **91**, 109.
Liedenbaum, C. (1998). "Improved Electro-Optical Characteristics of Polymer Light-Emitting Devices." Presented at the 200th W. E. Heraeus seminar: Electroluminescence of Organic Materials, Bad Honnef, Germany.
Logdlund, M., and J. L. Brédas. (1994). *J. Chem. Phys.* **101**, 4357.
Marks, R. N., D. D. C. Bradley, R. W. Jackson, P. L. Burn, and A. B. Holmes. (1993). *Synth. Met.* **55**, 4128.
May, P. (1996). *Soc. Inf. Display Int. Symp. Digest of Technical Papers* **27**, 192.
McElvain, J., H. Antoniadis, M. R. Hueschen, J. N. Miller, D. M. Roitman, J. R. Sheats, and R. L. Moon. (1996). *J. Appl. Phys.* **80**, 6002.
Miyaguchi, S. (1998). "OLED Full-Color Passive-Matrix Display." 9th International Workshop on Inorganic and Organic Electroluminescence, Bend, Oregon.
Nakada, H., and T. Tohma. (1996). "Applications of Organic LEDs: 256 × 64 Dot Matrix Display." Paper presented at 8th International Workshop on Inorganic and Organic Electroluminescence, Berlin.
O'Brien, D. F., M. A. Baldo, M. E. Thompson, and S. R. Forrest. (1999). *Appl. Phys. Lett.*, **74**, 442.
Pai, D. M. (1970). *J. Chem. Phys.* **52**, 2285.
Papadimitrakopoulos, F., K. Konstandinis, T. M. Miller, R. Opila, E. A. Chandros, M. E. Galvin. (1994). *Chem. Mater.* **6**, 1563.

Park, Y. W., M. A. Druy, C. K. Chiang, A. G. MacDiarmid, A. J. Heger, H. Shirakawa, and S. Ikeda. (1979). *J. Polymer Sci. Polymer Lett.* **17**, 195.
Parker, I. D. (1994). *J. Appl. Phys.* **75**, 1656.
Parthasarathy, G., P. E. Burrows, V. Khalfin, V. G. Kozlov, and S. R. Forrest. (1998). *Appl. Phys. Lett.* **72**, 2138.
Parthasarathy, G., G. Gu, and S. R. Forrest. (1999). *Adv. Mater.* **11**, 907.
Perçin, G., T. S. Lundgren, and B. T. Khuri-Yakub. (1998). *Appl. Phys. Lett.* **73**, 2375.
Pope, M., and C. E. Swenberg. (1982). *Electronic Processes in Organic Crystals.* (Oxford University Press, Oxford).
Pschenitzka, F., and J. C. Sturm. (1999). *Appl. Phys. Lett.* **74**, 1913.
Salaneck, W. R., and J. L. Brédas. (1996). *Adv. Mat.* **8**, 48.
Salaneck, W. R., S. Strafström, and J. L. Brédas. (1996). *Conjugated Polymer Surfaces and Interfaces* (Cambridge University Press, Cambridge).
Schoo, H. F. M., and R. J. C. E. Demandt. (1998). *Phillips J. Res.* **51**, 527.
Scott, J. C., J. H. Kaufman, P. J. Brock, R. DiPietro, J. Salem, and J. A. Goitia. (1996). *J. Appl. Phys.* **79**, 2745.
Scurlock, R. D., B. Wang, P. R. Ogilby, J. R. Sheats, and R. L. Clough. (1995). *J. Am. Chem. Soc.* **117**, 10194.
Sheats, J. R., H. Antoniadis, M. Hueschen, W. Leonard, J. Miller, R. Moon, D. Roitman, and A. Stocking. (1996). *Science* **273**, 884.
Shen, Z., P. E. Burrows, V. Bulović, S. R. Forrest, and M. E. Thompson. (1997). *Science* **276**, 2009.
Shimoda, T., S. Seki, H. Kobayashi, M. Kimura, S. Miyashita, R. H. Friend, J. H. Burroughes, and C. R. Towns. (1999). International Conference on Science and Technologies of Advanced Polymers, ICAP99, Yamagata, Japan, July 26–30.
Sze, S. M. (1981). *Physics of Semiconductor Devices*, 2nd ed. p. 251. (Wiley, New York).
Tang, C. W. (1994). "Organic electroluminescent multicolor image display device." U.S. Patent #5 276 380.
Tang, C. W. (1996). *Soc. Inf. Display, International Symposium, Digest of Technical Papers* **27**, 181.
Tang, C. W., and S. A. Van Slyke. (1987). *Appl. Phys. Lett.* **51**, 913.
Taylor, R., Z. Shen, and S. R. Forrest. (1995). *Topical Meeting on Organic Thin Films for Photonics Applications*, Paper MB-5, Portland, Oregon (September 11–14).
Tian, P. F., P. E. Burrows, V. Bulović, and S. R. Forrest. (1997). *Proceedings of Materials Research Society 1997 Fall Meeting*, Boston, MA.
Vaeth, K. M., and K. F. Jensen. (1997a). *Adv. Mater.* **9**, 490.
Vaeth, K. M., and K. F. Jensen. (1997b). *Appl. Phys. Lett.* **71**, 2091.
Vestweber, H., J. Pommerehne, R. Sander, R. F. Mahrt, A. Greiner, W. Heitz, and H. Bassler. (1995). *Synth. Met.* **68**, 263.
Vincken, M. H., ed. (1998). Special Issue on Polymer Light-Emitting Devices. *Philips J. Res.* **51**, 461–533.
Wakimoto, T., R. Murayama, K. Nagayama, Y. Okuda, H. Nakada, and T. Tohma. (1996). *Soc. Inf. Disp. Dig.* 849.
Warta, W., R. Stehle, and N. Karl. (1995). *Appl. Phys. A* **36**, 163.
Wu, C. C., J. C. Sturm, R. A. Register, J. Tian, E. P. Dana, and M. E. Thompson. (1997). *IEEE Trans. Electron Devices* **44**, 1269.
Xia, Y., E. Kim, X.-M. Zhao, J. A. Rogers, M. Prentiss, and G. M. Whitesides. (1996). *Science* **273**, 347.
Yan, M., L. J. Rothberg, F. Papadimitrakopoulos, M. E. Galvin, and T. M. Miller. (1994). *Phys. Rev. Lett.* **72**, 1104.
Yang, Y., and A. J. Heeger. (1994). *Appl. Phys. Lett.* **64**, 1245.

CHAPTER 2

Thin Film Electroluminescence

Regina Mueller-Mach and Gerd O. Mueller

COMMUNICATIONS AND OPTICS RESEARCH LABORATORY
HEWLETT-PACKARD LABORATORIES
PALO ALTO, CALIFORNIA

```
I.   INTRODUCTION . . . . . . . . . . . . . . . . . . . . . . . . .    27
II.  PHENOMENOLOGY . . . . . . . . . . . . . . . . . . . . . . . .    31
III. BASIC MECHANISMS IN THIN FILM ELECTROLUMINESCENCE . . . . . . .   41
     1. The Simple Model . . . . . . . . . . . . . . . . . . . . .    44
     2. Electrical Properties . . . . . . . . . . . . . . . . . .    45
     3. Optical Properties . . . . . . . . . . . . . . . . . . . .    55
     4. Luminescence Properties . . . . . . . . . . . . . . . . .    60
     5. Efficiency . . . . . . . . . . . . . . . . . . . . . . . .    68
     6. High-Field Electronic Transport . . . . . . . . . . . . .    73
     7. Design Rules . . . . . . . . . . . . . . . . . . . . . . .    77
IV.  THIN FILM EL DEVICES . . . . . . . . . . . . . . . . . . . .    80
     1. Direct View Displays . . . . . . . . . . . . . . . . . . .    81
     2. AMEL . . . . . . . . . . . . . . . . . . . . . . . . . . .    89
     3. Nondisplay Applications . . . . . . . . . . . . . . . . .    91
V.   MEASURING TFEL . . . . . . . . . . . . . . . . . . . . . . .    93
     1. Measurements on Test Areas . . . . . . . . . . . . . . . .    93
     REFERENCES . . . . . . . . . . . . . . . . . . . . . . . . .   101
```

I. Introduction

Thin film electroluminescence (TFEL) has a long history, which for quite some time was wishful thinking only. It addressed two goals:

- Large-area flat panel displays
- Large-area light sources.

The second goal was inherited from other forms of high field electroluminescence — powder electroluminescence (EL) of all kinds generated the fiction of walls of soft cold light, radiating with hitherto unknown efficiency from everywhere. Very soon after the first successful steps toward TFEL were made, however, it became obvious that any use in general lighting would require much higher efficiency to compete with conventional incandescent bulbs or even fluorescent lamps.

The common part of the goals — large area — had one direct consequence: the technology could not rely on single-crystalline material. Phosphors, luminescent solid-state materials, in powder form had been the first response to requests for large area. Since the discovery of electroluminescence by Destriau (1936), powder phosphors had been explored and to some extent even used for so-called cold lamps. Either driven by DC or AC voltages, they reached only rather limited lifetimes, which strongly depended on drive level or light output required. Even large information content displays have been manufactured for some years based on pulse driven (DC) powder screens. At present the only application of AC powder electroluminescence is in specialty lamps (e.g., for emergency guiding in airplanes, cinemas, etc.) Backlighting of LCDs with powder EL lamps has been developed, and it is believed it will become successful at some point.

Thus, while lighting became a domain for powder EL, thin film EL is now considered as a display technology. Other applications are possible, but not commercialized. The truth is that TFEL has reached a market volume of some $100 million, and defended it against increasing pressure from improved and very reasonable LCDs since 1996. This market is flat panel displays for mainly military, medical, and industrial applications, where the ruggedness, extreme temperature range, and 179° viewing angle are highly desired and honored even in terms of price. Possibly avionics applications requiring direct-sunlight readability will emerge. The stumbling block for TFEL was the inability to meet the demands for full-color in time (i.e., in 1995). Now that it is available on a reasonable brightness and efficiency level, markets have been taken by other technologies, and very soon a newcomer — the field emission display (FED) — will impose new challenges that TFEL must compete with. TFEL is now serving niche markets, where one or some of its unique advantages are a *conditio sine qua non*.

The birth of TFEL, as we understand it today, was a paper of Inoguchi *et al.* (1974) from Sharp Central Research Labs, given at the meeting of the Society for Information Display in 1974. It described the basic phenomena, presented almost unbelievable lifetime data, and indicated a wealth more multiplexability than anybody at that time was dreaming of. There are earlier roots of all the single features, which had to come together in Inoguchi *et al.*'s work to be so successful. Thin film EL of ZnS:Mn had already been described by Vlasenko and Popkov (1960) and the impact excitation of luminescent centers by hot electrons was claimed by Khang (1968). The success of Inoguchi *et al.* played a very visible part. He was invited by Pankove to write a book (Inoguchi and Mito, 1977) for the Springer series "Topics in Applied Physics." Not as visible for some time were the projects started in all major electronic companies worldwide to obtain firsthand judgment of the new technology. There was a great deal of skepticism as Chen and Krupka at Bell Laboratories reported (1972), that there is good reason not to expect any reasonable efficiency and/or brightness from much the same materials and structures. The reports by Inoguchi *et al.* and results obtained everywhere indicated that the drive voltages had to be in the 200-V range; too high for the transistors, everybody thought at the time. IBM, which had quite some activities in plasma displays and in TFEL in those days, decided to use TFEL as a screen technology for a storage CRT (Howard and Alt, 1977), addressing pixels by the electron beam and keeping them lit in a special "memory" TFEL mode. Sharp, on the other hand, started the development of driver ICs to address the TFEL matrix in an a-line-at-a-time mode, which is the standard method today. In 1983 the first 6-in. diagonal 320 * 240 pixel (quarter-VGA) panels were introduced by Sharp. About that time, IBM abandoned its TFEL program because of difficulties with the maintenance of the memory mode and with the electric stability of a basically unpatterned large screen area. Many other companies, whose efforts surfaced much later, had similar experiences — electrical stability and long-term life were problems. Planar Systems, Inc., was spun off by Tektronix in 1982 to commercialize the successful development made at this company. It is known as Planar America, Inc., and Planar International, based in Finland, the market leader. While Sharp and Planar are the big guys, several small companies are making very specialized TFEL products. A start-up in Canada, Westaim, demonstrated very promising color TFEL displays in a "thick-film–thin-film" technology. Luxell, also based in Canada, is specializing in the highest contrast under direct sunlight, featuring a proprietary thin film black layer technology.

The first Sharp product used ZnS:Mn as a phosphor material, a material that had had some use as a yellow-emitting CRT screen. It is the workhorse of TFEL, and an every-thing-goes material. It has been and is the most efficient, and has only one disadvantage, that it is yellow only, and tricks have to be used to make it multicolor. In Subsection 5 in Section III, we discuss the reasons for this unique position. It delivers a crisp, high-contrast image, it can be driven with gray scale, and the largest panels manufactured, by Planar, were 19-in. diagonal workstation monitors, monochrome 1024 * 768. The U.S. Army disclosed that they have about 50,000 TFEL displays at work, mainly in tanks. At present a variety of monochrome displays is available from Planar (Fig. 1, see color insert) and Sharp. Multicolor displays (Fig. 2, see color insert) are now also manufactured by Planar and Sharp based on ZnS:Mn, filtering red and green from the yellow.

Several colors, depending on the number of gray shades in each color, can be generated. To find a good material for blue, the third basic color, has been a rather tedious process, and does not yet seem to be complete. We are going to explain why it is not as easy as taking a well-established blue CRT phosphor, incorporating it into a TFEL structure, and that's it. There are now hopes on SrS:Cu,Ag after looking for quite some time into SrS:Ce and $SrGa_2S_4$:Ce. Full-color displays are on the market (Fig. 3, see color insert), but neither brightness nor color purity are more than acceptable. We come back to the two concepts applied (Subsection 1 of Section IV) and to the materials used (Subsection 4 of Section III).

A TFEL display is by far the simplest physical and mechanical design of all flat panel displays that one can imagine. It consists basically of one (glass) substrate and some thin films, sandwiched between two systems of electrode stripes, one on the substrate and one on top of the thin film stack. Only the electrodes are patterned (an artist's view is given in Fig. 4); it shares this simplicity with some organic LED displays.) This is true exactly for the monochrome display and for one type of the color display, which is called color-by-white or color-by-filtered-white (discussed in part b of Subsection 1 in Section IV). A protective plastic coating or a glass plate, glued to the top, completes the device. If one compares this simplicity with the complexity of an LCD or even that of a plasma display or FED, it is amazing that so little R&D money in comparison to the other technologies has been put into TFEL. Perhaps everybody thought that being so simple, it would come for free.

The mechanism is about as simple as the physical outline, as we will show, in spite of the fact that no rigorous theory exists, and no material forecast has been developed. The device is a pure capacity below a threshold voltage; above it, it becomes lossy because of the energy consumed by the light generated. It must be driven by alternating voltages, and the light emitted

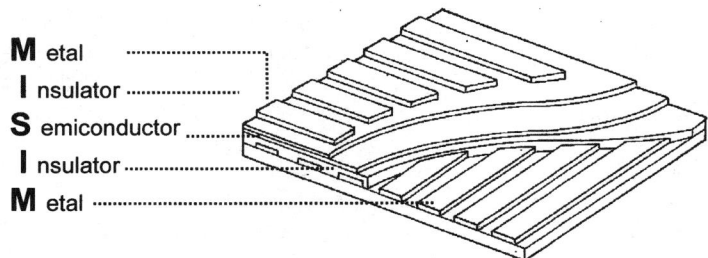

FIG. 4. Schematic of the basic structure of a TFEL device—an MISIM stack on a substrate, usually glass.

contains a strong component of twice the frequency of the drive. The feature differentiating all TFEL-able materials from usual phosphors is their high-field electronic transport mechanism. The electrons can attain high kinetic energies from the field almost as in a vacuum. Therefore, the correspondence to an all-solid-state field-emission-display is not far-fetched. In any one of those special phosphors electrons are emitted and accelerated, until they (impact) excite a luminescence center, which might well be the same as in the phosphors of thin or conventional CRTs. And there is no tip making, gate structuring, focusing grid preparation, vacuum baking and sealing, gettering, cathode poisoning, and envelope breaking. The expense is in the higher driver cost, and in (much) less overall efficiency. We argue that the P22 phosphors, which everyone sees every night on the TV screen, unfortunately are unable to do this job—the green and the blue because of field quenching of the luminescence, and the red, because the acceleration of the electrons is not good enough.

Much less has been published about TFEL than about optoelectronic III–V compounds, or LEDs, or even OLEDs, but it is completely impossible to be comprehensive. Therefore the reader is referred to good review articles and books, and, of course, to original papers as cited here or in the review articles (Smith, 1981; Mach and Mueller, 1982; King, 1992; Mach and Mueller, 1991, Mach, 1993; Mueller, 1993b; Tornqvist, 1996, 1997; Rack and Holloway, 1998; an excellent book written by Ono (1995) and Chapter 3 of this volume.

II. Phenomenology

A TFEL device consists of at least two thin films sandwiched between electrodes. It is driven by alternating voltages, in all practical devices by bipolar pulses. Normally three thin films, two dielectric films sandwiching the active or semiconductor or phosphor film make the device reliably working. Figure 4 shows a schematic drawing of the device structure— MISIM. Light is generated as soon as the pulse amplitude exceeds a threshold value. Figure 5 shows a typical brightness versus voltage $B(V)$ characteristic. Notice that the brightness or radiance is plotted on a logarithmic scale (Fig. 5 gives the data presented by Inoguchi *et al.* in 1974, including the first aging data published). The same data, replotted on a linear scale, would closely resemble the curve in Fig. 22 in Subsection 2 of Section III, after a rather curved portion, corresponding to the steep part of Fig. 5, a gently rising almost linear portion. These two parts clearly indicate the advantages of TFEL for matrix addressed displays: The very high

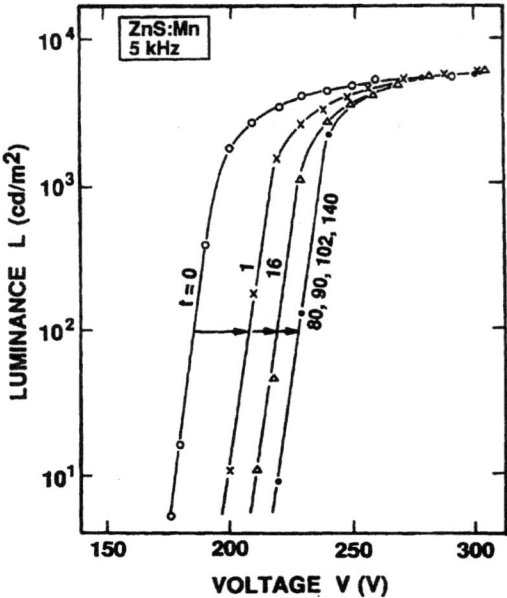

FIG. 5 Brightness versus (peak) voltage, showing the threshold behavior (in logarithmic scale), allowing for multiplex ratios of 1:1000 (after the historical paper of Inoguchi and Mito, 1974).

nonlinearity — four orders of magnitude change within 5 to 10 V amplitude change — allows for multiplex ratios of more than 1000. The gentle increase in brightness above this threshold allows for gray shades by amplitude modulation of an adding voltage beyond a selected voltage equal to the threshold value. Figure 5, however, also demonstrate the drawback of TFEL — the high voltage level, far beyond complementary metal-oxide semiconductor (CMOS) capabilities.

TFEL devices respond to voltage changes with light emission. Near to the leading edge of an applied pulse voltage U, conductive or dissipative current I, begins to flow, accompanied by the emission of light B, as shown in Fig. 6. Depending on the material properties, and especially on the type of luminescent center, the time dependence of the emission varies by orders of magnitude, very slowly decaying for Mn^{2+}, and fast, almost replicating the dissipative current, for Ce^{3+}. (Some special features near the trailing edge of the voltage pulse are discussed in Subsection 2 of Section III).

TFEL devices respond to changes in polarity: Applying two successive pulses of the same polarity produces at the leading edge of the second pulse

2 THIN FILM ELECTROLUMINESCENCE   33

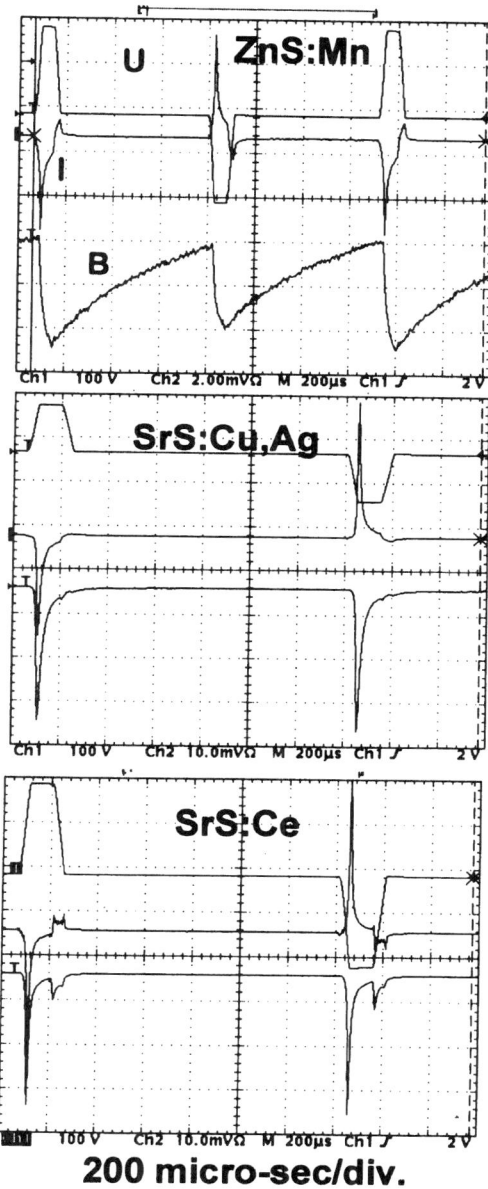

FIG. 6. Time dependence of (top to bottom) applied voltage U, dissipative current I, and brightness B, for the most important TFEL materials (top to bottom) ZnS:Mn, SrS:Cu,Ag, and SrS:Ce. The applied voltage is a bipolar pulse train.

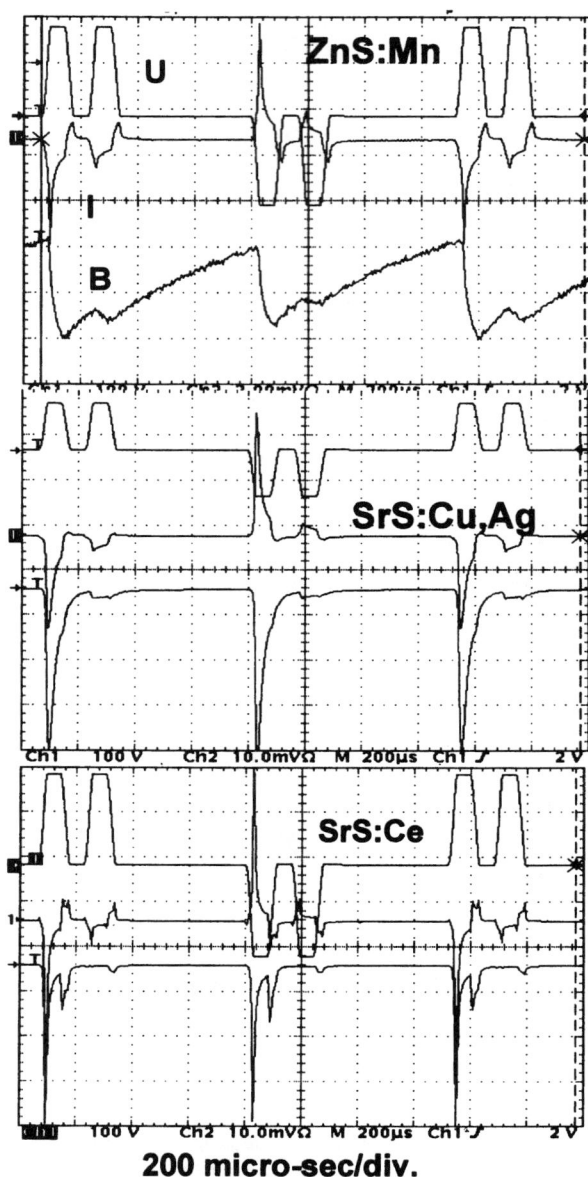

FIG. 7. Time dependence of (top to bottom) applied voltage U, dissipative current I, and brightness B, for the most important TFEL materials (top to bottom) ZnS:Mn, SrS:Cu,Ag, and SrS:Ce. The applied voltage is a train of biplor double pulses, demonstrating vanishing reponse in the second pulse of equal polarity.

only a small fraction of the light produced by the first one (Fig. 7). This memory of the previous type of polarity can last as long as hours, and in special cases, for days. Again there are quantitative differences between the materials, but no exceptions to this principle.

It is not necessary to apply pulses; sinusoidal voltages have been widely used to establish the phenomena and to select the right materials in the early days of TFEL research. In this case, the rise of the brightness is gentler, and the current persists over a longer time and does not reach comparable values (Fig. 8).

In summary, light emission occurs if a voltage is ramped beyond a threshold value, and only if previously no voltage of the same polarity and the same amplitude has been applied.

So far we covered the phenomenology of single pixels or test areas. As stressed in Section I, from the beginning the aim of TFEL was to make displays. Displays of some complexity are either matrix addressed or "active-matrix" driven. In the latter case, the active matrix supplies drive voltages to each pixel by being matrix addressed itself. Matrix addressing (Fig. 9) uses a sequence of phase-shifted (positive) voltage pulses applied to a system of rows to select (prepare) all pixels in one row at this pulse time for light emission. Light emission occurs if and only if at this time the respective column is driven negative. As all other rows stay unselected at this time, the column pulse does not cause light emission in any other row. This scheme is called *a-line-at-a-time* address mode. As TFEL acts only on polarity changes, a "reset" pulse of opposite polarity must be applied to all pixels at the end of a frame (a lit pixel lights again at this time). Alternative to the reset pulse scheme, every second frame must use reverted polarity for all actions (only one light pulse per frame).

To make a color (multi- or full-color) display, groups of pixels must emit different colors. Here we mention just the names for the techniques explained in Section IV; color-by-white is the most attractive method in TFEL, but patterned color, and even time-sequential color are possibilities that have been pursued.

As mentioned earlier, the steepness of the $B(V)$ response curve (Fig. 5) allows for very high multiplex ratios, which means that high numbers of lines (rows) can be sequentially selected and the brightness information written into its pixels without encountering crosstalk. Of course, there is electrical crosstalk in a matrix with appreciable capacities between each and every pixels. However, it does not translate into electro*optical* output because of this high nonlinearity.

There is purely optical crosstalk in TFEL devices. A very essential part of the TFEL phenomenology is light guiding. The light generated by the action of an applied voltage above threshold is only partly emitted through

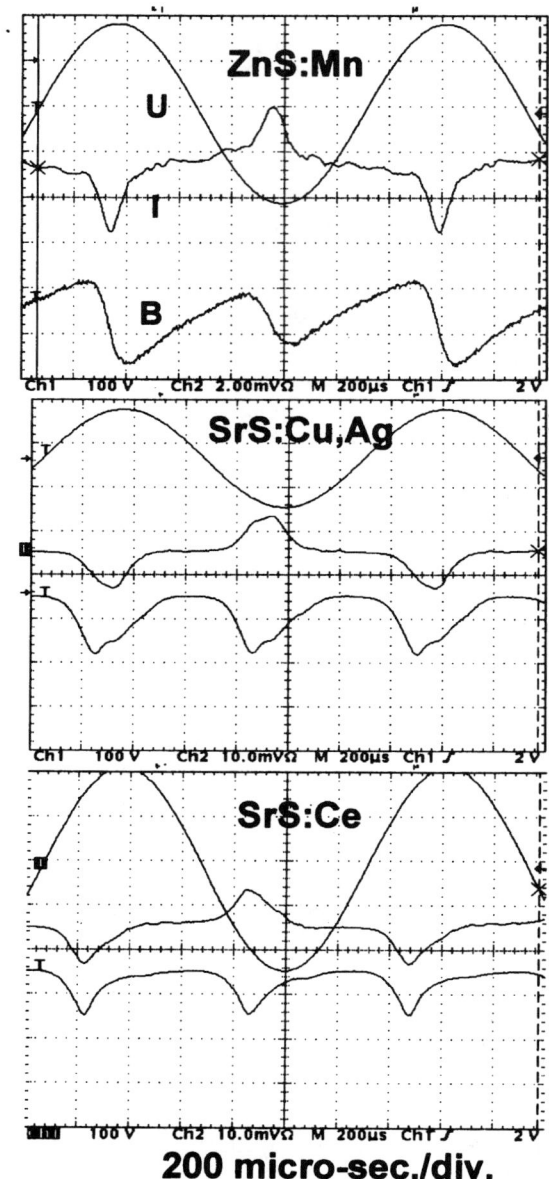

FIG. 8. Time dependence (top to bottom) of applied voltage U, dissipative current I, and brightness B, for the most important TFEL materials (top to bottom) ZnS:Mn, SrS:Cu,Ag, and SrS:Ce. The applied voltage is a sine wave. Notice the much broader current pulses, compared to pulse drive.

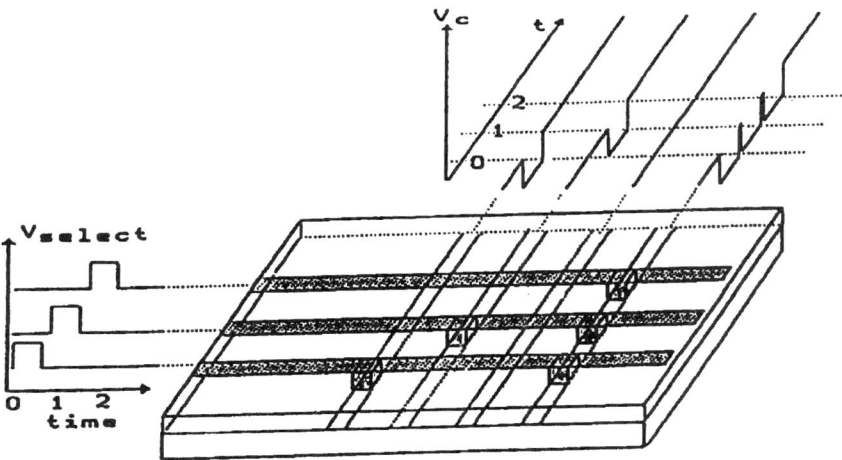

FIG. 9. Address scheme for passive matrix a-line-at-a-time TFELD device.

any of the transparent electrodes. The consequence of light guiding in the substrate plane is the brightly emitting edges of the glass substrate. Guided light can be traced by the (usually) small part of the light that is scattered out of the light guide and into air. Looking at a device with only 1 pixel lit, one observes light intensity decaying with increasing distance. This is best seen from the "backside," meaning the side on which there are nontransparent electrodes. A view of the backside of a display structure with only 1 pixel lit is shown in Fig. 10a. Every optical inhomogeneity (e.g., electrode stripe edges) acts as a scatterer and is clearly observable (as are mirror images from the glass surface).

As discussed in more detail in Subsection 3 of Section III, about 90% of the generated light is guided in the substrate plane. Only 10% is "used" by the viewer. However, the same concept of scattering-out that is usable for tracing can be used to make displays brighter. A little wash-out of edges accompanies the roughness. Figure 10b gives an example, although it is very visible only in the right part (log scale) of the figure.

In Fig. 11 2 pixels out of displays with rather different roughness are compared. The scattered light is followed along the Al electrodes. A "rounding" of the top and a tail on both sides are clearly visible. This can be used to quantify the scattering (Mach et al., 1988a). Roughness can be used to enhance brightness by up to 400% (Mach et al., 1990b) (see Subsection 3 of Section III for a more detailed discussion). Besides the smoothing of the otherwise crisp image a (unwanted) consequence of

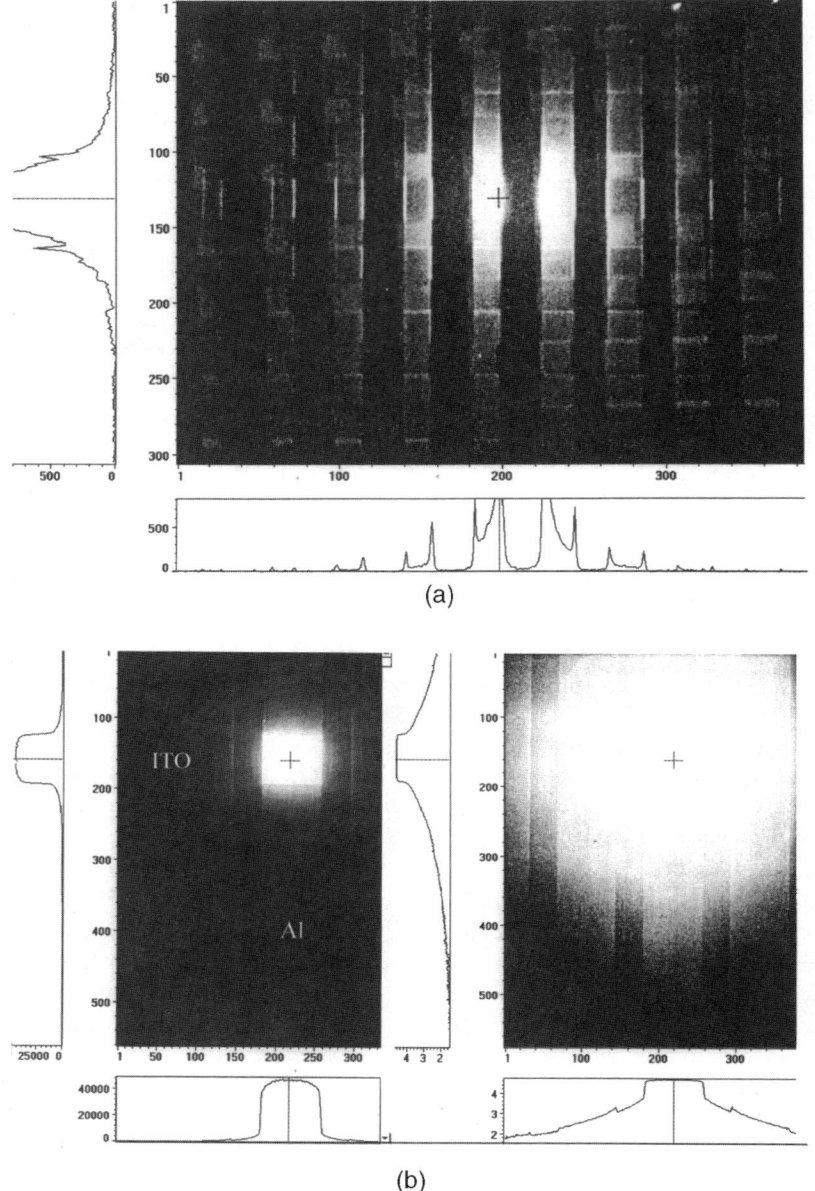

FIG. 10. Demonstrating the scattering of guided light by different roughness of the active film: (a) back view of a pixel in a smooth film of a matrix; (b) front view with high roughness, (left) linear brightness scale, and (right part) logarithmic scale.

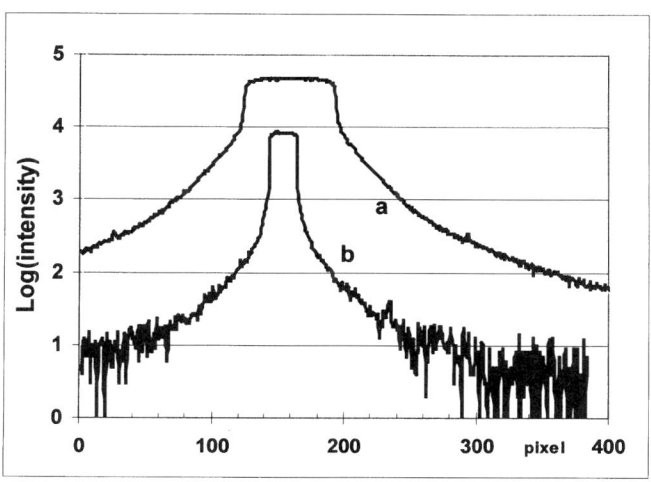

Fig. 11. Demonstrating the scattering of guided light by different roughnesses of the active film: front view of a pixel: curve a, low roughness and curve b, high roughness.

roughness is the increase in diffuse reflection, which reduces contrast. Therefore a very balanced use of scattering should be made, depending on the specific application.

Every thin film stack with thicknesses of the individual films being smaller or comparable to the wavelength of the light handled, gives rise to optical interference. Subsection 3 of Section III gives a more thorough account of the effects. The most visible is a slight color change with viewing angle. Much less pronounced than in LCDs and only rarely detectable with the naked eye, it can be willingly enhanced (microcavity effects, see Chapter 9) or suppressed by roughness. One of the most attractive features of TFEL is the viewing angle of almost 180°. In a rough sample, the light is distributed almost Lambertian, that is, its intensity decays as $\cos \Theta$ with increasing angle, Θ, against the normal (see Fig. 12a).

In TFEL *color* depends uniquely on the phosphor material, understood as the combination of host and dopant. For a given host, different dopants offer mostly dramatic alterations from blue to red, for instance. On the other hand, a given dopant might change its emission wavelength with the crystal field of the host. However, the variety of materials that have obtained (or are even likely to obtain) practical importance is very limited. As mentioned several times, the workhorse of TFEL—ZnS:Mn—emits in the yellow. Its spectrum—from a rough sample—is shown in Fig. 12a. A "smooth" sample exhibits angular dependence of the spectra (Fig. 12b), which,

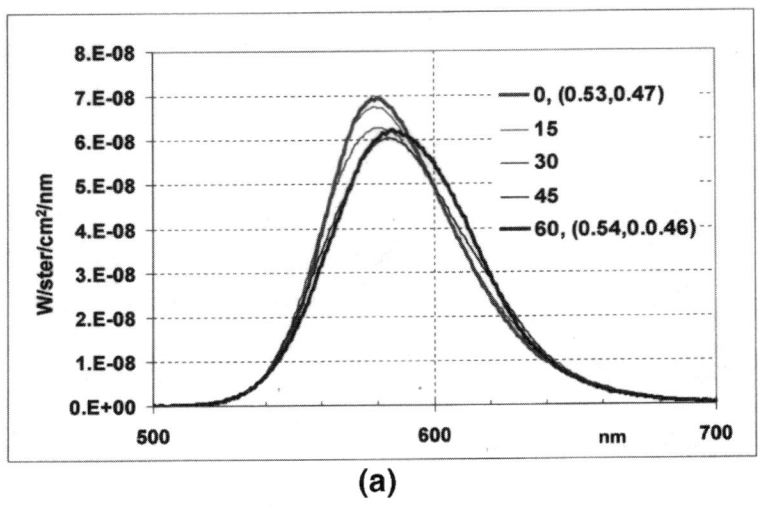
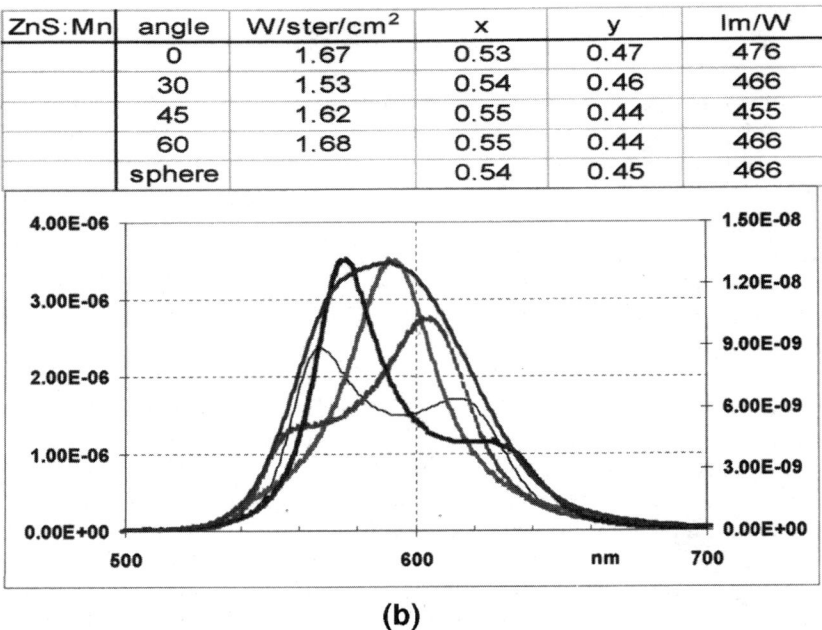

FIG. 12. Spectral distributions of the emission measured under different angles (given in degree in the legend) of (a) a rough ZnS:Mn thin film–thick film TFEL device of Westaim (sample by courtesy of Xing-Wei Wu from Westaim, Inc.) and of (b) a smooth ZnS:Mn sample (in-house); the distribution measured into a sphere is also given (right scale).

however, does not dramatically change the appearance, reflected by the color coordinates and the luminous equivalent, shown in the table at the top of the figure. Other dopants can be incorporated into the ZnS lattice (only in relatively limited concentrations, however). Figure 13 shows the spectra of the two most useful rare earth ions, RE^{3+}, which have $4f-4f$ dipole forbidden transitions; narrow lines in comparison to Mn^{2+}. Tb^{3+} is a good green emitter, which has found some limited military application, where interference with night goggles is a concern (Mikami *et al.*, 1987; Haerkoenen *et al.*, 1990). Ho^{3+} has been advocated as a rather good "white" emitter especially for AMEL use (Mueller-Mach *et al.*, 1997).

A long struggle to achieve blue emitters of good enough efficiency and longevity has involved mainly SrS, at first doped with Ce^{3+}, and more recently with Cu^+ or Cu/Ag^+ (Soininen, 1998; Sun, 1998; Sun *et al.*, 1997; Jones *et al.*, 1998; Park *et al.*, 1999). Figure 14 shows some spectral distributions. Clearly, Cu–Ag has the best color coordinates, without being perfect. The spectra are too broad to be used without filtering. The SrS:Cu–Ag spectra are influenced by the concentration of both dopants (see Figs. 38 and 39 in Subsection 4 of Section III).

Strontium-thiogallate, $SrGa_2S_4$ doped either with Ce^{3+} or Eu^{2+} has been investigated (Fig. 15), without finding really broad use (Bennalloul *et al.*, 1998; Barrow *et al.*, 1993) (see Section IV). The green spectrum is almost perfect from a saturation and color coordinate point of view. The blue one is too broad, and its lumen equivalent is rather low. Therefore, for a certain time calcium-thiogallate doped with Ce was under discussion (see Fig. 58 in Subsection 1 of Section IV), and was used in the dual-substrate exerciser of Planar, because of its better luminous equivalent. For a detailed review of the materials problems involved, see Chapter 3.

Summing up the phenomenological description of TFEL devices and technology, one can say that TFEL devices have excellent potential for high information content displays, because of virtually unlimited addressability, full color potential, extreme viewing angle and contrast. They are extremely thin, light weight, and as rugged, as any glass plate (or better). They are potentially cheap, except for their drive electronics.

III. Basic Mechanisms in Thin Film Electroluminescence

Most surprisingly, only very few materials are capable of generating efficient thin film electroluminescence. Discovered on ZnS:Mn, this material turns out to still be the most efficient. Besides ZnS, doped with rare-earth ions, SrS and CaS doped with rare earths, especially Ce^{3+} and Eu^{2+}, and

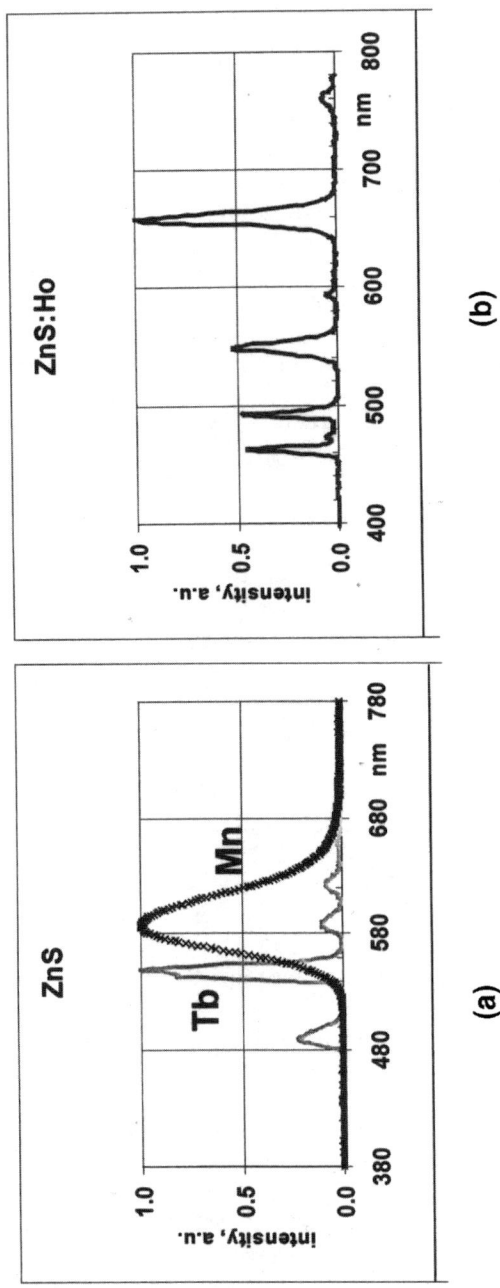

Fig. 13. Typical TFEL spectra of (a) ZnS:Mn and ZnS:Tb and (b) ZnS:Ho.

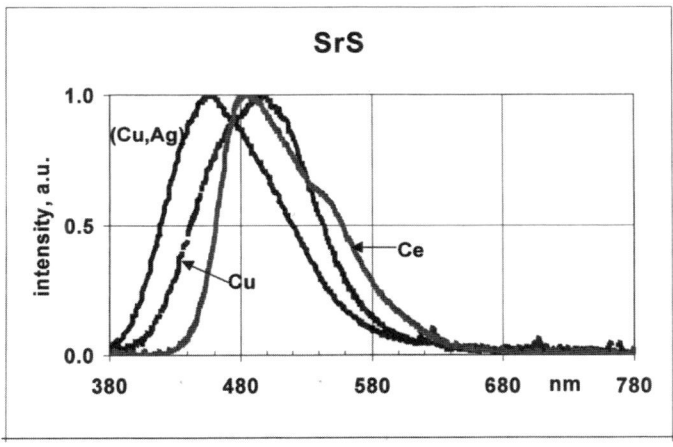

FIG. 14. EL spectra of SrS doped with Ce^{3+}, Cu^+, and Cu^+, Ag^+. (The latter two samples are by the courtesy of Sey-Shing Sun from Planar Systems, Inc.)

noble metal ions, constitute the group of "proven" TFEL materials. ZnSe has been shown to exhibit electron transport properties, which are very similar to those in sulfides (Mach et al., 1984), but it has no better dopant incorporation properties than ZnS. The interest in this material has, therefore, ceased. More recently, another group of sulfides, the thiogallates — MGa_2S_4, M = Sr, Ca, Mg, Ba — attained quite a bit of interest. Thus, one of the most basic notions seems to be that sulfides and selenides,

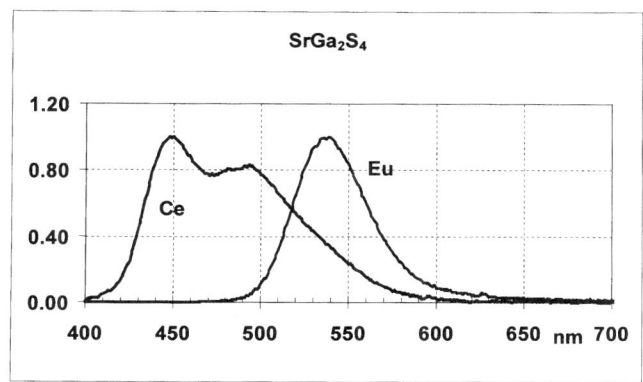

FIG. 15. Spectra of Sr-thiogallate, $SrGa_2S_4$, doped with Ce^{3+} and Eu^{2+} respectively.

and only those, are capable of doing well in TFEL. Many oxides have been tried over the years, and a renaissance of interest in oxides seems to be beginning because of the better dielectrics that are now available (Kitai, 1998; Kitai et al., 1997; Minami, 1998a, 1998b). It would be a highly prestigious task to determine the reasons for the predominance of the chalcogenides in some detail. Some speculations and brief arguments on this point are given later.

1. THE SIMPLE MODEL

General agreement exists on the "simple model" of the TFEL mechanism (Fig. 16). The figure depicts valence and conduction band edges versus a space coordinate perpendicular to the substrate plane. The dielectrics or insulators (I) are assumed to have higher gaps and lower dielectric constants than the phosphor or semiconductor (S) film (which is preferable). The depicted situation is that of a high electric field applied to the structure, which is electron accelerating toward the right side of the drawing. The basic assumption is that electrons are tunnel-injected into the S-film from states in the interface between dielectric and S-film. They are accelerated in the conduction band of the S-film and attain kinetic energies, which entitle them to impact excitation of dopants in the S-film.

The electric properties of TFEL structures at voltages above threshold can easily be interpreted as field ionization and avalanche breakdown. If this

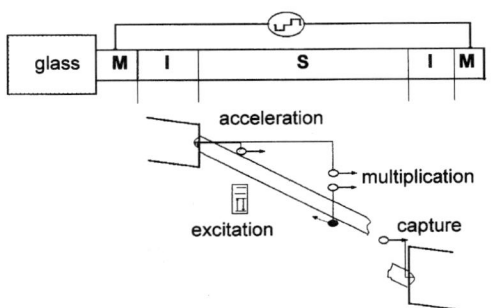

FIG. 16. The simple model of TFEL, showing a schematic of the dielectrics (*I*), and phosphor (*S*) films between electrodes (*M*). The lower part of the figure shows valence and conduction band edges in a high-field situation (with no space charge accumulated), and the main electron processes—acceleration, multiplication, excitation of dopants, and capture—are indicated.

field ionization and carrier multiplication are due to band-to-band excitation of electron–hole (e–h) pairs, we conclude that the kinetic energy of the impacting electrons is by far larger than the bandgap. How much larger depends on the band structure, which governs the balance of energy *and* momentum between the impacting and the generated electron, and the generated hole. From cathodoluminescence efficiency values of the respective materials one is used to think of about thrice the gap energy to be used in the generation of e–h pairs. Theoretical calculations of Duer *et al.* (1998), however, seem to indicate that a massive multiplication sets in even before the electrons attain 1.5 times the gap energy.

After having excited dopants, and generated new e–h pairs, the electrons and their descendants arrive at the interface of the S-film with the right (momentary anode) dielectric, and are trapped in the same kind of states from which they tunneled. The holes either drift to the left (cathode) side or are trapped in defect states of the S-film. Reversal of the voltage starts the process in the opposite direction of electronic travel. The excited dopants return to their ground state, hopefully radiatively.

2. ELECTRICAL PROPERTIES

Adopting this simple model has some immediate consequences: First, the threshold for charge transfer *and light generation* is set by the depth of the interface states, *if and only if* this threshold field E_t is higher than the field at which the electrons can attain appreciable energy from the field. We know that ohmic conduction is one in which energy input from the field is balanced by the loss to phonons. And here we deal with an energy increase to at least 2.5 eV, which is about the excitation energy of most of the relevant dopants. The second threshold field E_a, characterizes the field at which essential heating of charge carriers starts. Second, if the depth of the interface states in the two different interfaces is very different, so that the tunnel threshold is higher or lower than E_a, respectively, we expect a very different efficiency in the two polarities. If for the (momentary) cathode interface $E_t < E_a$, the efficiency will be low, as electrons are transferred (consuming energy) but not heated enough to excite. A much lower initial brightness in this polarity is the consequence. This very simple concept was experimentally confirmed as early as 1983 (Mueller and Mach, 1983; Mach and Mueller, 1986) (Fig. 17) by using UV excitation to generate electrons (e–h pairs) below the tunnel-threshold for electrons of the electroluminescent (EL) device in the dark. A similar technique to generate carriers below the tunnel threshold by UV excitation was later on used by several groups to study the motion of electrons and holes in high fields (e.g., Corlatan and

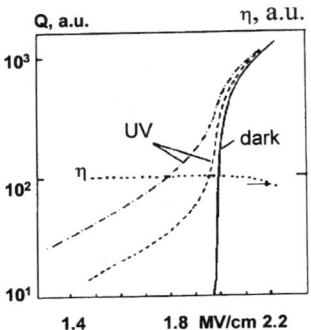

Fig. 17. Early experimental result (Mueller and Mach, 1983); Mach and Mueller, 1986), showing the modification of a typical transferred charge Q versus voltage (field strength) by incident UV light (two different intensities). Below the "acceleration" threshold the photoexcited carrier drift "ohmic" (dark) threshold is *not* set by availability of carriers.

Neyts, 1994). Later "asymmetric samples" (ITO–SiON–ZnS:X–Al), where the upper dielectric is missing, proved valuable to achieve carrier injection in one polarity at lower threshold fields than in the other polarity. Figures 18 and 19 show examples for different dopants. In both cases, the metal contact injects electrons at too low an electric field for efficient acceleration, leading to a lower efficiency or even different spectral distribution in this polarity (Mueller-Mach et al., 1996, 1999). Rather striking evidence is the vanishing blue emission of Tm^{3+}, when "early" (low-field) injection of electrons does not allow for high enough electron energy to excite more than the infrared level. The concept of the two thresholds was also successful in the explanation of degradation phenomena (Fig. 20) of ALE-deposited ZnS:Mn (using $ZnCl_2$ as a precursor) by several groups over the years (Mach et al., 1988b; Neyts and Viljanen, 1994; Kononets et al., 1996). The efficiency of these devices decreases with time, as in one polarity the electrons are injected at too low an electric field after some driving. This forced the manufacturers to look for different Zn precursors (Soininen et al., 1997).

Based on the mentioned "avalanching," the electric behavior can be modeled. The generally used equivalent circuit is shown in Fig. 21. The threshold of carrier multiplication is obviously mirrored by the Zener voltages of the back-to-back diodes, and the voltage drop on the S-film (C_s) remains constant as soon as the capacitively divided total voltage exceeds it — "field clamping." The current–voltage characteristic of the (ideal) Zener diode is a delta function at its threshold voltage; the current shoots up just at the Zener voltage. How close can a real TFEL structure get to this

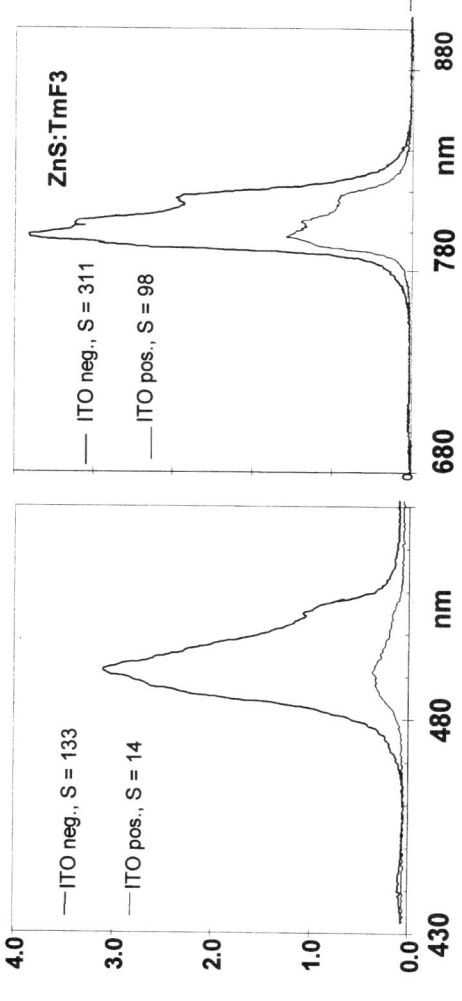

FIG. 18. An asymmetric sample of ZnS:TmF3 with cooler electron distributions for ITO$^+$, than for ITO$^-$, as shown by the smaller ratio of 485/790 nm emission: S is the light sum over the respective band.

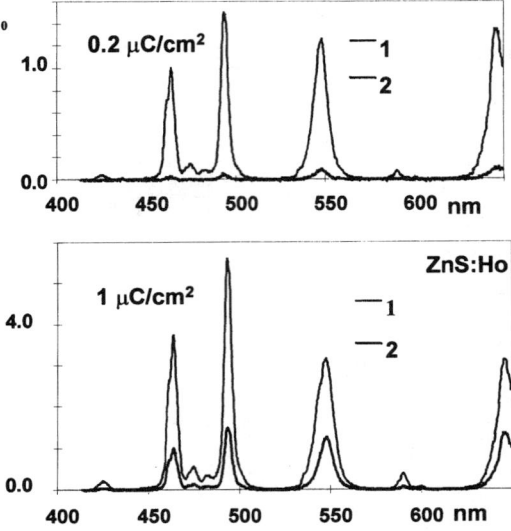

FIG. 19. Spectra of an asymmetric ZnS:Ho^{3+} sample, showing "early" (low-field) electron injection in polarity 2: in the upper graph at 0.2 μC/cm^2 the difference is more pronounced than at 1 μC/cm^2 (lower graph).

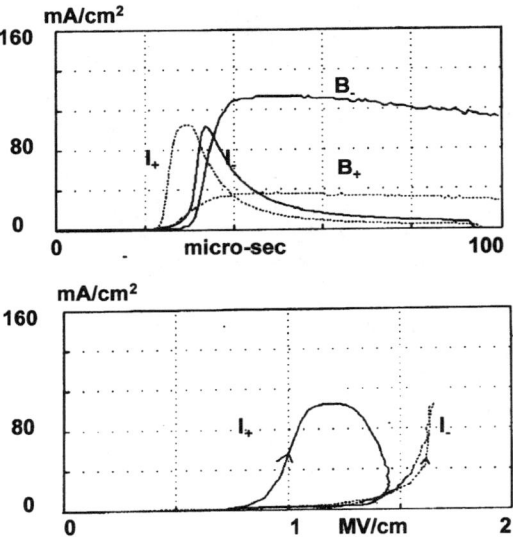

FIG. 20. Asymmetric ZnS:Mn sample (ALE) showing (a) "early" injection for AL$^+$, resulting in less brightness and (b) N-type current–voltage characteristic, beginning at 0.9 MV/cm, while the other polarity exhibits rather "normal" S-type behavior (after Mach et al., 1988b).

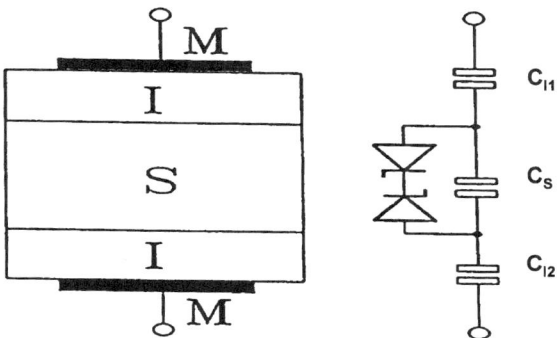

FIG. 21. The back-to-back Zener diode model and its correspondence in the film stack (left).

idealization? Of course, it is impossible to measure inside the structure (e.g., the voltage drop on the S-film); however, as we elaborate on in Section V, it is easy to measure the transferred charge. The transferred charge versus voltage $Q(V)$, curve of the equivalent circuit is just the straight line of Fig. 22. The dotted curve is a measured TFEL characteristic. Fitting can easily be done in the linear part, which is

$$Q = C_i^2/(C_i + C_s) * (V - V_{th})$$

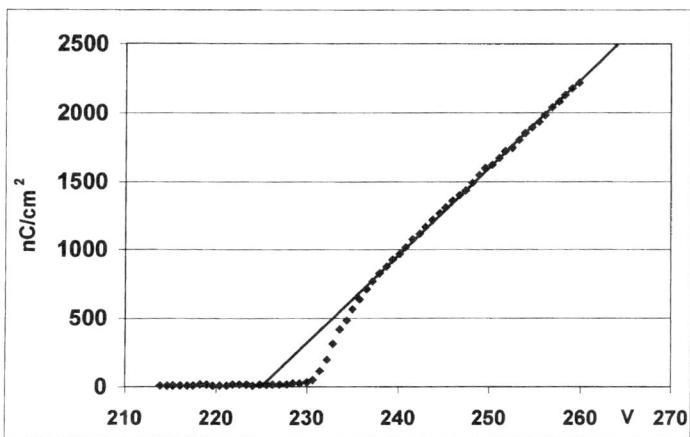

FIG. 22. Transferred charge versus voltage $Q(V)$ of a ZnS:Mn TFEL device, compared to a back-to-back Zener circuit for a two-parameter fit in the linear part (solid line).

yielding values for C_i and C_s, if C, the total capacity measured below threshold, is also put into the arithmetic. The deviation at the kink is what one expects from the soft onset of tunneling and/or multiplication at threshold. It is not really soft, 5 V (or a 3% increase of total field) is the range to adjust to breakdown behavior. It is this field range in which the brightness increases by four orders of magnitude (Fig. 5). (The fit is a neat destructionless determination of the thickness of the active film; Mueller *et al.*, 1988a; Mueller, 1992).

Having obtained by the fitting procedure described, the values of the capacities, one can try to calculate from measured values of applied voltage $U(t)$ and the dissipative current $j(t)$, the voltage on the S-film, U_s:

$$U_s(t) = \alpha * U(t) - 1/C_i * \int^t j(t') \, dt' + \text{const} = \alpha * U(t) - Q(t)/C_i + Q(T)/C_i/2$$

The integration runs from a time $t = 0$ before the leading edge of the pulse to a time t. The integration constant has been determined from continuity with the following opposite polarity half-period. The average field F_s is easily obtained by dividing through the thickness of the S-film ($d_s = C_s/$ known permittivity). To become independent of irrelevant device parameters as the area A, we divide the current j by it and speak about current density $I = j/A$ in amperes per square centimeter. Field clamping is shown in Fig. 23b, which gives $F_s(t)$ for a series of applied peak voltages V that correspond to transferred charges between 0.5 and 3.2 $\mu C/cm^2$ through the sample. Figure 23a shows the measured current densities $I(t)$ together with an example of the voltage pulses. It is evident that by the action of the polarization field $F_s(t = 0)$ the onset of current flow moves "down" the ramp of the leading edge of the voltage pulse.

Plotting $I(t)$ versus $F_s(t)$, we obtain results as shown in Fig. 24, which was taken on SrS:Cu$^+$, the most promising "new" material; this is shown together with the time dependencies from which it was derived. A rather good resemblance with the Zener diode is obvious. Further examples in Fig. 25 are on ZnS:Mn, SrS:Ce, and CaS:Eu (Mach and Mueller, 1991).

The examples of Figs. 24b and 25 relate to different phosphor materials. The same material, however, can show very different behaviors in dependence on the interface of the special sample. Let us call the examples of Figs. 24b and 25 "S-type." One different, extreme case was given for the positive polarity of the degraded ALE$^+$ sample in Fig. 20. Let's call it "N-type." Intuitively one is inclined to explain the S-type behavior as a consequence of space charge storage, which increases the cathode field, even when the average field decreases (and, of course, it is the average field, which we have

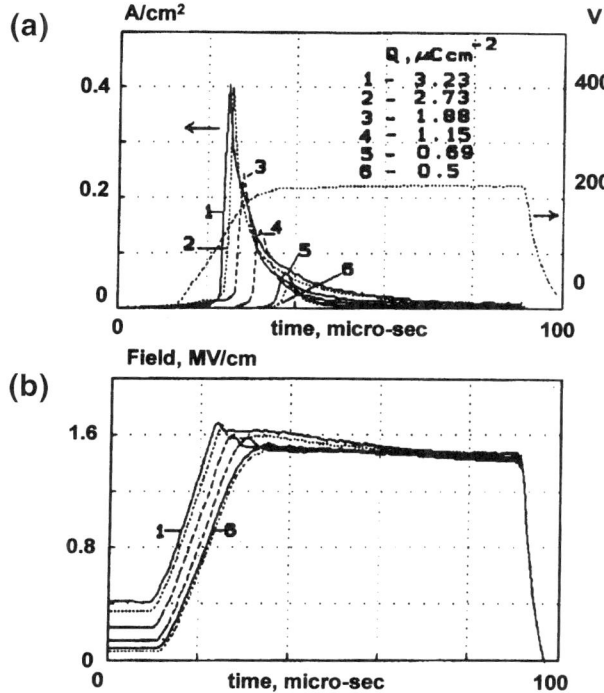

FIG. 23. (a) Time dependencies of applied voltage (only one shown), and dissipative current; the numbers at the current curves refer to the transferred charges given in the legend. (b) From the respective voltages and currents, the time dependence of the average field in the S-film has been calculated, nicely illustrating "field clamping" (after Mach and Mueller, 1991).

calculated). With the same kind of hand-waving arguments, one can ascribe N-type characteristics to interface state exhaustion; even an increasing field cannot sustain the current. This is no derivation or theoretical explanation, it is just heuristic, but did prove rather meaningful when manipulating the interface or analyzing the defect structure of the volume.

Strangely enough, this method of analyzing the electric "conduct" of TFEL structures has not achieved much popularity; most papers use the analysis of capacity versus time or voltage. It certainly contains the same information, and possibly because of the training in SPICE models, it got to some self-explaining intimation (e.g., Wager and Keir, 1997).

As read from the last few graphs, the current density appears high, of the order of 1 A/cm^2, but as the carrier velocity is of the order of the thermal velocity, the underlying electron concentration is fairly low. It is by no

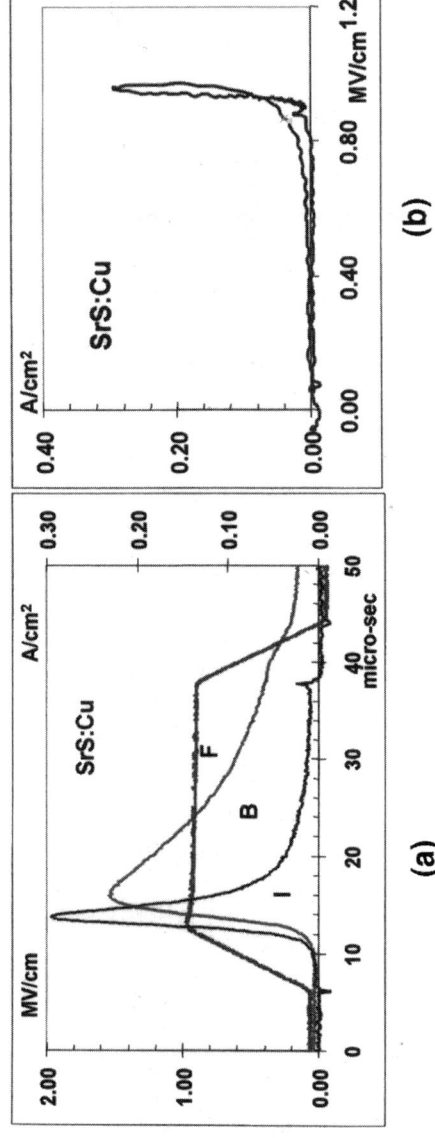

FIG. 24. SrS:Cu TFEL: (a) time dependence of voltage, current, brightness, and average field in active film and (b) current density versus average field (sample by courtesy of S.-S. Sun from Planar Systems, Inc.)

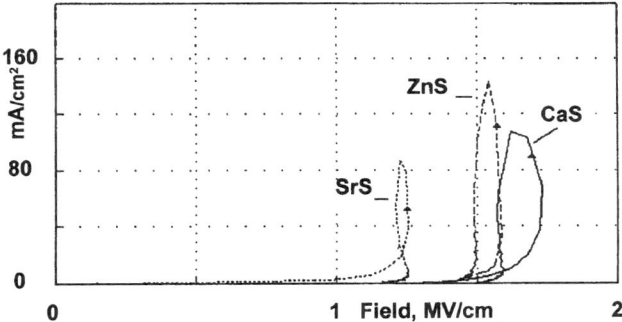

FIG. 25. Current density versus average field for samples of SrS:Ce^{3+}, CaS:Eu^{2+}, and ZnS:Mn (after Mach and Mueller, 1991).

means spatially constant, but will increase exponentially from cathode to anode as long as the field can be supposed to be constant. However, by the carrier multiplication itself space charge is developing, which in turn reduces the field and decreases multiplication. So, in principle a self-limited multiplication process is most likely. No heuristic classical model is available in the literature as far as we know. The parameters entering such a model would necessarily be the dependence of a multiplication probability — multiplication length — on the local field, the hole drift length before trapping, and a starting space–charge distribution, which reproduces after one full cycle in the stationary case. To obtain stationarity at all, some assumptions about electron–hole recombination must go into the model — space charge cannot be accumulated indefinitely. The simplest concept is the presumption, that electrons meet the trapped holes of the last half-cycle during current flow in the opposite polarity pulse (excluding trailing edge current at this time). In fact a so-called "blue flash" (actually it is a rather broadband, "white") was observed very early in ZnS:Mn (Smith, 1981; Douglas et al. 1993) and even in undoped ZnS. It is clearly visible in Fig. 26 but it has exactly the length of the (dissipative) current pulse. It is followed by the slowly decaying emission of the dopant. This timing makes it more likely to be due to intraconduction band transitions of kinetic electrons.

One "odd" electric feature, which has resulted in a wealth of literature, is "trailing-edge current." First detected was the accompanying trailing edge emission. Both are closely coupled, but current being the *sine qua non*. Figure 27 gives an example. Most likely the interface states at the (momentary) anode do not accept the electrons rapidly or "deeply" enough to immobilize them and to prevent their counterflow under the action of the

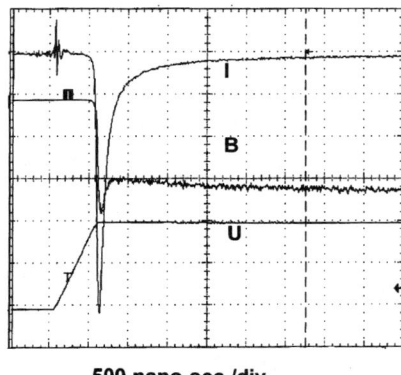

500 nano-sec./div.

FIG. 26. Digitizer screen images of (top to bottom) dissipative current (*I*), brightness (*B*), and applied voltage (*U*) showing the so-called "blue flash," which closely follows the current and is riding on the slowly varying Mn emission. Notice the time scale of 500 ns/div.

space–charge field. While prominent in SrS and CaS, trailing edge features have only been observed in ZnS:Mn samples, which were specially prepared.

Quite some discussion has been devoted in the literature to "space charge," differentiating between "static," "permanent" and "dynamic," one or the other being thought to relate to ZnS and SrS, respectively (Neyts, 1994). It is obvious from the previous considerations that space charge is

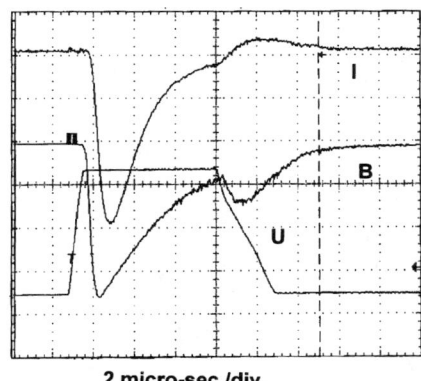

2 micro-sec./div.

FIG. 27. Trailing edge features of SrS:Ce^{3+}: time dependence of (top to bottom) dissipative current (*I*), brightness (*B*), and applied voltage (*U*), showing at the trailing edge of the voltage pulse a "countercurrent" and an additional brightness peak.

generated and destroyed in each half-cycle of the applied voltage. It is very unlikely, however, that a space-charge-free situation can be developed during normal operation. How large the amount of persisting space charge after "normal" switch-off of the applied voltage is, relative to the maximum amount of space charge during operation, will depend not only on the specifics of the sample but also on the "normal" switch-off procedure and possibly the leakage of the insulators. There is some similarity with the almost forgotten "hysteresis" or "memory" devices (Howard et al., 1982) based on heavily doped ZnS:Mn, which attracted much attention in the early 1980s. A probable distinction between ZnS and SrS (or, better to say, SrS:Ce, as no extensive studies on undoped SrS are known) is the different defect structure. A high concentration of donors in SrS:Ce^{3+} seem to produce a much more homogeneous space charge background than in usual ZnS:Mn, even during a first ramp-up of the voltage.

3. Optical Properties

Two essentials govern the optical behavior of a TEFL stack: an interference filter, and its active material with a refractive index >2.2, which causes severe light guiding. Chapter 4 covers in some detail the special effects related to film thicknesses comparable with the light wavelength on emission. So we take another viewpoint, adding some practical considerations to this discussion. The first part of this section is devoted to light guiding, the second to interference.

a. Light Guiding and Scattering

Figure 28 illustrates what portion of the light generated in the center of a cube has a chance to exit through the faces. Using the refractive index of ZnS:Mn, which is 2.3, this is about 10% of the light impinging on the respective face, or more precisely, the light that hits this surface under an angle of less than 25.8°. All other rays are totally reflected (and might still travel around in the cube, if they have not found another face to exit or have been absorbed finally). The escaping rays are redistributed into the half-space, 2π, by refraction. (Thus, almost nothing enters any secondary optics of limited acceptance cone). The considerations become really involved if the cube becomes a slab with tiny dimensions perpendicular to one large plane surface, which is the usual geometry even of a pixel (width at least one order of magnitude larger than thickness). The 10% of the light impinging on the (top) surface is also 10% of the light generated, if the back surface is

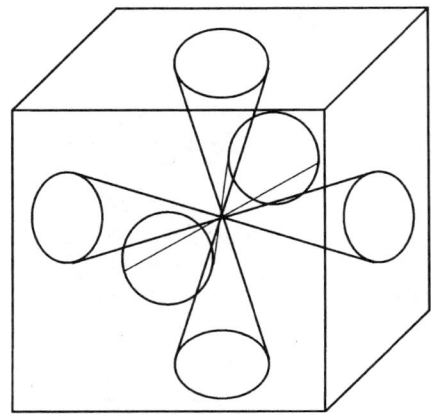

FIG. 28. Schematic of the escape cones of light from a cube of high-refractive-index material.

perfectly mirrored. Thus, 90% of the generated light is guided in the plane of the substrate. For a complex structure, such as the TFEL stack, this does not mean that all light is guided in the active film, but a percentage is guided in the dielectrics (depending on its index), and, for example, 15% (of the generated) travels in the glass (of index 1.53). This estimate neglects interference effects.

Noting for the following discussion of efficiency that the light extraction or light outcoupling efficiency is about 10% for a back-mirrored structure; 90% of the total is traveling in the substrate plane. In a careful analysis of different ZnS:Mn samples, it was shown that light extraction can be increased to about 40% by making the (any) surface rough (Mach et al., 1988a, 1990b). Adding to the facts on spectral distributions given in Section II, spectra of a "white" TFEL device made by Westaim in their proprietary thick film–thin film technology are given in Fig. 29. They were measured under the angles indicated, using a telespectroradiometer. Integrating the flux over the whole visible range gives the angular dependence of the emission intensity, as displayed in Fig. 30 (see Section V for the measuring technique used). A Lambertian distribution would show up as a straight line in this graph. The rather complete lack of interference effects in the spectra and the angular distribution are a strong hint toward massively enhanced light extraction, caused by the rough ceramic substrate.

A group of people from Westinghouse had the great idea to make use of the light guided in the substrate plane. Kun et al. (1986) invented the "edge

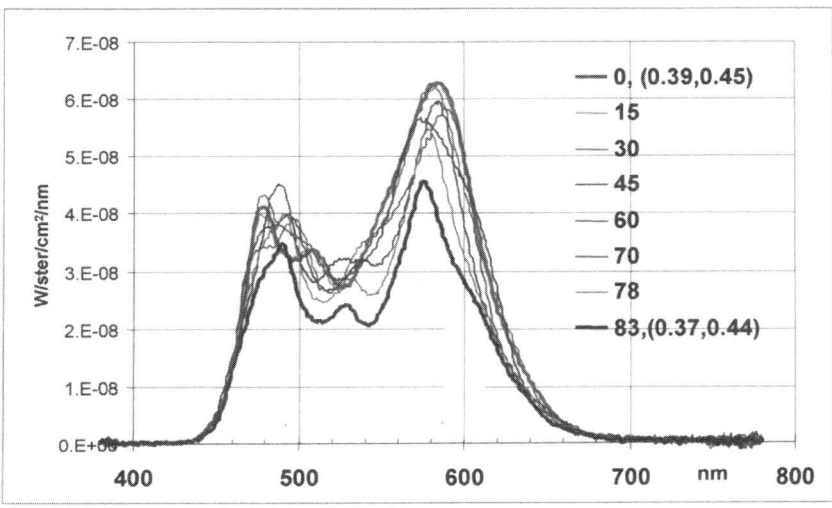

FIG. 29. Spectra of a "white" TFEL stack on a rough ceramic substrate (sample by courtesy of Xing-Wei Wu from Westaim, Inc.) under different angles, showing almost perfect Lambertian behavior of the spectrally integrated intensity (see Fig. 30).

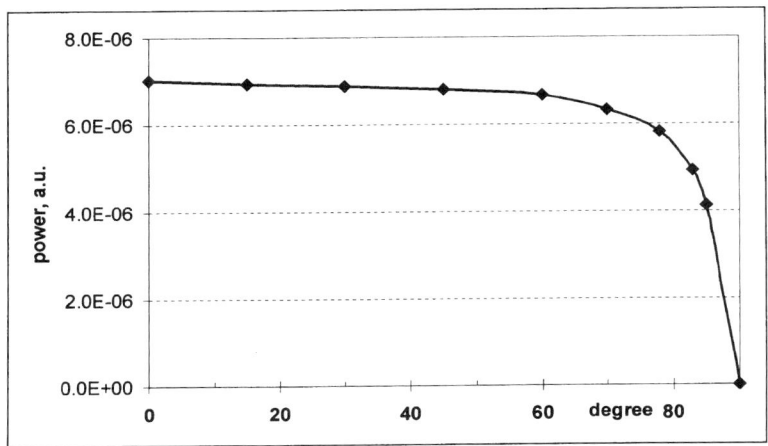

FIG. 30. Integrals over the spectra shown in 29 versus viewing angle; a Lambertian characteristic would be a constant in this mode (see Section V).

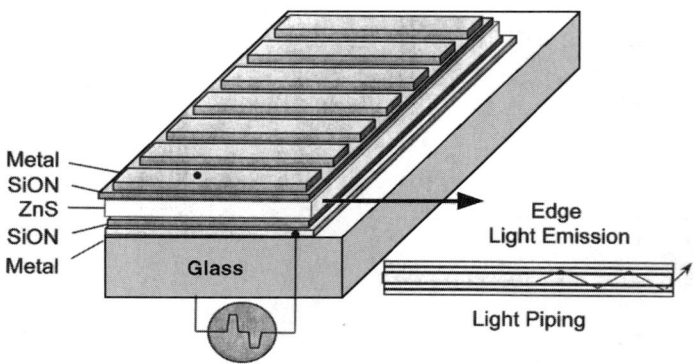

FIG. 31. Edge-emitting TFEL structure for use as exposure device for electrophotographic printers. The device (after Kun et al. 1986) uses the light guided in the film plane.

emitter." As discussed in more detail under devices, the basic idea was to sum over all the guided light under an elongated pixel that travels in this direction, and cause it to exit at an edge perpendicular to the long dimension (Fig. 31). The summing, of course, must be considered with some care: light generated far from the emitting edge sums logarithmically, as the intensity law for guided light is $1/r$, where r is the distance traveled. Nearer to the edge, the exit cone must be taken into account, and nearer means less than about 5 pixel widths from the edge. This discussion assumes that only the upper electrodes are structured to form the long pixels, and no separation of the pixels by cutting through the active film — trenching — is attempted. Thus, light guiding occurs only in the plane (spreading it into 2π) not along the pixel. To improve the situation, trenching has been undertaken, but etching develops the polycrystalline structure of the active film (Fig. 32) and leads to high scattering losses. A better way to achieve one-dimensional guiding is the separation of light production (in an intact film) and guiding, which can occur in a "dome" structure above each pixel (Figs. 33a and 33b). Making these light-piping domes from a material, which is absorption-free, can allow for much longer summing pixels, than any other concept (Mueller, 1996).

b. Interference Effects

If scattering does not account for an appreciable part of the intensity, interference effects will be clearly visible in any spectrum taken into a limited angular field. Vlasenko et al. (1970) were the first not only to mention this influence but to explain and try to exploit it. Examples for this difficulty in

FIG. 32. The polycrystalline nature of the ZnS film renders the trenching of the edge emitter pixels useless as scattering from the rough surfaces introduces losses that are too high.

measuring on TFELD are discussed in Section V. In principle, all measurements from which an efficiency is to be derived should be taken using an integrating sphere, or summing over a closely spaced net of measurements taken with a gonioradiometer. This is very necessary if the aim is a material optimization. Figure 34 shows simulations of interference effects. An energy-constant spectrum is assumed, and the percentage of the light that would be radiated without interference effects into 4π, emitted into a forward direction, is plotted versus wavelength. If the aim is to optimize forward radiance or luminance; of course, just this quantity should be tested. Enhancing interference effects by the use of microcavities can result in patterns that are very difficult to realize otherwise. Figure 35 shows the results of applying a dielectric mirror of some complexity to a TFEL device utilizing the 800-nm emission of ZnS:Tm^{3+} and emitting this sharp line almost totally into a hollow cone of 50° opening and halfwidth of 8° (Mueller *et al.*, 1996, 1997).

Interference effects can be used to achieve high contrast by enhancing the absorption of a very thin metal film in a dielectric stack (Dobrowolski *et al.*, 1992). Luxell demonstrated a 3:1 contrast ratio in direct sunlight based on this concept. The claimed advantage is that the absorption is only enhanced in the visible (or any other wanted spectral region), so that the IR and UV are still reflected, improving the thermal balance.

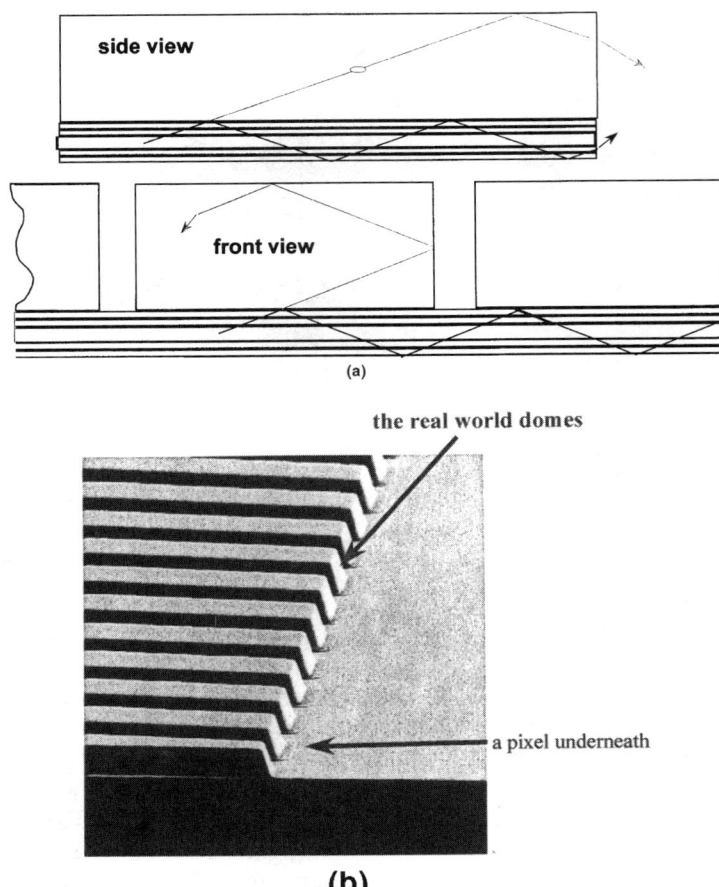

FIG. 33. (a) Light guides made from smooth surface glassy material (domes) on top of the pixels enables one-dimensional light guiding and summing over the entire pixel length. (b) scanning electron microscopy (SEM) micrograph of a (piece of) a 300 dpi "domed" printbar.

4. Luminescence Properties

Luminescence centers used in TFEL devices (TFELDs) were introduced in the "simple model" as levels lying outside the band structure of the active material. Most of the luminescence processes discussed in the other chapters of this volume are recombination processes between free electrons and free holes, eventually trapped into some shallow defects (donor–acceptor pair recombination). TFEL makes use primarily of innershell transitions of

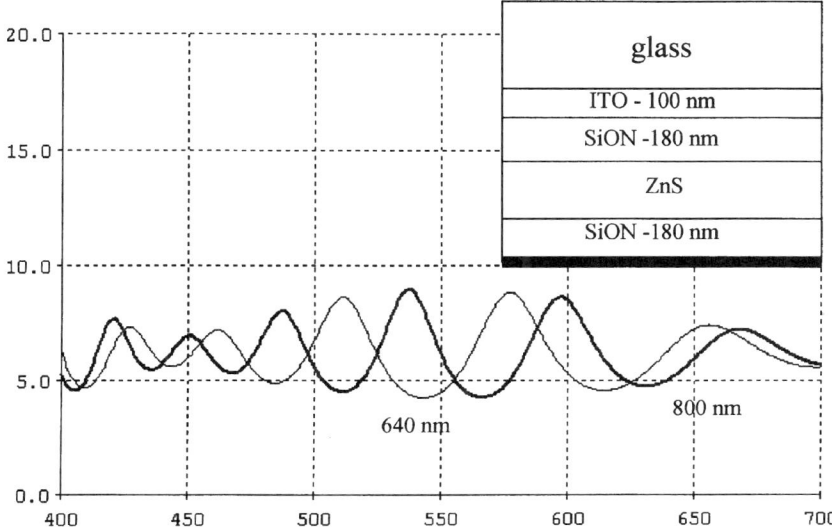

FIG. 34. Interference effects modulating the emission of an assumed perfect white spectrum: viewing angle, 0; back electrode, Al, 180-nm SiON, 800- or 640-nm ZnS, 180-nm SiON, 100-nm ITO, 7059 glass.

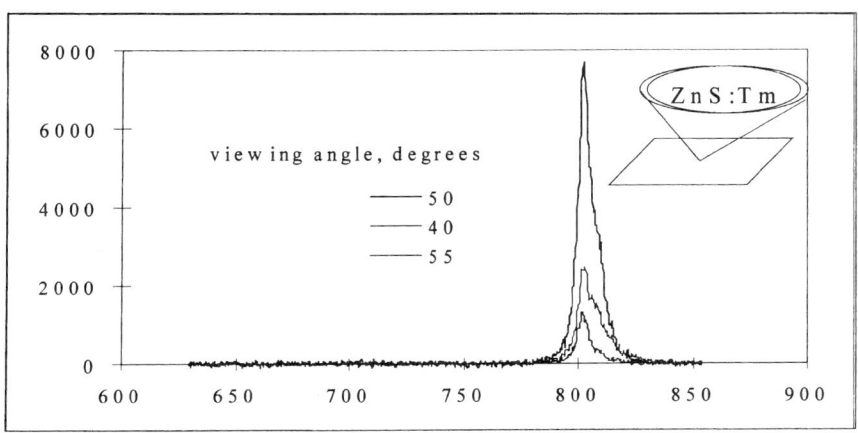

FIG. 35. Hollow cone of light emission from a ZnS:Tm microcavity EL (MiCEL) device.

impact-excited electrons. These electrons are not supposed to leave their atomic site, and therefore, for instance, do not contribute to the current. The most used and best understood material for TFELDs, ZnS:Mn, is an excellent example for the kind of luminescence transitions encountered. Mn^{2+} has five electrons in its $3d$ shell, the two s electrons are used in establishing bonds, which are of the same nature as the Zn^{2+}-compound-forming ones. A coulomb interaction — impact — of any of the very kinetic conduction electrons, moving in the electric field of about 1.5 MV/cm, excites one of the $3d$ electrons into an excited state *within the same ion*. The nature of the excited state is well characterized mainly from photoluminescence experiments (Busse *et al.*, 1976, 1979). As long as the main quantum number does not change, the transition is parity forbidden, and hence the excited state is long-lived (1.5 ms). A level scheme of Mn in ZnS and Cu in SrS (Payne *et al.*, 1984) is shown in Fig. 36. Phonon interaction accounts for most of the width of the emission band of ZnS:Mn, which can be described well by Huang–Rhys theory.

Very early in the TFEL history the question surfaced as to why not to dope the ZnS with acceptors, so successful in cathodoluminescence of CRTs (P22 phosphors), as for example, Ag^- or Cu^-. Some attempts to do so ended in completely unsuccessful TFEL devices (but were partly successful in powder EL devices, where the fields are not as high). The explanation is field quenching. The high electric field (> 1 MV/cm) ionizes the holes from the acceptors, as shown by Gumlich (1981). The power of field quenching has even been demonstrated on deeply bound excitons in ZnS:Te. Te forming an isoelectronic defect in ZnS, can rather tightly bind excitons. With

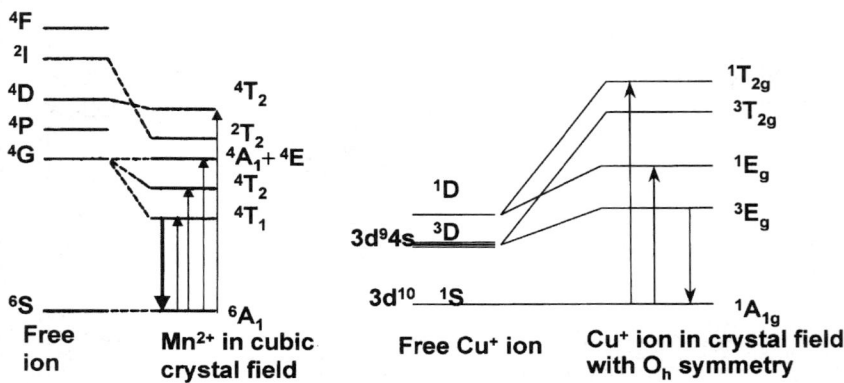

FIG. 36. Energy level schemes of Mn^{2+} in ZnS (after Busse *et al.*, 1976) and Cu^+ in SrS (after Payne *et al.*, 1984) in the crystal fields given.

increasing Te concentration, the binding occurs to pairs, triples, and even quadruples of Te. Binding energies as high as 0.7 eV have been reported (Iseler and Strauss, 1970; Heimbrodt and Goede, 1986). If one attempts to use this material in TFEL not only no EL emission is observed, but even photoluminescence is heavily quenched by fields well below the charge transfer threshold (Fig. 37) (Mach et al., 1991).

Thus, only inner shell transitions of electrons, well shielded against external electric fields, can be used in TFEL. This limits the selection to $3d$ transition metals such as Mn and noble metals such as Cu and Ag with unfilled $4d$ shells. The latter came into the game rather late, but from the PL-excitation spectra (Figs. 38 and 39) as reported by Park et al. (1999) and Jones et al. (1998), it is now rather clear that their role is completely different from their role in the P22 phosphors. The counterplay of Ag,Cu in EL is not completely understood at the moment and neither is the temperature quenching of these emissions. The alternative to d-metals are rare earths (REs), mainly as RE^{3+}, which except for Ce^{3+}, show only $4f-4f$ transitions (Kobayashi et al., 1985), which are parity (dipole) forbidden, and long-lived (Fig. 40; Mueller-Mach et al. 1999). Some can be doubly charged, for example, Eu^{2+}, which has Ce^{3+} dipole-allowed $5d-4f$ transitions. The dopants showing dipole-allowed transitions are crystal field sensitive, and so their emission shifts to different energies in dependence on the host (Fig. 41) (Kobayashi et al., 1985; Eichenauer et al., 1996; Yamashita, 1987).

The dipole-allowed $CaS:Eu^{2+}$ shows a rather interesting red emission, as does $SrS:Ce^{3+}$ in the blue. Field quenching has been suspected for both materials (Ando, 1992; Tanaka et al., 1989). This seems to be in contrast to what has been stated about inner shell transitions. However, in the case of SrS and the two dipole-allowed REs, the ions appear to have relatively

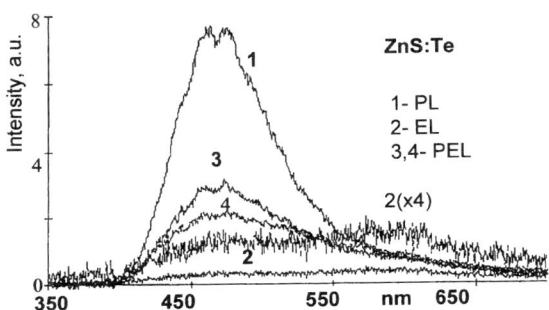

FIG. 37. Field quenching of photoluminescence and electroluminescence in $ZnS:Te^{2-}$ (after Mach et al., 1991).

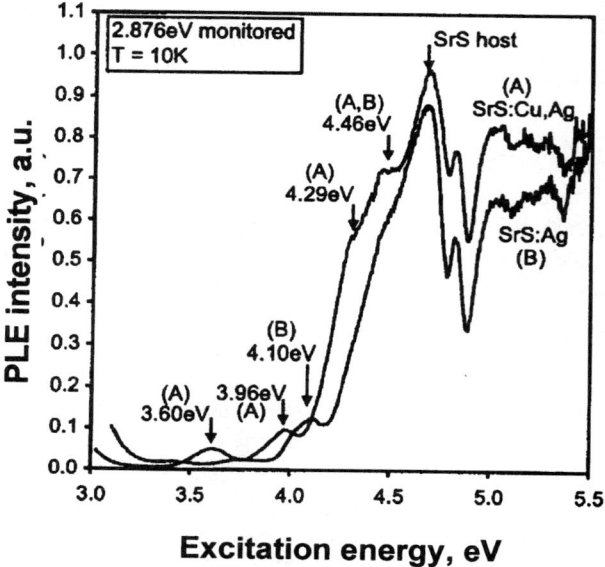

FIG. 38. PL excitation spectrum of the 431 nm emission of (A) SrS:Cu,Ag and (B) SrS:Cu at 10°K. (after Park et al., 1999).

strong interaction with extended (band) states. While the relative position of the Mn^{2+} ground state to the bands of ZnS is rather uncertain, and numerous trials to see any excitation into band states failed, Ce^{3+} and Eu^{2+} in SrS or CaS behave completely different. Ando (1992) proposed to explain his field quenching experiments by the assumption of field ionization of excited electrons into the conduction band of the host. Additional experiments made it likely that really the excited (5d) states of both ions are located within some tenths of an electron volt below the conduction band edge. A crystallinity dependent conduction band tailing, reducing the effective separation of the excited state of the dopant from the band explained a strong variation of the strength of the field quenching (FQ) from sample to sample (see also Mach et al., 1990a).

An interesting experiment has shown that the FQ is directly related to current and transferred charge. Applying a voltage to the TFEL sample and shining a very short laser light pulse on it, the sample responds with strong photoluminescence (PL) if the wavelength is tuned to an absorption band of the dopant. If the light pulse is short enough and the excitation is in the lowest absorption band, the decay curve of the PL reveals the internal quantum efficiency (QE) of the excited transition (see Section V). Figure 42

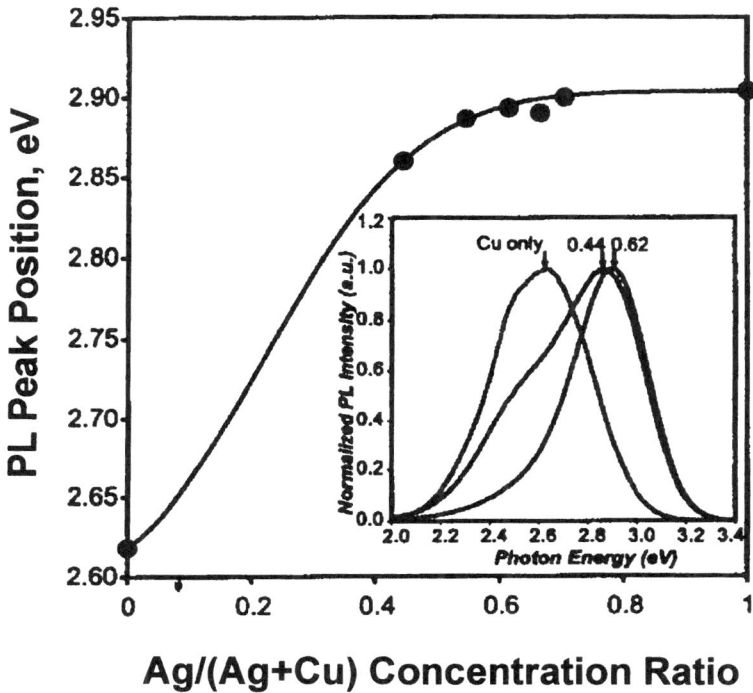

FIG. 39. Shift of peak emission with increasing Ag concentration in SrS:Cu,Ag samples; inset shows some spectra to illustrate the gradual decrease of the Cu peak around 475 nm with increasing Ag concentration (after Park et al., 1999).

shows results on CaS:Eu. Besides the decrease of QE with increasing voltage, strong photocurrents are observed, which correlate well with the decrease of the light sum. While the peak photocurrent is difficult to measure (mismatch of the sample impedance and the cable), the integrated photocurrent, the photocharge Q_p is easy to determine. Both quantities are plotted together with the quantum efficiency η_{lum} in Fig. 43 versus applied voltage for a moderate CaS:Eu sample and a good SrS:Ce sample (Mach and Mueller, 1993; Mueller, 1993a, 1994).

Some of the field-quenched dopants will remain ionized until electrons become available, which then are trapped and decay radiatively, emitting the typical dopant spectrum. This notion is consistent with the observation that the spectra in the so-called trailing edge emission are really the typical spectra. However, there are quantitative discrepancies: more emission is recorded than the deficit due to FQ can account for (Mueller, 1993a, 1993d, 1994).

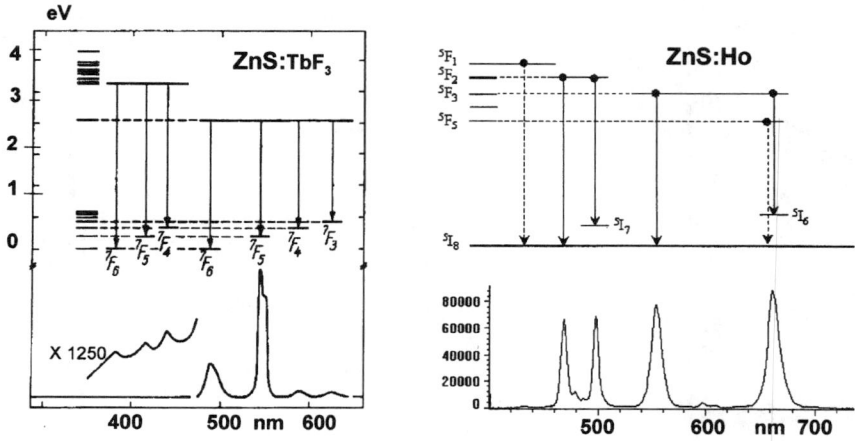

Fig. 40. Energy level schemes and related spectra for ZnS:TbF$_3$ (after Kobayashi et al., 1985) and ZnS:Ho^{3+} (after Mueller-Mach et al., 1999; this level schema for Ho is controversial at the moment).

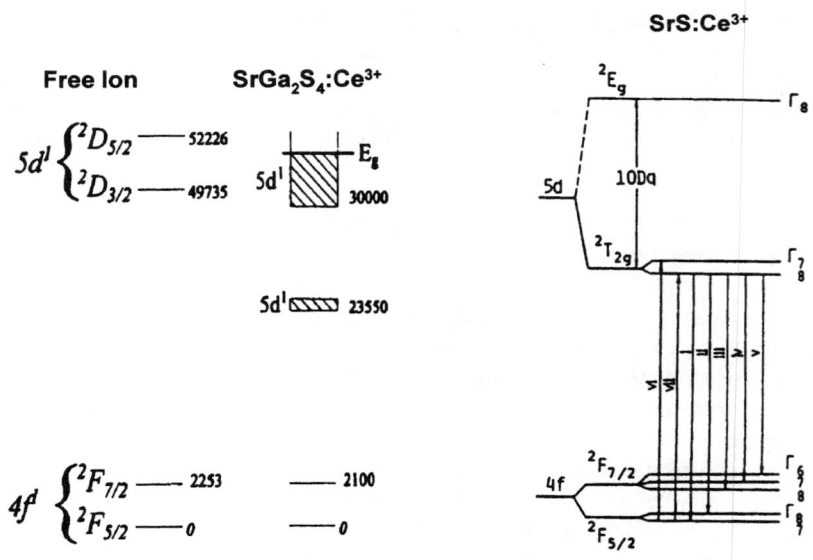

Fig. 41. Energy level schemes of Ce^{3+} in SrGa2S$_4$ (after Eichenauer et al., 1996) and SrS (after Yamashita and Mishitsuji, 1987), respectively.

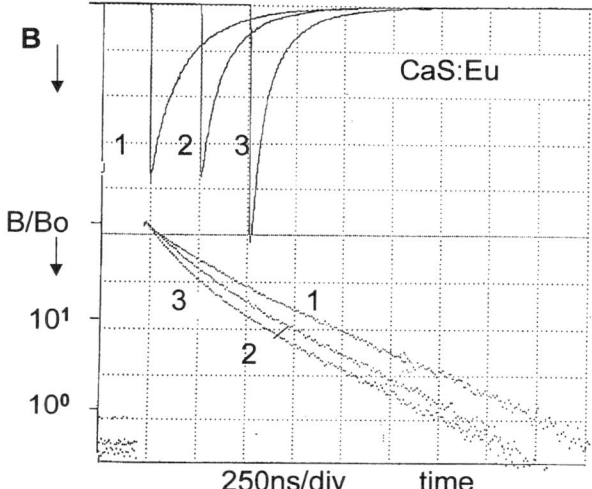

FIG. 42. Field quenching of the Eu^{2+} emission in a CaS TFEL structure: brightness versus time, excited by a laser pulse of 450 nm, upper part on a linear scale, shifted arbitrarily on time axis (increasing downward), lower part same data on a logarithmic scale, normalized to initial brightness B_0: curve 1, no field; curve 2, well below EL threshold; and curve 3, slightly above threshold (after Mach and Mueller, 1993).

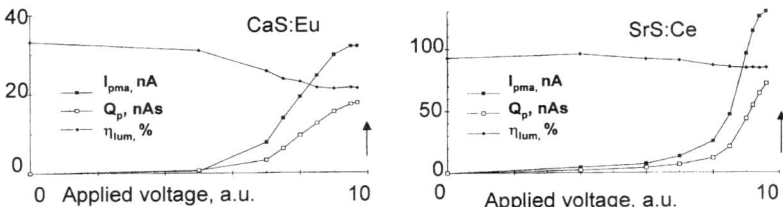

FIG. 43. Field quenching and ionization in TFEL samples of (a) $CaS:Eu^{2+}$ and (b) $SrS:Ce^{3+}$; uppermost curve, quantum efficiency in percent; peak current, I_{pma}; transferred charge, Q_p in nanoamperes per second, all versus applied voltage, given in arbitrary units. The steep rise in Q_p indicates the EL threshold; saturation takes place in the region of field clamping, indicated by the arrow.

5. Efficiency

One of the crucial parameters in almost every application is efficiency. Even in nonbattery operated devices, the efficiency must be high enough to prevent excessive heat production. Besides this, from a modeling point of view, the proof of validity is in the derivation of good numerical estimates and of the dependencies of the efficiency values on material and device parameters.

In a complex physical system, the trial of a factorization of the efficiency into probabilities of independent processes is often a good idea. The *simple model* (Subsection 1 of Section III) lends itself to this procedure. Starting with light extraction or optical outcoupling of the generated photons, we have already noted that this is a very unlikely process. Without active help, only 10% of the photons make it through a smooth plane-parallel stack geometry. We note $\eta_{out} = 0.1$.

Of the excited dopants, only a fraction return radiatively to the ground state; this fraction we designate with η_{lum}. This strongly depends on the dopant concentration because of concentration quenching.

Only a fraction of the electrons energetic enough to excite a dopant, will do so, some will miss the target or excite other species, for example, e–h pairs, by impact on valence band electrons. The fraction that will excite the right dopants are called η_{exc}. Perhaps not all the electrons attain enough energy to excite the preferred dopants; the fraction hot enough are called η_{hot}.

Thus, we end up with a product of four (Mach and Mueller, 1984b)

$$\eta_q = \eta_{hot} * \eta_{exc} * \eta_{lum} * \eta_{out}$$

Let us try to break down this product a bit further. The excitation probability of centers or dopants by any one of the electrons will be proportional to the concentration of the centers, N (cm^{-3}). It is usual and reasonable to introduce an impact cross section σ (cm^{-2}). Here $1/(N*\sigma)$ is the impact length for excitation, and the total drift length d_s, divided by the impact length gives the impact probability. Thus, $\eta_{exc} = d_s * N * \sigma$. We can measure the efficiency η, for instance, in watts per watt. This reminds us of the dimension of η_q, which we were talking about until now: it was something like a quantum efficiency, $\eta_{q'}$ = the fraction of electrons transferred, which can excite, which really excite, which radiatively deexcite, and finally come out as photons (it can be > 1). Thus, to convert into power units, we must multiply by the quantum energy $h\nu$, and divide this by the energy input, which is $e * E_{th} * d_s$. And we get

$$\eta = \eta_q * h\nu/(e * E_{th} * d_s) = h\nu/(e * E_{th} * d_s) * \eta_{hot} * \sigma * N * d_s * \eta_{lum} * \eta_{out}$$

The general behavior as a function of dopant concentration can be derived from this formula. Expecting that the luminescence efficiency is constant at low dopant concentrations N, and all other parameters do not depend on N, we get the initially straight line increase with N (Fig. 44). At a higher concentration, the so-called concentration quenching sets in and, if η_{lum} decreases more strongly than linear with N, will bend the curve downward. Hurd and King (1979) and later Benalloul et al. (1990) and others undertook some systematic studies of this dependence, $\eta_{lum}(N)$ and really found this behavior for ZnS:Mn. Based on the same methodology, and including a variety of different deposition methods, Mach et al. (1990b) could compare directly to this figure (Fig. 45) and gave a general account of the efficiency of ZnS:Mn: experimental values for the efficiency, η, η_{lum}, N [by secondary ion mass spectrometry (SIMS)], and E_{th} and the selection of *smooth surfaced samples* ($\eta_{out} = 0.1$), permitted a quantitative determination of

$$\sigma * \eta_{hot} = 4 \times 10^{-16} \, cm^2$$

At the same time optimum doping conditions were determined to be $N = 2 \times 10^{20} \, cm^{-3}$, with a luminescence efficiency $\eta_{lum} = 0.4$.

This is as far as one comes without any reasonable doubt, and error bars, which are less than $\pm 20\%$. (Above a concentration at which the impact length becomes comparable with the free-fall acceleration length, the kink

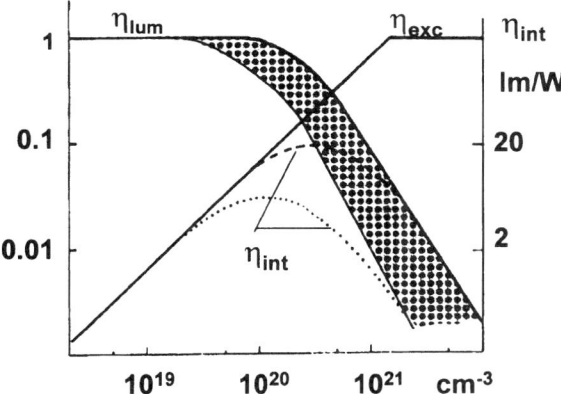

FIG. 44. Factorization of the efficiency of TFEL. Schematic showing on the left scale the luminescence efficiency η_{lum} decreasing from 1 at the concentration quenching limit; the excitation efficiency η_{exc} increasing linearly with the dopant concentration; the scatter region between η_{int} and η is caused by the outcoupling efficiency, which depends on the morphology of the films.

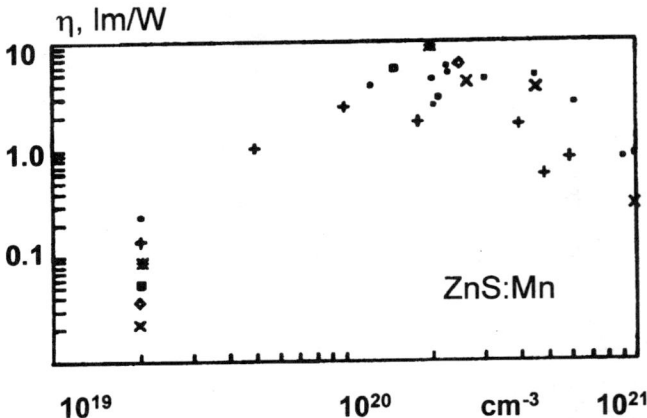

FIG. 45. Experimental values of the efficiency of ZnS:Mn TFEL samples versus their Mn concentration for comparison with Fig. 44 (after Mach et al., 1990b).

in η_{exc} of Fig. 44, a shift of the threshold voltage occurs).

The breakdown of the product $\sigma * \eta_{hot}$ into its factors has raised quite some debate in the past years. If only the "lucky electrons" (Ono, 1995, and references therein) can "run away" and attain sufficient energy (so that, e.g., $\eta_{hot} = 0.01$), one has some difficulties to explain the other factor: How can the Mn center have a cross section of $\sigma = 4 \times 10^{-14}$ cm², a value that is 200 times larger than its geometrical cross section? Assuming a reasonable cross section just slightly above the geometrical, $\sigma = 4 \times 10^{-16}$ cm², leads us to $\eta_{hot} = 1$, or "all electrons are hot." In this case, the impact length or mean free path between exciting impacts in an optimum doped ZnS:Mn sample is 1.2×10^{-5} cm, which compares to an absolute minimum acceleration length to 2.5 eV of 2×10^{-6} cm for ballistic or loss-free acceleration (in an average field of 1.5×10^6 V/cm). In many other cases—low field injecting metal contacts, for instance, which yield much lower products $\sigma * \eta_{hot}$, the fraction of hot enough electrons is probably really much below 1 (see Figs 18–20).

A comparably reliable database is not available for any of the other materials that would justify making similar evaluations. As obvious from Table I, in which literature data have been converted from the lumens per watt usually given to physically meaningful watt/watt values, the total span of efficiencies is not very large, but ZnS:Mn is the best. The most likely reason is its large impact excitation cross section, which together with the concentration at which concentration quenching begins (and which does not change very much), determines the optimum value. For values obtained for further materials see Chapter 3.

TABLE I

EFFICIENCIES OF THE MOST COMMON EL MATERIALS

	Luminous Equivalent					
	lm/W*	lm/W	mW/W	x	y	Reference
ZnS:Mn	5	466	10	0.54	0.45	Mach et al. (1990a)
SrS:Ce,CuBr2	1.3	323	4	0.21	0.4	Mach et al. (1990b)
SrS:Ce	1	333	3	0.2	0.38	Liu et al. (1998)
SrS:Ce,Cl,Mn,Ag	1.9	386	5	0.27	0.5	Velthaus (1997)
SrS:Ce	0.7	398	1.8	0.3	0.5	Soininen (1998)
SrS:Cu,F	0.22	203	1.1	0.15	0.23	Sun (1998)
SrS:Cu,Ag	0.2	150	1	0.17	0.16	Sun (1998)

*Measured external values.

The most uncertainty in any conversion of the measured value of (external) efficiency into internal efficiency is introduced by the light extraction factor. A careful comparison of data taken on ZnS:Mn samples prepared in different technologies, made a spread by a factor of four most likely. This includes samples that were made rough in an attempt to maximize this partial efficiency (Mach et al. 1990b). But there is no doubt that, for instance, the samples on ceramic substrates of Westaim (see Fig. 55 in Subsection 1 of Section IV for the schematic, and Figs. 12a and 29 for spectra), which were not available in 1990, compare favorably with the roughest samples of this "old"' paper.

For ZnS:Mn one other point must be raised: Its efficiency is very dependent on the drive level—or even better to say on its brightness level. The efficiency decreases rather dramatically as soon as a relatively well understood cross-relaxation mechanism allows two neighboring excited Mn ions to transfer the excitation of one to the other, exciting the latter to a second excited state, about 4.2 eV above the ground state. This latter transition as well as the decay is dipole allowed. By reabsorption in the host this double excitation energy is lost (Mach and Mueller, 1984a; Mueller et al. 1988b). Other materials also show drive-dependent efficiencies. The mechanism responsible in the Mn case is very unique, but other cross-relaxation processes are thinkable. More likely, however, these are space–charge effects, and thermal ones.

Strangely enough, the literature does not contain a thorough discussion of ways to improve the efficiency. The careful reader will have asked herself already, where does the 95% of the energy input go, which is not (even internally) converted into light? One possible or even probable answer is:

into the radiationless recombination of electron–hole pairs, unintentionally generated (Mueller, 1993a, 1994). SrS:Ce, which outputs even less light than ZnS:Mn, can eventually tell us something about the direction to pursue. Measuring the field quenching of the Ce^{3+} emission and the trailing edge emission (Mueller, 1994), in the same geometry one can compare intensities — one missing and one "additional." As it is the same geometry, one can regard the missing intensity as a measure of the number of ionized Ce centers in arbitrary units. The trailing edge emission measures the same number in the same arbitrary units, if the explanation we offered is correct and complete. Doing the experiment as described, one gets a much larger trailing edge emission than the quenched intensity can account for. The following explanation has been offered: Not only the electrons ionized form the Ce centers might drift back into the bulk, but, of course, also the normal transferred ones, if the interface allows for such. But other than in the ZnS:Mn case mentioned, they recombine at least partly radiatively, emitting the characteristic Ce^{3+} band. How can this be? Of course, most of the electrons arriving at the anode originated from e–h pair generation, leaving behind trapped holes. Thus, the recombination will be with those holes. If this recombination is able to transfer into the Ce band, we have this recovery or recuperation of "lost" energy, which we long for. Thus, very likely we must complete our discussion of trailing edge features with the remark that trailing edge current can be accompanied by emission, even if *no* ionized centers have been generated. The precondition is that the recombination of the back-traveling electrons with trapped holes is transferring energy into the wanted emission band. Codoping with suitable hole traps could improve the efficiency.

Another possibility, not explored yet, is to avoid the loss instead of recovering losses. This means to avoid the generation of e–h pairs. The way to suppress e–h pair generation is to limit the thickness of the S-film to one acceleration length to 4 eV, which is not sufficient for multiplication, but enough to excite every useful dopant. The problem is the crystallinity, which in general improves only after the first 50 to 100 nm to acceptable values.

In this context, the increasing interest in oxides should be mentioned. No indication of avalanching in oxide phosphors has been published. As one immediate consequence, multiplexing is very limited, because of the very soft electrooptic characteristics, $B(V)$, and $Q(V)$. However (Minami 1998a, 1998b), relative good efficiency values up to 10^{-3} W/W have been reported (Fig. 46). The required voltages are, in general, much higher, starting to reach usable luminance values in the 400- to 500-V range. These characteristics do favor applications of oxide phosphors in single (low count) pixel devices, for instance, for advertisement.

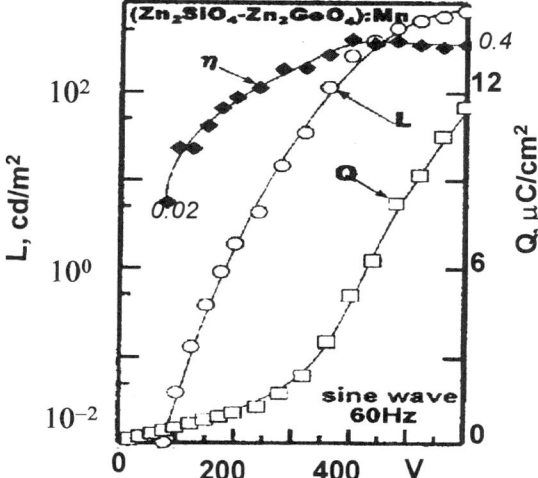

FIG. 46. Luminance L, transferred charge Q, and luminous efficiency η versus applied voltage for a zinc silicate germanate $Zn^2(Si,Ge)O_4$:Mn TFEL sample (after Minami et al., 1998b).

6. HIGH-FIELD ELECTRONIC TRANSPORT

It is a challenge to solid-state physics to understand electron transport at fields, which exceed the usual "hot electron" phenomena region (Gunn effect etc.) by about an order of magnitude. The only other area in which the interest in fields in megavolts per centimeter arose at about the same time was the electric breakdown of transistor gate oxides. A group at IBM (DiMaria and Fischetti, 1987) has done extensive work on this subject, and some universities joined the effort. The experimental situation in this context is, however, very much different, as is the material. While the TFEL structure gives the unique opportunity to stabilize the electric breakdown by virtue of the two sandwiching dielectrics (even one distributed series capacity will do), the experiment on gate oxides — SiO_2 mainly — has been done in a DC mode, with any breakdown being catastrophic. And, of course, it is hardly possible to find a dielectric with higher breakdown field than SiO_2. Thus, the TFEL gives a unique opportunity to study the high-field transport in the materials, which show the typical breakdown characteristics, as shown in Fig 25.

a. Theoretical Studies

The theoretical treatment starts out from the methods proven in the investigation of high-field conduction in GaAs. The balance between energy gain from the field and energy loss to the phonon determines the kinetic energy of the carrier within a band structure, which is assumed as basically unchanged with respect to the zero-field case. The problems lie in the derivation–estimate of the coupling constants between electron and the different phonons. Usually, some of them must be treated as fitting parameters, which are determined in specially designed experiments. The first results for fields above 1 MV/cm indicated for ZnS that there is no chance to excite Mn to EL (Brennan, 1988), in somewhat striking disagreement with the industrial success of this material already at this time. Further refinement of the theory (Duer *et al.*, 1998, and citations therein) brought the energy distribution of the electrons closer to the observation of excitation of levels at about 2.2 eV (Fig. 47), but it took until very recently that the possibility of carrier multiplication and e–h pair creation became theoretically "real." Calculations, using the best available band structures seem to indicate, that the e–h pair creation sets in even below 1.5 times the energy gap. While "good" band structure data are available for ZnS, much less accurate data for SrS or CaS hinder the calculations, and make even the establishing of differences between the materials difficult. No attempts are known, to consider changes in the band structure by the high fields as a possibility.

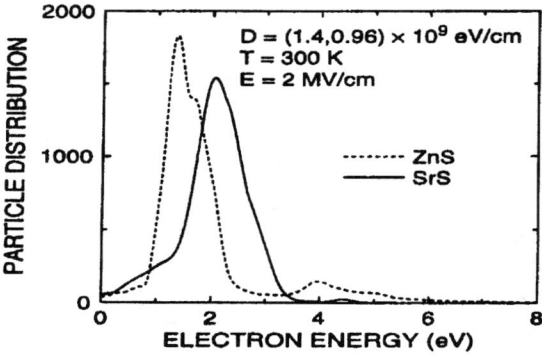

FIG. 47. Theoretical electron energy distribution in ZnS and SrS under the action of a field of 2 MV/cm (after Duer *et al.*, 1998).

b. Electron Energy Distribution

The TFEL structure by itself offers the opportunity to measure the electron energy distribution (EED). Going from a dielectric-sandwiched S-film to an S-film stacked on a dielectric (Fig. 48.), covered by a very thin electrode, one obtains the chance to emit electrons into a vacuum. Electron energy analysis in a vacuum is a methodology that is well established; the three-grid analyzer has little room for error. A very important feature of electron transmission through metal films is well known from Auger analysis — the transmission increases if one goes from about 30 eV down to 1eV. Metal films act as low-pass filters for electrons in this energy range. Having this in mind, a TFEL structure without a top dielectric and with an Au electrode below 10 nm thickness was used as an electron-emitter into vacuum (Mueller *et al.*, 1990, 1991; Mach, 1992). This concept was pursued by two other groups (Kitai *et al.*, 1991; Okamoto and Kobayashi, 1994) as a possible replacement of Spindt tips for field emission displays. Here we are interested only in the results and their bearing on the mechanism of TFEL. Figure 49 compares the dissipative current $I(t)$ (upper part) with the vacuum current $I_{vac}(t)$ at the respective counter voltage U given. Using standard procedures (time averaging and differentiation with respect to the counter voltage) these data were condensed into EEDs, as shown in Fig. 50. There the energy E is just the equivalent to the counter voltage with no corrections made for electron affinity and work function differences between the Au top electrode and the (Au covered) tungsten grids of the analyzer. Two sets of data are given in Fig. 50, one for a transferred charge of 0.1 and the other for 0.6 $\mu C/cm^2$. This dependence has been analyzed in more detail,

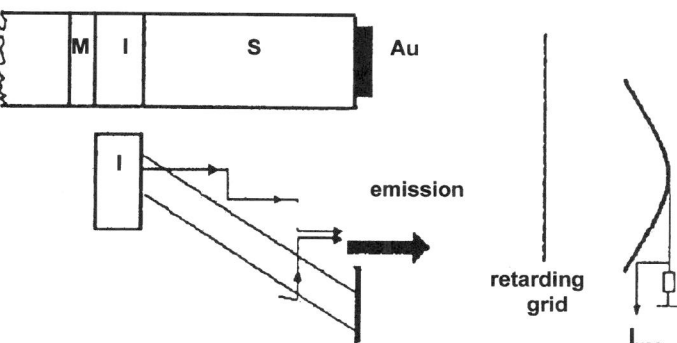

FIG. 48. Schematic of the experimental arrangement to measure the energy distribution of vacuum emitted hot electrons from ZnS (and ZnSe).

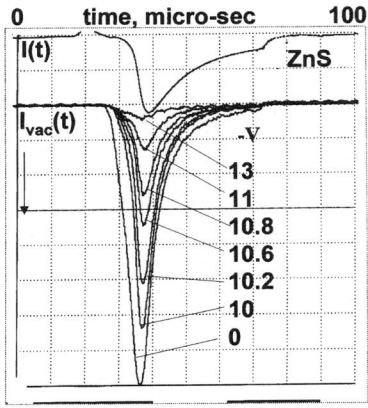

FIG. 49. Time dependence of (upper part) dissipative inner current $I(t)$ and (lower part) vacuum current $I_{vac}(t)$ at the counter voltages given (in volts) (after Mach, 1992).

as shown in Fig. 51. With increasing transferred charge the I_{vac} curves at a counter voltage of 1 V shrink to smaller widths. A tentative explanation being the accumulation of space charge in the anode near region "cooling" the electrons.

Independent of the very details, the striking result is the peak of the EED of the vacuum emitted electrons at about 11 eV (Fig. 50), which is not far from three times the gap (3.7 eV). To look for a suspected systematic trend with the gap energy, the same experiments were performed on ZnSe structures. The peak at about 7.5 eV holds the same relationship to the

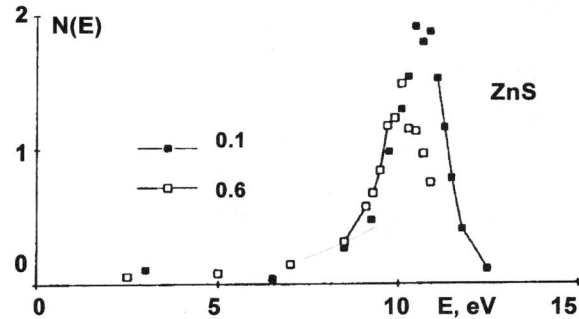

FIG. 50. Electron energy distributions $N(E)$ of the vacuum emitted carriers as evaluated from Fig. 49 for two transferred charge values, 0.1 and 0.6 $\mu C/cm^2$, respectively.

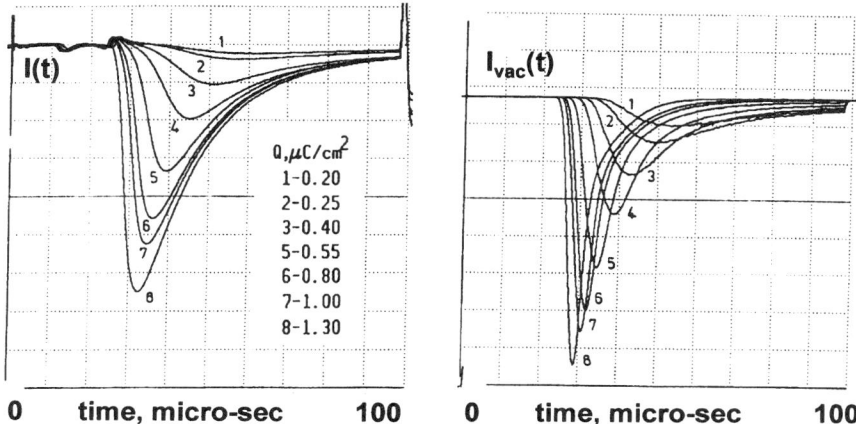

FIG. 51. Time dependence of (a) dissipative current $I(t)$ and (b) of vacuum current $I_{vac}(t)$ at 1-V counter voltage, $I_{vac}(t)$ at the various transferred charge values given in the inset. The vacuum current becomes narrower in comparison to the inner current with increasing charge transfer = increasing space charge buildup.

smaller gap (2.8 eV) of ZnSe. No independent confirmation of these exciting results has been attempted to our knowledge, but Okamoto and Kobayashi (1994) were able to show that electrons can also be emitted into the vacuum from SrS TFEL-devices.

One hand-waving explanation of these EEDs is the almost free-fall acceleration at the high field, until multiplication as ultimate energy loss mechanism sets in. A pile-up of carriers at this threshold of steeply increasing multiplication probability would occur and give rise to rather low electron counts at lower energies. The experimental outcome is in full disagreement with the theoretical results, which, however, refer to a different, highly nonstationary situation.

7. Design Rules

The parameters to play with in the design of a TFEL device are the film thicknesses, the material for the dielectric films, and eventually the dopant concentration of the S-film, assuming that its material is predetermined by the choice of the emission color. The parameters, which must be known, are drive frequency and voltage capability of the column drivers, as well as the brightness expectation.

The capacities and the threshold field of the S-film material in contact with the chosen dielectric determine the transferred charge–voltage characteristic of a TFEL pixel or test area.

$$Q(U) = \alpha * C_i * (U - U_{th}) = \alpha * C_i * (U - d_s * E_{th}/\alpha), \qquad \alpha = C_i/(C_i + C_s)$$

From that one gets, neglecting any changes of efficiency with drive

$$B(U) = \eta * p(U) = \eta * 2f * Q * U_{th} = \eta * 2f * E_{th} * d_s * C_i * (U - d_s * E_{th}/\alpha)$$

One must take into account that d_s and α are not independent of each other, as $C_s = \varepsilon\varepsilon_0/d_s$. A graph in Fig. 52 shows the dependence of input power p (equal to the brightness, in units of η), on d_s for given values of $C_i = 40$ nF/cm², $f = 60$ Hz, $E_{th} = 1.5 \times 10^6$ V/cm, and the applied voltage U as parameter. A set of $p(U)$ curves with d_s as the parameter of most influence is shown in Fig. 53. Of course, all these simulations do not correctly depict the turn-on region nearest to the threshold (see Section V). If η is dependent on drive level or input power, or as in the case of ZnS:Mn on the brightness itself, the schematic plots hold for $Q(V)$, not for $B(V)$. However, note that the design consideration should start out from $Q(V)$.

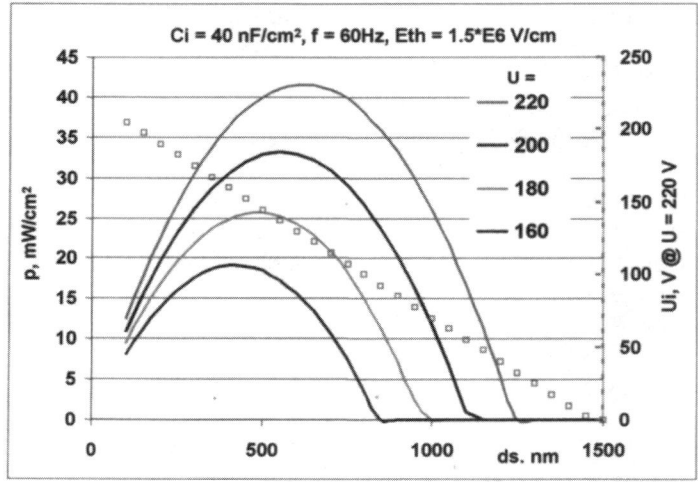

FIG. 52. Input power p versus phosphor film thickness d_s for the applied (peak) voltages V given in the legend; constant parameters: insulator cap. 40 nF/cm², threshold field $1.5 * 10^6$ V/cm, 60 Hz drive. Also given (right scale) voltage over the (two) insulator films versus d_s at 220 V applied voltage.

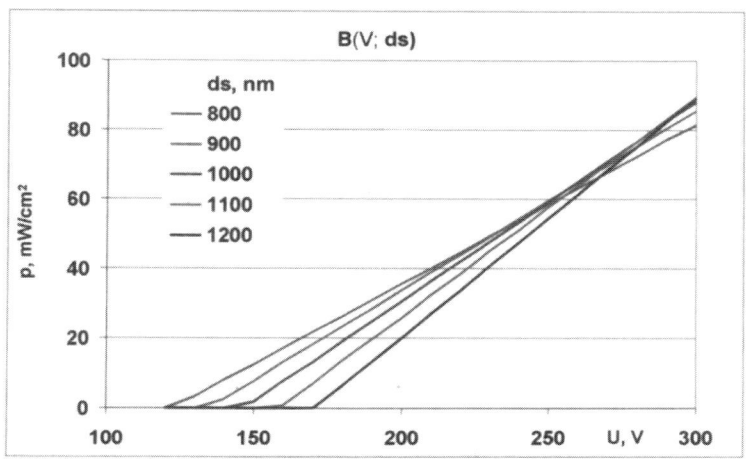

FIG. 53. Input power p versus applied voltage for different phosphor film thicknesses, given in the legend with the same constant parameters as in Fig. 52.

Let us come back in this context to the often used unit B_v, which designates the luminance at V volts above threshold; very commonly B_{40} is given, as most display applications operate with maximum column driver voltages of 40 V. If this value is used as a criterion in a process optimization, there is a high risk that changes are attributed to the efficiency η, but what really changes is the thickness of the phosphor film d_s. This is accompanied by a change of the capacitance, and of the threshold voltage, as shown in Fig. 54. Should the threshold change but the capacitance remain constant, a variation of the threshold field is the most likely reason. Changing process temperatures usually results in thickness changes, as the sticking coefficient of some of the components is strongly temperature dependent. Thus, B_{40} is a convenient way to eliminate the problems encountered in a constant-voltage comparison, but possibly hides important design changes. And by no means it can reveal dependencies of the efficiency.

One of the most important design rules for TFEL devices is to keep the dielectrics as thin as possible, or better, get the highest possible insulator capacitances C_i. This is also demonstrated in Fig. 54. The luminance (B_{40}) increases for any thickness of the active film linearly with C_i, while the threshold voltage (right axis) decreases (nonlinearly). This was pointed out by Morton as early as 1987. Howard (1977) published a figure of merit for the TFEL dielectrics that relates the charge at breakdown field of the dielectric ($\varepsilon_0 \varepsilon_I * E_b$) to the maximum transferable charge of the device.

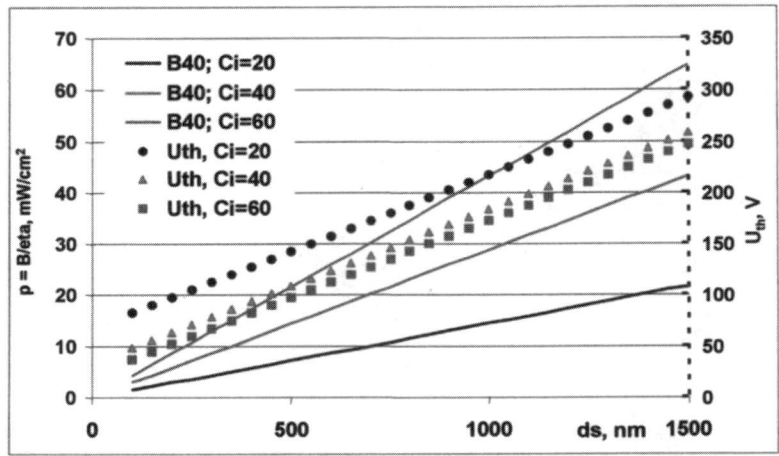

FIG. 54. Input power p_{40} or luminance B_{40}, at 40 V above threshold versus phosphor film thickness for different insulator capacitances, given in the legend; also given (right scale) the respective threshold voltages (row driver voltages).

Usually the practical limit is set by particulates or roughness of the top surface of the active film, which causes breakdown even below the "intrinsic" breakdown field. As is obvious from Fig. 53, the second rule is to choose the active film thickness d_s as high as consistent with the capability of the row (select) drivers.

IV. Thin Film EL Devices

Sharp Corp. of Japan was the first industrial supplier of TFEL displays. Introduced in 1983 and driven by an in-house manufactured driver IC, their first displays used ZnS:Mn as the active material. The size was about that of a postcard, quarter-VGA (see Section I). Second to market was Planar Systems, USA, a spin-off of Tektronix. Their first product was very close to the one offered by Sharp, and differentiating to different market sectors began later on. The big hope of both manufacturers to find access to the growing laptop computer production did not materialize, partly because of initial technical difficulties, but mostly because of higher prices and lower efficiency. The benchmark for the mass markets was liquid crystal displays (LCDs), becoming more and more available from more and more manufacturers at that time. The relatively slow but steady market penetration by

TFEL displays (TFELD) was driven by the special features they had to offer: As mentioned, no visual distortions in dependence on viewing angle, high contrast, grisp images, and a wide operating temperature range. The U.S. military was a great supporter of the technology that could easily meet military ruggedness specs; recently announced, more than 50,000 TFELD are working in U.S. Army applications. Another great niche market for TFELD were medical instruments, especially patient condition surveillance in hospitals. There the "flatness" and the high viewing angles were most valued. Without putting too much emphasis on the order, a third important application field was and is industrial control and measurement. Again, the harsh environment of some of the application spots, together with a suitable form factor, favor TFELD over other display types. Relatively late in the short history of TFELD, the transportation industry developed an interest in them.

1. DIRECT VIEW DISPLAY

a. Monochrome Displays

The initial 10 years of industrial production of TFELD were based solely on monochrome, yellow-emitting ZnS:Mn devices. Some good examples are shown in Fig. 1 in Section I. The improvements over the years aimed mainly at contrast and brightness. Much effort was expended for some time to increase efficiency. The initial determination toward the battery-operated instrument market was still a driver. Progress was made mainly by reducing the capacitive power losses, described in some detail in part c of this subsection. Standard products of Planar are shown in Fig. 1 together with some specs. The diversification of monochrome products is best followed by looking on the producers" Web pages. Most noticeable is the increase in contrast and at the same time decrease in power consumption over time.

Going beyond this "integral" contrast, a new technology for direct sunlight readability is being commercialized by Luxell, Canada. Its working principle is described in Subsection 3 in Section III (Li *et al.*, 1994).

A breakthrough with respect to brightness (and efficiency) was released in 1994 by Westaim. Their new thick–thin film technology (Wu *et al.* 1994, 1996; Doxsee and Wu, 1997; Liu *et al.*, 1998) allowed for high brightness because of much better light extraction. In the meantime, it has also been successfully applied to full-color displays. The basic structural changes in this technology (Fig. 55) are

- Use of a ceramic substrate with thick film electrodes and through-connections to the drivers

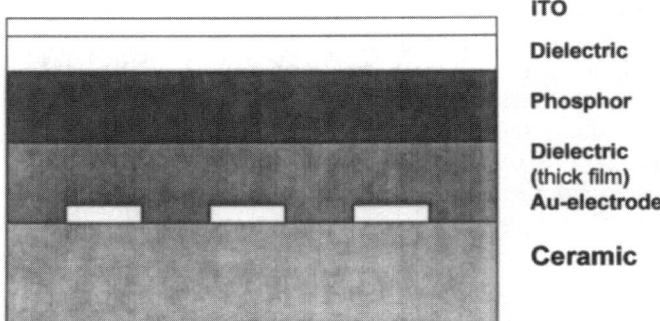

FIG. 55. Schematic of the thick film–thin film structure featured by Westaim, Inc., Canada.

- A thick film, high-permittivity dielectric
- A thin film phosphor film, a very thin top dielectric, ITO top electrodes, laser trenched into the row system

The major advantages are the very high scattering by the rough ceramic "reflector," increasing the light extraction efficiency, and the possibility to anneal to high temperatures. The latter feature is most valuable for blue phosphors. The very high insulator capacitance allows also for very high power input at, for example, 40 V above threshold (see Subsection 7 in Section III).

b. *Multi- and Full-Color*

In many applications, the availability of at least one second color is much wanted. Highlighting, warning, and alerting are examples of the reasons. As outlined in Section II almost all colors can be obtained by using different rare earth dopants in ZnS. However, good efficiency and maintenance are not easily achieved. But most importantly, the only very little changes to existing monochrome technology necessary, gave the multicolor ZnS:Mn version the lead. The principle is shown in Fig. 56. The structure is inverted. The metal electrodes are put down on the substrate, ITO (or any transparent electrodes) are deposited on top of the stack, and they are topped by a system of color filters, assigning every second row (or column) to one of the wanted colors. In this special case, green and red filters are applied. The emission spectrum of Mn^{2+} is wide enough to allow for this split into two colors, as shown in the figure. Even without intensity modulation four colors

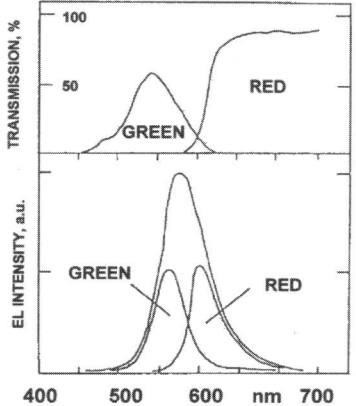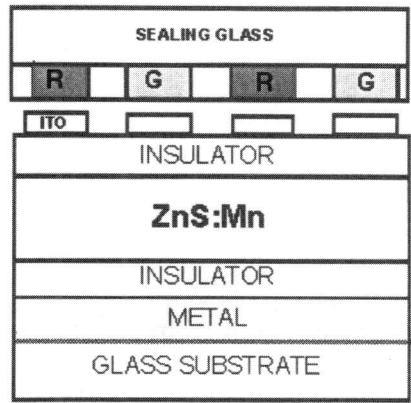

FIG. 56. Operating principle and schematic of the structure of the multicolor TFEL displays (after Okibayashi et al., 1991).

become possible; gray shading allows for much more. First announced in 1991 (Okibayashi et al., 1991) and described in some detail by Sharp, it is manufactured by Sharp and Planar. Figure 2 shows a multicolor TFELD of this type from Planar (Haaranen et al., 1992). This multicolor scheme deviates appreciably from the conventional way of thinking about color displays, which is derived from the way the CRT does it. In the CRT phosphors, which emit different colors, in general, red, green, and blue, are placed in certain patterns on the screen. Three electron beams hitting either of them, generate color dots side by side, and the observer's eye averages spatially, to "see" a mixed color. The same impression could be generated if the phosphor were a white-emitting one, and the RGB components generated by a filter mosaic. To our knowledge nobody proposed this for a CRT. In a TFELD, however, patterning of the active material(s) would mean three lithographic processes to remove the first deposited thin film from the pixel spots, which the second material should hold, and so on. The borders between the different materials are as all the active material exposed to electric fields of some megavolts per centimeter in operation. This was at least one of the considerations, which led to the proposal of "color-by-white" (Tanaka et al., 1988). In the schematics of Fig. 57 the two different approaches, patterned phosphor and patterned filters, are shown side by side. There is no doubt that the patterned filter design is much easier to manufacture. However, it will do, if and only if the efficiency of the white phosphor is good enough to allow for the light-absorbing filtering. In each

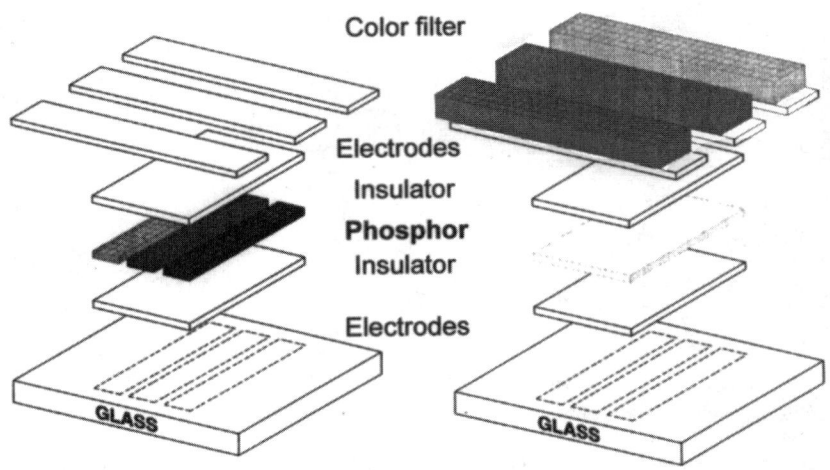

FIG. 57. Full-color display concepts: left, patterned phosphor; right, color-by-white or by (patterned) filter.

lit pixel, about three times the amount of light has to be generated, as two-thirds are absorbed in the filter.

There were steps along the way that should be briefly mentioned. A partial use of the patterned phosphor approach was attempted by Planar in the so-called dual-substrate display (King, 1994; Barrow et al. 1993). Figure 58 shows the principle: On one substrate green and red emitting pixel side by side are made from ZnS:Tb and ZnS:Mn, with the latter filtered to remove its green content. On a second substrate, a monochrome blue emitting display was deposited. The active material being a calcium-thiogallate doped with Ce^{3+}. An optical contact between the two substrates put face to face was established by filling the space in between with suitable oil. Of course, all the electrodes of the upper display (the ZnS one) must be transparent. It is not known at what price this display type was offered nor whether any have been sold. The approach, however, proved feasibility of several important steps, which became important for the further development of the field.

In parallel to this development several groups worked on the color-by-white display. There is no white-emitting phosphor available, therefore multilayered films or stacks of phosphor films had to be used. The combination of the red and green containing spectrum of ZnS:Mn, and either SrS:Ce or more recently SrS:Ag,Cu for the blue, cover an appreciable part of the visible, as Fig. 59 shows. Nire et al. (1992) was the first to

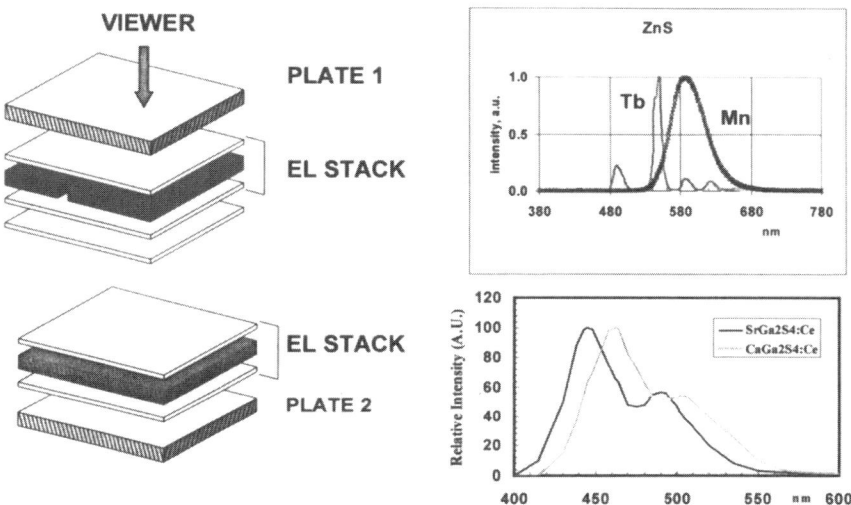

Fig. 58. Design of the Planar "dual-substrate color exerciser": Plate 1 carries red = filtered ZnS:Mn, green = ZnS:Tb; Plate 2 carries blue = $CaGa_2S_4$:Ce, the gap is filled with oil.

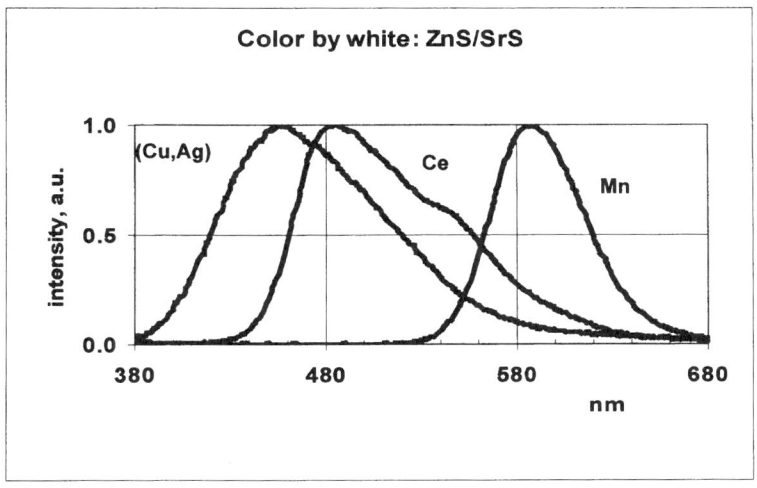

Fig. 59. Spectra of "ingredients" of latest color-by-white versions.

demonstrate a successful version in 1992. It consisted of an active layer composite $ZnS:Mn^{2+}|SrS:Ce^{3+}|ZnS:Mn^{2+}$; the individual films were deposited on top of each other by multisource deposition and sandwiched between dielectric films. The stack without filters had a luminance of 125 cd/m² at 60-Hz drive. The mixed white, after being filtered and recomposed, showed a luminance of 2.1 cd/m² at 20 V above threshold. The prototype with 213 × 200 pixels on 4.4′ diagonal used black stripes between the color filters, which reduced the fill factor to 14% per color, but gave excellent contrast. Thus, in spite of the low luminance, the appearance was also excellent.

The problem, as in many other studies of the blue-emitting SrS:Ce was the difficulty in getting reasonable amounts of blue light out of it (Chapter 3). So it took another 5 yrs until Planar International brought the first color-by-white full-color display to market. Figure 3 shows it. It is a 4.8-in. diagonal quarter-VGA display, meant mainly for demanding measurement applications. The principle structure and choice of materials is still the same as used by Nire (Toernqvist, 1996; Soininen, 1998). The main specifications are given in Fig. 3 and the color coordinates are shown in Fig. 60. To

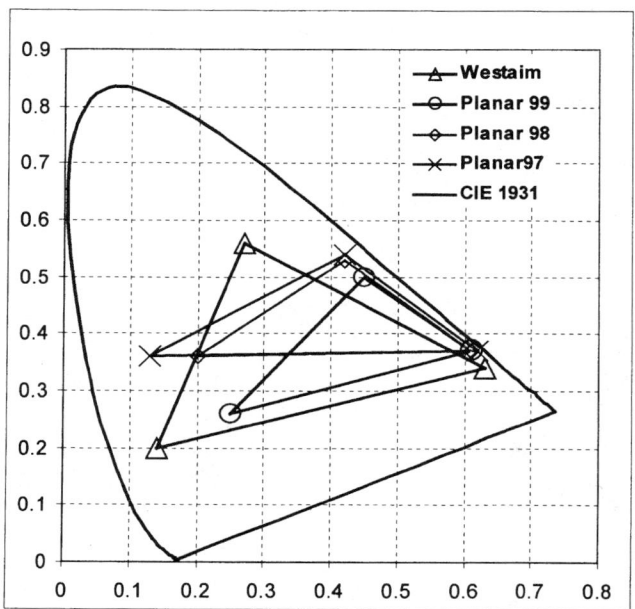

FIG. 60. Comparison of color space covered by the latest three versions of Planar and a prototype of Westaim; (x, y) data from Soininen, 1998, and Doxsee and Wu, 1997.

increase the brightness, the drive frequency was raised to 240 Hz, and 45 cd/m² of mixed white were achieved. The contrast ratio is also excellent: 2:1 at 40,000 lx of ambient light on the panel. Figure 60 also shows the color coordinates of former panels from Planar and one prototyp developed by Westaim (spectrum and panel design see Fig. 61 in the color insert, and Fig. 62). Figure 61 gives the specs published in 1997 (Doxsee and Wu, 1997). The secret of Westaim seems to be the very good blue reached in SrS:Ce by their special technology.

The next step is certainly under way. A so-called "true white QVGA" display with the same dimensions was announced by Planar (Soininen, 1998), which is also based on a two-film active layer, but incorporating SrS:Cu instead of SrS:Ce. The improvement in color coordinates becoming accessible is also indicated in the color diagram of Fig. 60. In fact SrS:Ag,Cu is the great hope for breaking the *blue barrier* (see also Subsection 4 in Section III), so pertaining to TFEL over its life. The progress is most visible in the color space expansion, however a saturated blue and consequently a "real" white are still in the future. First projections based on SrS:Cu,Ag are included in Fig. 60 as "Planar 99," but quite an improvement can be expected.

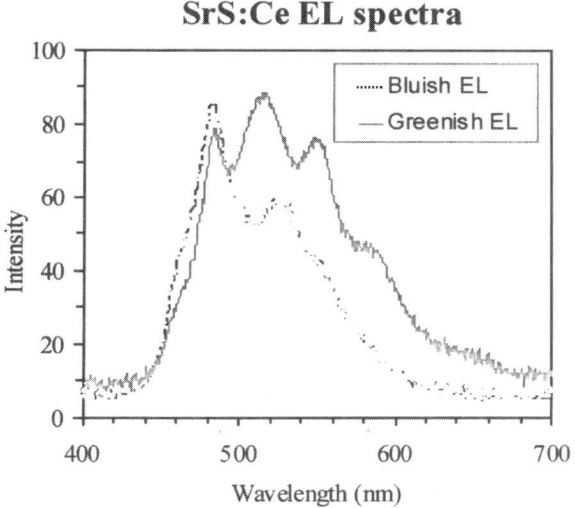

FIG. 62. Spectra published by Westaim about different SrS:Ce results in bluish and greenish (after Liu *et al.*, 1998).

c. *Limiting Factors*

Before reviewing more devices, some general remarks should be made to put achievements and expectations into perspective. TFEL was in the 1980s thought of as a technology for portable devices. Light weight, the lightest of all flat panel technologies, high contrast, eventually up to direct sunlight readability, and high efficiency were claimed as advantages. In fact, all but the last one are true advantages. In the case of efficiency, the picture is more complex. One has carefully to discern between "pixel efficiency" and "display or screen efficiency." As discussed in some detail in Subsection 5 of Section III, the pixel efficiency (i.e., the efficiency measured on an unstructured test area) can be as high as 5 lm/W. This is a value of which manufacturers of back-lit LCDs only dream, even the OLED folks will find it acceptable, and FEDs will eventually approach or surpass it slightly. The screen efficiency, however, is much lower; without special power-saving measures it is at most one half of the pixel efficiency, if the screen is reasonably large. The reason for this is the fact that the capacitance of the film stack is high, typically 10 nF/cm^2 or even higher, and the drive voltage is in the 100- to 200-V range. This means that for a drive frequency of 60 Hz, 120 times in a second the total capacity of about 6000 nF (assuming a 14-in. diagonal display) has to be charged and discharged to say $U = 160$ V. Each time $C*U^2 = 0.1$ W will be consumed in the resistances involved. So, the total losses, even without lighting up any of the pixels, are in the order of 10 W. This is the price to pay for the *possibility* to generate light. Paying the same amount of power again, and using the 5 lm/W pixel efficiency as an example, the whole area can be lit to 250 cd/m^2, which is a pretty good brightness. In a crude approximation, one can estimate where the sweet design spot lies: making the capacitance smaller (for instance by increasing the dielectric thickness) drives the operating voltage higher; as the lost power goes with its square, this is the wrong direction. Going thinner, reduces the voltage, increases the breakdown risk, but what is more important, reduces the total number of available emitting dopants. This can be made up for by increasing the drive frequency, and that, in turn, increases the total losses. No real optimization has been published, but very advanced driving schemes have been discussed by King (1992) and are partially realized in Planar products. One of the most convincing approaches starts from the thought that the energy stored in a capacitor can be recovered during the discharge cycle. Every oscillator does this. Simply put an inductance in series, and you win. However, the problem lies in the fact that we have oversimplified. The capacitance, which any one of the select drivers sees, depends on the number of column drivers firing. Thus, this image content-dependent capacity makes it extremely difficult to recover the energy in inductances. As King pointed out, stepping the voltage pulses up can also save energy, as 2 times $C*(U/2)^2$ is smaller than $C*U^2$. Driver complexity and compatibility with gray shading of those approaches has not been discussed in the literature. In general, there seems to be a much smaller pitch for battery-operated applications now than in the early days of TFEL. Another remark should be allowed: Sarcastically, the lower the pixel efficiency, the less important is the switching penalty. And the example with the 5 lm/W is the very best, which one can get for monochrome yellow, at least up to now (see Subsection 5 of Section III for a discussion of pixel efficiency).

The decreased ambition for battery-operated applications of the main makers of TFELD—Planar and Sharp—can also be seen from another perspective: Sharp is the top maker of AM-LCDs and Planar Systems has acquired Standish Industries and entered the LCD market. Thus, both are able to offer the best solution for any application.

2. AMEL

As pointed out in Section I, TFEL is possibly the flat panel technology with the least problems in multiplexing up to high numbers of rows. "Active matrix," because it is known for overcoming the multiplexing problems of LCD, is intuitively tied to this issue. So the question arises, why active matrix (AM) in TFEL? Well, AM is an excellent technique to reduce the number of connections to the addressing system. As is now done in plasma displays, the driver circuits can be mounted on the border of the display substrate. But this is clearly a second-best solution. A real drastic reduction of the connections can only be expected from an AM. The second advantage of an AM is the reduction in power consumption by isolating nonaddressed pixels. Both issues count most positively in a small display, as, for example, are needed for head-mounted displays. The scheme of Fig. 63 illustrates that besides an appropriate Si chip, the structure is very simple. On a (well-planarized) top surface of the AM chip a dielectric-sandwiched active material is deposited, which is topped by a blanket ITO electrode. The AM connects with pixel electrodes to the TFEL stack, each one of them isolated from all others. The further path is through a switchable transistor to ground. If this metal-oxide semiconductor (MOS) transistor is conducting, the AC voltage applied between ground and the ITO top electrode drops on the TFEL stack; if it is nonconducting, the portion over the stack is below threshold. The capacitor C in relation to the stack capacity controls the fraction of the voltage dropping on the stack in an pixel-off condition. Thus, the transistor does not have to stand all the drive voltage, but only the voltage above threshold, necessary to produce the wanted luminance.

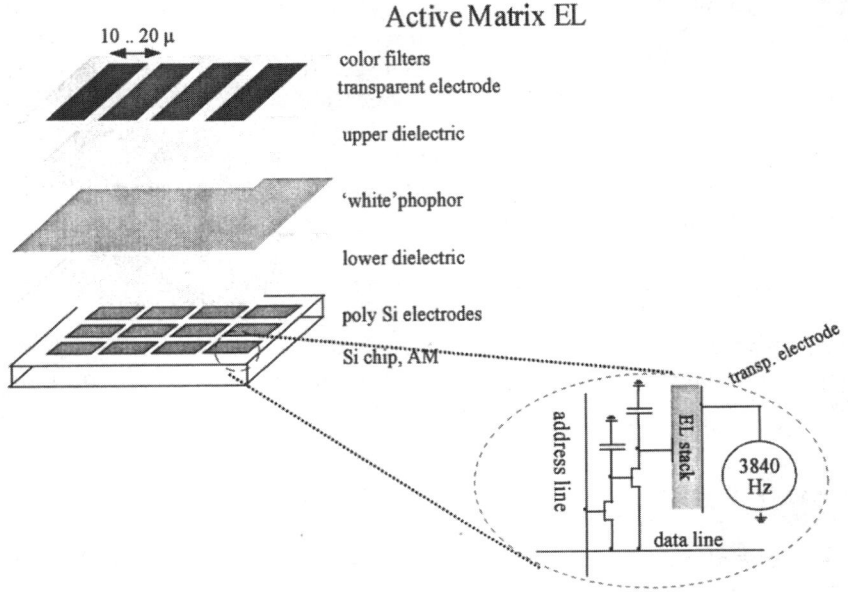

FIG. 63. Schematic design and schematic of pixel circuit in Planar's active matrix EL display.

Planar, together with a consortium comprising Sarnoff Corp. and AlliedSignal, has demonstrated 1280 × 1024-pixel AMEL devices with a pixel pitch of 12 μm and a data rate of 100 MHz on each of the eight data channels (Hsu et al., 1998; King, 1996, Arbuthnot et al., 1996; Khormai et al., 1994). This should be enough for a 60-Hz refresh rate at 6-bit gray shading. The gray shading is done by on-time modulation for a basic AC drive frequency of 10 kHz. Silicon-on-insulator (SOI) appears to be the best and adequate choice of technology for the AM to achieve those specifications. Pictures of the monochrome AMEL are shown in Fig. 64 (see color insert) and of a multicolor (by white) AMEL in Fig. 65 (see color insert). The conversion from monochrome to color is a change in active material, which can be much the same as in direct-view displays, but, of course, it is also a significant change in the addressing scheme. Reportedly (Hsu et al. 1998) the monochrome version could produce 875 cd/m^2, which should be enough to emulate fairly large virtual displays, depending on the projection optics. The power consumption of 0.9 W is still relatively high for battery-operated head-mounted displays, but possibly not the convergence point of the development. A combination with microcavities (see part b in Subsection 3 in Section III and in Chapter 4), which takes into account that the acceptance cone of the optics is about 10°, could increase the efficiency, if suitable luminescence centers were used (Mueller-Mach et al., 1996). However, a breakdown of the power consumption into "chip" and "stack" is not available at present.

After the technology has been more established, prices should come down rather rapidly, approaching the prices of of ASICs (on SOI) of the same area. The effort to achieve smaller and smaller pixels has begun.

3. NONDISPLAY APPLICATIONS

None of the nondisplay applications of TFEL has yet been commercialized. All of them consider TFEL as a light source with special features. Its attraction can come from either

- The arbitrarily shaped, large or small area, which can be *homogeneously* lit
- The light guiding and summing along a light path

Many attempts were made to introduce TFEL for back-lighting (e.g., of LCDs), but the relatively high cost and the competition of LEDs coupled to plastic light guides did turn these activities down.

a. Planar Light Sources

"Planar" here doesn't stand for the name of the company selling TFELDs, but has the generic sense of "plane surfaced" light source. In fact, slightly curved surfaces are also possible. This kind of light source has the advantage of being capable of rather high luminance–radiance. Values up to 50,000 cd/m^2 have been reported (Mueller-Mach and Mueller, 1996). As in all reported cases, glass or ceramic substrates have been used, thermal management is an issue, and those high values have been achieved only for time intervals below 1 ms and duty cycles below 1:50. The drive frequencies can be chosen as high as 100 kHz. On glass substrates, high-contrast ratios can be achieved without much effort, even higher ones by applying the techniques developed for avionics displays.

The TFEL light sources can be transparent, which can be a significant advantage over usual LED-lit plastic back lights, which must rely on some kind of light scattering out of the guiding plane for their operation. This inevitably reduces transparency and/or contrast.

b. Edge Emitters

The name coined by Kun, who also invented the device (Kun et al., 1986), describes its purpose very well. As mentioned in Section III, one of the big disadvantages of TFEL is the high refractive index of all "good" active materials. It allows for only about 10% of the generated light to leave the structure through one front face. The rest of the light is guided in the active film, the other films, and the substrate (if transparent). Breaking a TFELD and looking on the edge will demonstrate the action: the edge is much brighter than the face. This is especially true if looking into a "long" pixel ("long" refers to the direction perpendicular to the edge). The guided light for every point source in the active layer spreads radial into 2π but sums up along the length of the pixel. A group at Westinghouse prepared to use this idea for "print bars," replacing the scanning laser in an electrophotographic printer by a multitude of edge emitter pixel, imaged onto the drum by a lens array (Fig. 66). Some 5000 pixels would be enough to give a 600 dpi (dots per inch) printer the same principal advantages over a scanning light source as a matrix-addressed display over the scanning electron beam CRT: 5000 times longer dwell time; absolute, jitter-free alignment; and higher contrast. An impressive patent portfolio has been developed later on

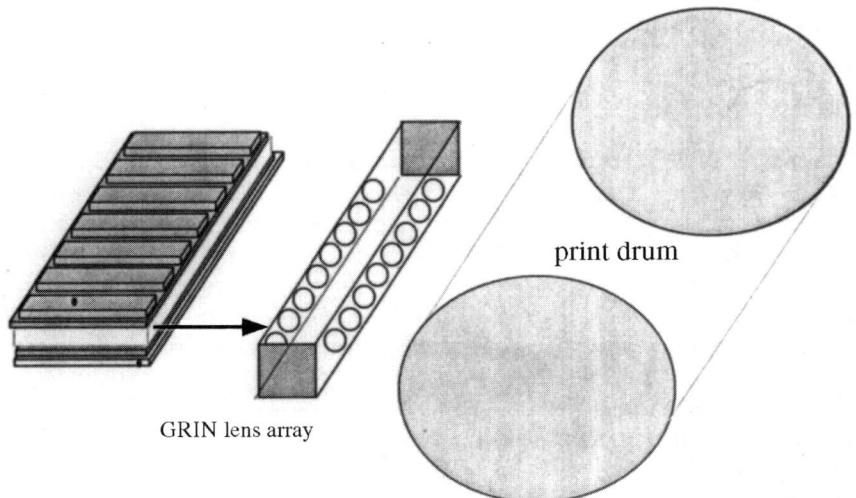

FIG. 66. Schematic of an electrophotographic printer design using an edge emitter array to write the information on the print drum: (left to right) edge emitter array (print bar), graded index (GRIN) lens array (SELFOC), and print drum.

by Hewlett-Packard (Mueller, 1996), but cost considerations prevented the device from entering the marketplace. One of the highlights of the development is shown in Fig. 67 (see color insert), showing the emission pattern of a 1200-dpi device with every other pixel addressed.

V. Measuring TFEL

Neither standardized methods nor even commercial equipment has emerged over the two decades of R&D on TFEL. As mentioned, many of the early investigations were just measurements of $B(V)$ characteristics, very often even with sinusoidal voltage drive. The main reason is that no bipolar pulse sources with output voltages of 200 to 300 V were available on the market. One solution of the problem was the application of push-pull drive: two pulse generators connected to the two electrodes of a sample fire with equal frequencies but phase shifted to each other, pushing and pulling the electrons from left to right and vice versa. Unfortunately this cheap and easy to implement method renders quite impossible some measurements we discuss in this section.

The measurements to be described here refer mostly to engineering samples or test areas, which are not capacitively meshed to other areas or pixels (as, e.g., in display matrices). They rely mainly on a well-defined capacity strictly determined by the known area. They are aiming for the understanding of the mechanism, the determination of material parameters of relevance, and the influence of deposition or aftertreatment parameters. A second category of measurements, not described here, is necessary to characterize the light emission from displays. The general features of interest in this field are the radiance–luminance, contrast, color (or spectra), and the uniformity of those values (Baron, 1994).

1. MEASUREMENTS ON TEST AREAS

a. Brightness

The most important parameter of a "light source" or any emitting device is its radiance, or if expressed in photometric units, its luminance. In the older literature brightness is often used if no special preference is given to either photo- or radiometric units. Radiance and luminance values are interrelated by the spectrum of the source and by the eye sensitivity V_λ or $V(\lambda)$. Repeating here textbook knowledge, the conversion between radiant power or flux, and luminous flux is most conveniently done by a quantity called "luminous equivalent" (LQ), which is given by

$$LQ = \int P(\lambda) * V(\lambda) \, d\lambda \Big/ \int P(\lambda) \, d\lambda$$

This brings us to the second very fundamental value of any light source — its spectrum. Measured in radiometric units, as watts per nanometer, we have used it in the preceding equation, $P(\lambda)$. As $V(\lambda)$ is expressed in lumens per watt, LQ is also given in lumens per watt.

b. *Spectra*

The use of the notion of a spectral distribution of power radiated by any source [as, e.g., $P(\lambda)$], does not directly relate to an unambiguous measurement procedure. The usual spectral measurement gear is designed to accept the emission from the source within a rather narrow "acceptance cone," given in the most straightforward case by the area of the spectrometer slit (or the coupling fiber bundle) and the distance from the source. So, in general something in units such as watts per steradians per nanometers are obtained. Going to a infinitesimal cone, we would obtain the *spectral radiant intensity*. The *spectral luminous intensity* is obtained by multiplying the spectral radiance by $V(\lambda)$. The spectral radiance is the primary quantity on which all colorimetric calculations are based, directly or in some instruments indirectly; that is without really obtaining the spectral radiance (three- or multifilter colorimeters etc.).

If the angular distribution is *Lambertian* (i.e., if it varies as $\cos \Theta$), the measured spectral radiant intensity multiplied by π gives the spectral power distribution or *spectral flux* $P(\lambda)$ in watts per nanometer. However, Lambertian distributions are hardly possible in TFEL, and this has caused much confusion in the literature. It is for this reason, that this "old tutorial stuff" is repeated here at some length. The reason for the problems encountered is *optical interference*. In Fig. 12b a rather typical spectrum of a ZnS:Mn sample is displayed. Radiance, measured under four different angles to the substrate normal into an solid angle of 0.25° is plotted versus wavelength. To give some indication of how well even for the (spectrally) integrated power a Lambertian relation holds, the integral values are given in the figures. This is no extreme case. Of course, the color coordinates and the luminous equivalents also depend on the viewing angle, as is evident from the values given in the figure.

In many cases, the information about "brightness (or color) normal to the surface"' is wanted. However, it should not be used in the same context as saying that TFEL has a viewing angle of 180° and is therefore superior to other display technologies. One should be extremely cautious in measuring radiance into a small fixed angular cone as a criterion for materials or device development: the angular distribution *changes* with any of the thicknesses of the constituent layers of a TFEL device. A simulation was shown in Fig. 34. A variation of 20% in the thickness of the active layer, which is easily encountered in an optimization series with respect to deposition temperature, has been assumed. By our experience, this is realistic.

The best way not to fool oneself during technological development is the measurement of flux into an integrating sphere. Figure 12b also gives the spectral flux into 2π for the sample measured under the different angles. A very "normal" ZnS:Mn spectrum is obviously obtained.

Some words for beginners concerning "imaging" or "tele spectroradiometers" (Pritchard type) in which the object is imaged on to a circular aperture (hole in a mirror), which is the entrance port to the instrument. The virtues of those instruments are that one sees where the measurement is taken from and the quantity measured is *radiance*, usually given in watts per steradian per nanometer per centimeter squared. Converting this reading by the eye sensitivity, we get *luminance* in candelas per square meter or lumens per steradian per square meter. (Some authors call this "brightness" as well.) The reading, however, is correct only if a uniformly emitting area of the test object fills the aperture completely. (Notice that the calibration is done in this way, and the aperture is imaged to the entrance slit of a spectrometer.) As long as this filling condition holds, a Lambertian source will give constant readings versus angle, as the radiant intensity decreases as the cosine, but the surface area, emitting into the aperture, increases as the cosine of the same angle. Figure 12b is a neat illustration; the wavelength integrated radiance is almost constant versus angle = Lambertian, but the spectra change significantly.

c. *Drive dependence*

The sample must be driven electrically. As outlined in the preceding section an AC voltage must be applied. In a crude approximation, brightness is proportional to the frequency, as each polarity change generates a light pulse. Slow decay relative to the period of the applied voltage might cause the impression of an only slightly modulated, "rippled" DC emission. If this happens, one should be cautious if the modulation becomes low, as the proportionality of brightness with frequency can be about to cease.

Pile-up of excitation occurs, as the decay before the action of the next pulse is not completed.

d. Input Power, Efficiency

The definition of the input power as given in textbooks of electrical engineering is

$$P = f * \int U(t) * I(t)\, dt$$

with $U(t)$ is the applied voltage and $I(t)$ is the total current, f is the frequency, and the integral taken over a full period is $1/f$. This prescription is not easy to implement, as the capacitive currents, which account for the adding-to-zero parts, may dominate (in the case of fast changing voltages), reducing the precision. Depending on the steepness of the current pulses a high dynamic range, and a high time resolution of the digitizing system must be used.

Much easier to implement is the measurement of transferred charge Q, which is the integral of the dissipative current. It is most easily obtained in a bridge circuit (upper inset of Fig. 68). The upper inset also shows the waveform as it appears on the sensing capacitor C_m. The amplitude of this waveform U_m gives

$$Q = U_m * C_m$$

The equivalent circuit — the back-to-back-Zener circuit of Fig. 21 — implies that above the threshold, the voltage stays clamped; there is no increase in the voltage drop U_s (the Zener voltage) at the semiconductor layer (C_s) (Fig. 23b). Therefore the power input into the semiconductor layer (and as the dielectrics are assumed to be ideal, into the device) is given as

$$P = 2 * f * U_s * Q_s = 2 * f * U_{th} * Q$$

with $U_s = \alpha * V_{th}$; $Q_s = Q/\alpha$; and $\alpha = C_i/(C_i + C_s)$, as given by electric engineering.

However, deviations occur, implying in fact deviations from the equivalent circuit. Especially, if there is a (trailing edge) current in the reverse direction, the charge Q is no longer an adequate variable to be multiplied by a clamped voltage (Mueller et al., 1988a; Mueller, 1993c) Figure 69 shows the close agreement between the energy/pulse $= \int U(t) * I(t)\, dt$ and $Q * V_{th}$

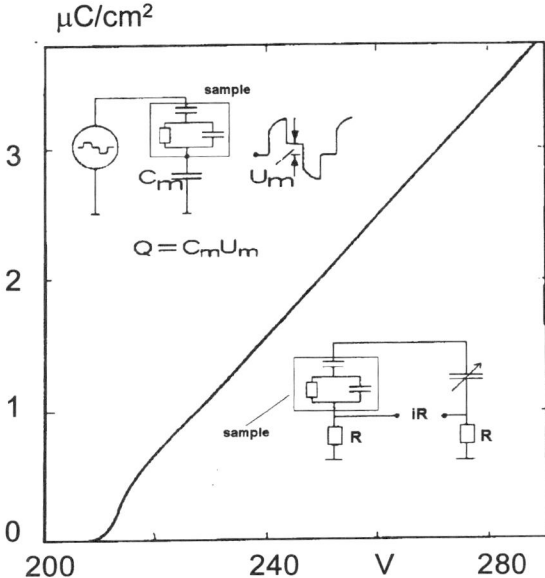

FIG. 68. Transferred charge Q versus (peak) voltage of an almost "ideal" TFEL sample; upper inset: measuring circuit for $Q = U_m * C_m$; lower inset: measuring circuit for the dissipative current $I(t)$ after zeroing by the adjustable capacitor below threshold.

for a ZnS:Mn sample, and the disagreement for an SrS:Ce, encountering trailing edge current.

Now, knowing the output — optical flux, measured in an integrating sphere — and the input electric power, determining the efficiency is as easy as a division of these two numbers.

e. *Partial Efficiencies*

As already outlined in Subsection 5 of Section V, splitting efficiency into partial efficiencies is a powerful method in optimization. The partial efficiency, which can be directly assessed, is the internal *luminescence efficiency* of the center (dopant). In the case of ZnS:Mn, it can even be measured in EL (Benalloul et al., 1990). The underlying notion is that the decay after excitation by a pulse, which is short in comparison to the decay time, contains all the information. During such a short excitation the same start situation for the spontaneous decay is reached in any sample; especially if it

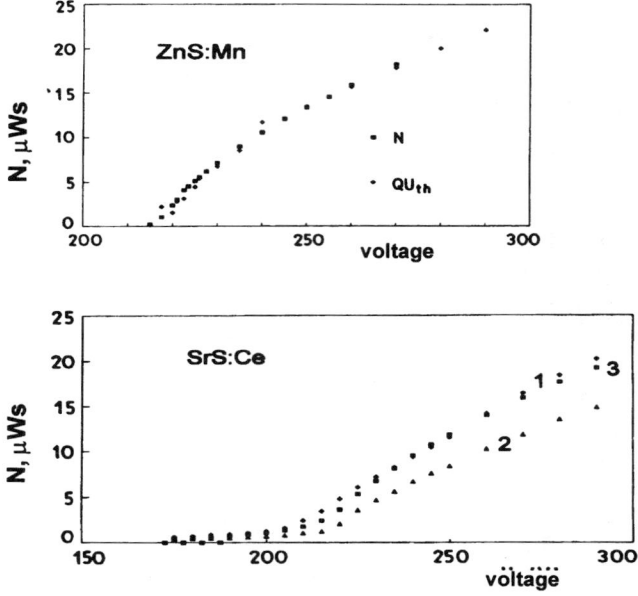

FIG. 69. Electrical energy input per period N versus (peak) voltage (a) of a ZnS:Mn sample, demonstrating agreement between the measurements of via transferred charge, and classical integral and (b) of an SrS:Ce sample showing discrepancies between the measuring methods related to the trailing edge current (after Mueller, 1993d).

is the same as in a fictive sample of 100% efficiency. This fictitious sample decays by definition exponentially with the radiative decay time τ_{rad}; the normalized integral (normalized to the amplitude) is equal to τ_{rad}. If the normalized integral of any real measurement is τ, τ/τ_{rad} is the luminescence efficiency. This method is elegant as it is independent of absolute intensity values. Not even any constancy of sensitivity over times longer than about five decay times is required. In ZnS:Mn, the decay time of the Mn emission is controlled by concentration quenching, and therefore a direct function of Mn concentration. As early as 1979 (Hurd and King, 1979) this method was calibrated and used to determine [Mn]. Not too high an excitation has to be used, however (Mach and Mueller, 1984a; see also Subsection 5 in Section III).

The method can also be applied in photoluminescence. It is essential to absorb the light into the lowest lying excited state and to use a short pulse (Mueller, 1993d). Dye lasers give this kind of pulses, even in the case of fast decaying (dipole-allowed) transitions. The method has been used extensively on Ce^{3+} and Eu^{2+} in SrS and CaS (Mueller, 1994; Mach and Mueller, 1993). A modification of this method has been used to characterize the widely suspected field ionization of the excited state of Ce^{3+} in SrS and Eu^{2+} in CaS (Figs. 42 and 43). It measures the change by the applied field in the normalized decay integral or light sum. As photoconductivity is induced, if field ionization occurs, electrons will be shifted to the dielectrics, if the experiment is done in a TFEL sample. An example obtained on CaS:Eu is shown in Fig. 42. A rate constant for the process, competing with the radiative decay is easy to evaluate, if the field-free case is strictly exponential. (Moving the photoexcitation pulse in time relative to the drive pulse one can, on the other hand, map the electric field in the sample.) Sensing the transferred charge in the usual way on the series capacitor shows directly the movement of the field-ionized electrons, and in principle allows the comparison of the number of "lost" photons with the number of moved electrons (multiplication) (Mueller, 1992, 1993d).

f. Photoluminescence

PL is a powerful tool to assess the quality of the semiconductor or phosphor film. Not only the emission spectra but moreover the excitation spectra of the emission line or band wanted in EL is worthwhile to investigate. Most materials exhibit direct excitation lines of the dopant's internal transitions, and show the host absorption, from which the dopant can be excited via quantum-mechanical transfer. An example for a well crystalline thin film sample of CaS:Eu^{2+} is given in Fig. 70. Bad samples

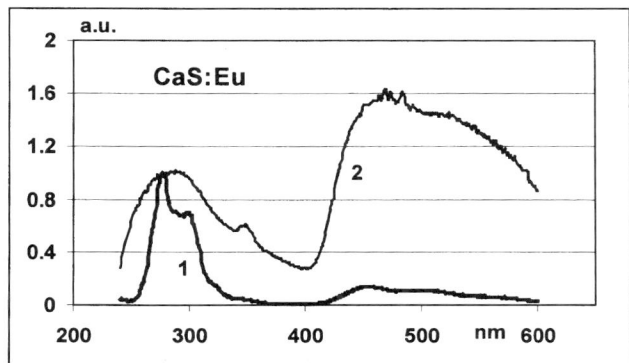

FIG. 70. Comparison of excitation spectra of CaS:Eu: (curve 1) thin film of good crystallinity and (curve 2) powder of poor crystallinity; the doping concentration is about the same.

(curve 2) very often show a rather smeared portion between the dopant and host excitation.

g. Space Charge

Very sophisticated experimental procedures have been described in the literature (e.g., Neyts, 1994) to determine the lasting changes introduced in samples by the application of pulse voltages above threshold, or simply by driving. One of the most sought after parameters is permanent or temporary space charge. As soon as the applied voltage exceeds threshold, carrier multiplication sets in and the creation of holes leads to space charge, either drifting toward the cathode or trapped on its way. As at least ZnS notoriously, but with some likelihood also the IIA–VI compounds, contains high concentrations of deep hole traps, space charge accumulation is probable. Most workers try to free the space charge by shining light on the sample; some apply high numbers of pulses just below threshold during or after illumination. In most cases, transferred charge is the measured quantity. How well the virginity of the sample is restored remains uncertain in most cases, as no easy-to-assess criterion of virginity is available. However, space–charge distribution in the sample is a very important issue, and every effort to determine it is highly welcome. It might be strongly connected to the fact that in many cases the nice linear $Q(V)$ shown in Fig. 22 is not found, but rather sublinear curves.

h. Rutherford Backscattering

RBS offers an excellent possibility to nondestructively analyze the success of the deposition processes in producing the designed structure. Except for ZnS:Mn, in which Mn is too similar to Zn, the dopant concentration, thickness, and composition of the individual films can be obtained. In most cases, the sensitivity of the method is just high enough to turn out the dopant concentrations with an 10% error bar. This is especially true for rare earths, which show up quite distinctly in an RBS spectrum.

i. Secondary Ion Mass Spectroscopy

SIMS is more suitable to analyze individual films for impurities and low dopant concentrations (and Mn in ZnS:Mn), but has the principle drawback of requiring calibration for each element, a real advantage of RBS that it does not possess. Also the depth resolution of SIMS is better than that in

RBS. It often reveals dopant concentration profiles, which are far from the wanted box shape.

j. Cathodoluminescence

Over the early years of TFEL research cathodoluminescence (CL) was used in the process of selecting candidate materials and acquiring hints for dopants with supposedly high efficiencies. However, analyzing the differences and similarities of TFEL and CL more closely, one comes to the conclusion that in some cases it has been useful, in some misleading. The most prominent example for the latter is $ZnS:Tm^{3+}$. In CL, it is a rather efficient emitter of blue around 480 nm. Numerous attempts to achieve an efficient emission in this line in TFEL, however, failed. ZnS:Tm gives a rather efficient EL at 800 nm, which, of course, is nothing to run for in the display business, but can be useful for other applications. Another example for "mislead" is the trial to get good TFEL out of oxide phosphors, which behave marvelously in CL, but because of their high electron masses are not able to supply enough energetic electrons for easy multiplexing (see Subsection 5 in Section III).

Acknowledgments

The authors gratefully acknowledge all the assistance, discussions, ideas, and contributions by their colleagues and coworkers at Central Institute of Electron Physics, Berlin; Edge Emitter Technology, Fremont, California; Hewlett-Packard Laboratories, Palo Alto, California; and by the numerous researchers worldwide with whom they had an intense exchange of samples and ideas over the last 18 years in this fascinating field.

References

Ando, M. (1992). In *Proc. 6th International Workshop on EL*, El Paso, TX, p. 85.
Arbuthnot, L., Mendes, S., Sproull, J. W., Aguillera, M., Aitchison, B., King, C. N., Dolny, G., Ipri, A., Hsueh, F. L., Stewart, R. G., Keyser, T., Schlesinger, S., and Becker, G., (1996), *SID96 Digest*, p. 374.
Baron, P. (1994). *1994 SID Seminar Lecture Notes*, F-5.
Barrow, W. A., Coovert, R. C., Dickey, E., King, C. N., Laakso, C., Sun, S. S., Tuenge, R. T., Wentross, R., and Kane, J. (1993). *SID93 Digest 24*, p. 761.
Benalloul, P., Benoit, J., Mach, R., Mueller, G. O., and Reinsperger, G.-U. (1990). 4th International Conference on II-VI Compounds 1989, Berlin, *J. Cryst. Growth* **101**, 989.
Bennalloul, P., Barthou, C., Benoit, J., Eichenauer, L., and Zeinert, A. (1993). *Appl. Phys. Lett.* **63**, 1954.

Benalloul, P. et al., (1998). *J. Alloys Compounds* **227**, p. 709.
Brennan, K. (1988) *J. Appl. Phys.* **64**, 4024.
Busse, W., Gumlich, H. -E., Meissner, T., and Theiss, D. (1976). *J. Lum.* **12–13**, 693.
Busse, W., Gumlich, H.-E., Geoffroy, A., and Parrot, R. (1979). *Phys. Status Solidi* **93**, 591.
Chen, Y. S. and Krupka, D. C. (1972). *J. Appl. Phys.* **43**, No. 10, 4089.
Corlatan, D. and Neyts, K. (1994). In *Proc. 7th International Workshop on EL*, Beijing, 1992, p. 211.
Destriau, D. (1936). *J. Chem. Phys.* **33**, 587.
DiMaria, D. J., and Fischetti, M. V. (1987). *Appl. Surface Sci.* **30**, 278.
Dobrowolski, J. A., Sullivan, B. T., and Bajcar, R. C., (1992). *Appl. Opt.* **31.28**, p. 5988.
Douglas, A. A., Wager, J. F., Morton, D. C., Koh, J. B., and Hogh, C. P. (1993). *Appl. Phys. Lett.* **63**, p. 231.
Doxsee, D. D., and Wu, X. (1997). In *3rd International Conference on Science and Technology of Phosphors*, Huntington, CA, p. 33.
Duer, M., Saraniti, M., and Goodnick, S. (1998). In *4th International Conf. on Science and Technology of Phosphors and 9th International Workshop on EL*, Bend, Oregon, p. 207.
Eichenauer, L., Jarofke, B., Mertins, H. C., Dreysig, J., Busse, W., Gumlich, H.-E., Bennalloul, P., Barthou, C., Bennoit, J., Fouassier, C., and Garcia, A. (1996). *Phys. Status Solidi A* **153**, 515.
Fouassier, C. (1996). In *Proc. 8th Workshop on EL*, Wissenschaft und Technik Berlin, p. 313.
Gumlich, H.-E. (1981). *J. Lum.* **135**, 795.
Haaranen, J., Toernquist, R., Koponen, J., Pitkaenen, T., Barrow, B., and Laakso, C. (1992), *SID92 Digest* **23**, 348.
Haerkoenen, G., Haerkoenen, K. and Toernqvist, R. (1990), *SID 90 Digest*, p. 232.
Heimbrodt, W., and Goede, O., (1986). *Phys. Status Solidi B* **135**, 795.
Howard, W. E., (1977). *IEEE Trans. Electronic Devices* **ED-24**, 903.
Howard, W. E. and Alt, P. M. (1977). *Appl. Phys. Lett.* **31**, 399.
Howard, W. E., Sahni, O., and Alt, P. M. (1982). *J. Appl. Phys.* **53**, 639.
Hsu, J., Hsueh, F. L., Stewart, R. G., Ipri, A., Connor, S., Arbuthnot, L., Sproull, W., Aguilera, M., Green, P., Nguyen, T., Keyser, T., Schlesinger, S., Becker, G., Kagey, D. (1998). *SID 98 Digest*, p. 949.
Hurd, J., and King, C. N. (1979). *J. Electron. Mater.* **8**, 879.
Inoguchi, T., and Mito, S. (1977). In *Topics in Applied Physics.* (Springer, Heidelberg), p. 222.
Inoguchi, T., Takeda, M., Kakahara, Y., Nakata, Y., and Yoshida, M. (1974). *SID74 Digest*, p. 84.
Iseler, G. W. and Strauss, A. J. (1970). *J. Lum.* **3**, 1.
Jones, T. C., Park, W., Mohammed, E., Menkara, H. M., Wagner, B. K., Summers, C. J., and Sun, S.-S. (1998). *Proc. 4th International Conf. on Science and Technology of Phosphors and 9th International Workshop on EL*, Bend, Oregon, p. 215.
Katiyar, M. and Kitai, A. H. (1992), *J. Lum.* **52**, 309.
Khang, D. (1968). *Appl. Phys. Lett.* **13**, 210.
Khormai, R., Thayer, S., Ping, K., King, C., Dolny, G., Dolny, G., Ipri, A., Hsueh, F., Stewart, R., Keyser, T., Becker, G., Kagey, D., and Spitzer, M. (1994). *SID94 Digest*, p. 137.
King, C. N (1992). *1992 SID Seminar Lecture Notes*, Vol. 1, p. M-6.
King, C. N., (1994). *1994 SID Seminar Lecture Notes*, Vol. 1, p. M-9.
King, C. N., (1996). "Active Matrix EL," In *Proc. 8th Workshop on EL*, Wissenschaft und Technik Berlin, 1996, p. 375.
Kitai, A. H., (1998). In *Proc. 4th International Conf. on Science and Technology of Phosphors and 9th International Workshop on EL*, Bend, Oregon, p. 199.
Kitai, A., Dalauc, N., and Zhizheng, H. (1991). *SID91 Digest* **22**, p. 440.

Kitai, A. H., Xiao, T., Liu, G., and Li, J. H. (1997). *SID97 Digest* **28**, 419.
Kobayashi, H., Tanaka, S., Shanker, V., Shiiki, M., Kunou, M., Mita, J., and Sasakura, H., (1985). *Phys. Status Solidi A* **88**, 713.
Kononets, Y. F., Toernqvist, R. and Vlasenko, N. (1996). In *Proc. 8th Workshop on EL*, Wissenschaft und Technik Berlin, p. 259.
Kun, Z. K., Leksell, D., Malmberg, F., Asars, J., and Brandt, G. B., (1986). *SID86 Digest* **86**, 270.
Li, L., Dobrowolski, J. A., Sullivan, B. T., Simpson, R., and Bajcar, R. C. (1994). *SID94 Digest* **25**, 140.
Liu, G., Xiao, T., Lobban, K., and Wu, X. (1998). In *Proc. 4th International Conf. on Science and Technology of Phosphors* and *9th International Workshop on EL*, Bend, Oregon, p. 179.
Mach, R. (1992). In *Proc. 6th International Workshop on EL*, El Paso, TX, p. 31.
Mach, R. (1993). In *Solid State Luminescence*, A. H. Kitai, Ed. (Chapman & Hall, London), p. 229.
Mach, R., and Mueller, G. O. (1982). *Phys. Status Solidi A* **69**, 11.
Mach, R., and Mueller, G. O. (1984a). *J. Lum.* **31 & 32**, p. 954.
Mach, R., and Mueller, G. O. (1984b). *Phys. Status Solidi A* **81**, 609.
Mach, R., and Mueller, G. O. (1986). *Proc. 18th ICPS*, Stockholm, p. 1587.
Mach, R., and Mueller, G. O. (1991). *Semicond. Sci. Technol.* **6**, 305.
Mach, R., and Mueller, G. O. (1993). In *Electroluminescent Materials, Devices, and Large Screen Displays*, Esther M. Conwell, Milan Stolka, and M. Robert Miller, Eds., *Proc. SPIE* **1910**, 48.
Mach, R., and Mueller, G. O. (1994). *Phys. Status Solidi A* **144**, 493.
Mach, R., Mueller, G. O., Schulz, G., Kalben, J. V., and Gericke, W. (1984). *Phys. Status Solidi A* **81**, 733.
Mach, R., Schrottke, L., Mueller, G. O., Reetz, R., Krause, E., and Hildish, L. (1988a). *J. Lum.* **40 & 41**, 799.
Mach, R., Mueller, G. O., Reetz, R., and Reinsperger, G.-U. (1988b). *SID88 Digest*, p. 23.
Mach, R., Mueller, G. O., Schnuerer, E., Selle, B., Ohnishi, H. (1990a). *Proc. 5th Intern. Workshop on EL*, Helsinki; *Acta Polytech. Scand.* **170**, 197.
Mach, R., Mueller, G. O., Schrottke, L., Benalloul, P., and Benoit, J. (1990b). In *Proc. 5th International Workshop on EL*, Helsinki; *Acta Polytech. Scand.* **17**, 161.
Mach, R., Reinsperger, G.-U., Mueller, G. O., Selle, B., Matzkeit, G. (1991). In *Proc. II-VI Conf. Tamano 1991*; *J. Cryst. Growth* **117**, 1002 (1992).
Mikami, A., Ogura, T., Tanaka, K., Yoshida, K., and Nakajima, S. (1987). *J. Appl. Phys.* **61**, 3028.
Minami, T. (1998a). In *Proc. International Conf. on Science and Technology of Phosphors* and *9th International Workshop on EL*, Bend, Oregon, p. 195.
Minami, T., Kuboto, Y., Miyata, T., and Yamada, H. (1988b). *SID98 Digest*, p. 953.
Morton, D. C., (1987). *SID87 Digest*, p. 345.
Mueller, G. O. (1992). In *Proc. 6th International on EL*, El Paso, TX, p. 102.
Mueller, G. O. (1993a). In *Proc. 13th International Display Research Conference*, Eurodisplay '93, Strasbourg.
Mueller, G. O. (1993b). In *Solid State Luminescence*, A. H. Kitai, Ed. (Chapman & Hall, London), p. 133.
Mueller, G. O. (1993c). *Phys. Status Solidi A* **139**, 271.
Mueller, G. O. (1993d). *Phys. Status Solidi A* **139**, 263.
Mueller, G. O. (1994). In *Proc. 7th International Workshop on EL*, Beijing, p. 7.
Mueller, G. O. (1996). In *2nd International Conference on Science and Technology of Display Phosphors*, San Diego, CA, p. 82.

Mueller, G. O., and Mach, R. (1983). *Phys. Status Solidi A* **77**, K179.
Mueller, G. O., Mach, R., Selle, B., and Schulz, G. (1988a). *Phys. Status Solidi A* **110**, 657.
Mueller, G. O., Neugebauer, J., Mach, R., and Reinsperger, G.-U. (1988b). *J. Crystal Growth* **86**, 890.
Mueller, G. O., Mach, R., Halden, E., and Fitting, H.-J. (1990). *Proc. XX Intern. Conference on the Physics of Semiconductors*, Tessaloniki, Greece, p. 2510.
Mueller, G. O., Mach, R., Reinsperger, G.-U., Schulz, G. (1991). In *Proc. Int. Display Research Conf.*, San Diego, CA, p. 815.
Mueller, G. O., Mueller-Mach, R., Alinsog, E., Lee, H., and Harrison, D. (1996). In *Proc. 8th International Workshop on EL*, Wissenschaft und Technik Berlin, p. 399.
Mueller, G. O., Mueller-Mach, R., Alinsog, E., Lee, H., and Harrison, D. (1997). *J. Lum.* **72-74**, 1002.
Mueller-Mach, R., and Mueller, G. O. (1996). In *Proc. 8th Workshop on EL*, Wissenschaft und Technik Berlin, p. 381.
Mueller-Mach, R., Mueller, G. O., and Nauka, K. (1996). In *2nd International Conference on Science and Technology of Display Phosphors*, San Diego, CA, p. 191.
Mueller-Mach, R., Mueller, G. O., Fouassier, C., Garcia, A., Fa, X. W., Zhong, G. Z., Zhao, L. J., and Sun, J. M. (1999). In *3rd International Conference on Science and Technology of Display Phosphors*, Huntington, CA, to be published.
Neyts, K. (1994). In *Proc. 7th International Workshop on EL*, Beijing, p. 359.
Neyts, K., and Viljanen, K. (1994). In *Proc. 7th International Workshop on EL*, Beijing, p. 28.
Nire, T., Matsonu, A., Wada, F., Fuchiwaki, K., and Miyakoshi, A. (1992). *SID92 Digest* **23**, 352.
Okamoto, S. and Kobayashi, K. (1994). In *Proc. 7th International Workshop on El*, Beijing, p. 194.
Okibayashi, K., Ogura, T., Terada, K., Taniguchi, K., Yamashita, T., Yoshida, M., and Nakajima, S. (1991). *SID91 Digest* **22**, 275.
Ono, Y. A. (1995). *Electroluminescent Displays*. World Scientific Publishing Co., Singapore.
Park, W., Jones, T. C., and Summers, C. J. (1999). *Appl. Phys. Lett.* **74.13**, p. 1785.
Payne, S. A., Austin, R. H., McClure, D. S. (1984). *Phys. Rev. B* **29**, 32.
Rack, P. D. and Holloway, P. (1998). *Mater. Sci. Eng. Repts.* **21**, 171.
Smith, D. H. (1981). *J. Lum.* **23**, 1–2, 209.
Soininen, E, (1998). In *4th International Conf. on Science and Technology of Phosphors* and *9th International Workshop on EL*, Bend, Oregon, p. 165.
Soininen, E., Haerkoenen, G., and Vasama, K. (1997). In *3rd International Conf. on Science and Technology of Phosphors*, Huntington, CA, p. 105.
Sun, S.-S. (1998). In *Proc. 4th International Conf. on Science and Technology of Phosphors* and *9th International Workshop on EL*, Bend, Oregon, p. 183.
Sun, S.-S., Dickey, E., Kane, J., and Yocom, P. N. (1997). *IDRC Proc.*, p. 301.
Tanaka, S., Yoshiyama, H., Nishiura, J., Ohshio, S., Kawakami, H., and Kobayashi, H. (1988). *SID88 Digest* **19**, 293.
Tanaka, S., Yoshiyama, H., Nishiura, J., Ohshio, S., Kawakami, H., and Kobayashi, H. (1989). *Springer Proc. in Physics*, Vol. 38, p. 56.
Toernqvist, R. (1996). In *Proc. 8th International Workshop on EL*, Wissenschaft und Technik Berlin, p. 279.
Toernqvist, R. (1997). *SID97 Digest* **28**, 855.
Velthaus, K. O. (1997). In *3rd International Conf. on Science and Technology of Phosphors*, Huntington, CA, p. 81.
Vlasenko, N. A., and Popkov, Z. (1960). *Opt. I Spectrosk.* **8**, 39.
Vlasenko, N. A., Zynyo, S. A., and Pukhlii, P. (1970). *Optics and Spectr.* **28**, 66.

Wager, J. F., Douglas, A. A., and Morton, D. C. (1992). In *Proc. 6th International Workshop on EL*, El Paso, TX, p. 92.
Wager, J. F. and Keir, P. D. (1997). *Annu. Rev. Mater. Sci.* **27**, 223.
Wu, X. (1996). In *Proc. 8th International Workshop on El*, Wissenschaft und Technik Berlin, p. 285.
Wu, X., Bailey, P., Carkner, D. E., Doxsee, D. D., Foo, K., Slade, S. L., Smy, W. M. (1994). In *Proc. 7th International Workshop on EL*, Beijing, p. 227.
Yamashita, N., and Michitsuji, Y. (1987). *J. Electrochem. Soc.* **134.11**, 293.

CHAPTER 3

Materials in Thin Film Electroluminescent Devices

Markku Leskelä, Wei-Min Li, and Mikko Ritala

DEPARTMENT OF CHEMISTRY
UNIVERSITY OF HELSINKI
HELSINKI, FINLAND

I. INTRODUCTION 107
II. DEVICE STRUCTURES AND CHARACTERISTICS 108
III. FILM DEPOSITION TECHNIQUES 111
IV. TFEL PHOSPHORS 112
    1. *Requirements for the Phosphors* 112
    2. *Dominant Materials* 117
    3. *Emerging Materials* 136
    4. *Materials of Minor Importance* 139
    5. *Color by White* 152
V. OTHER MATERIALS 154
    1. *Substrates* 154
    2. *Transparent Electrodes* 156
    3. *Metal Electrodes* 157
    4. *Insulators* 158
VI. CONCLUSIONS 165
    REFERENCES 166

I. Introduction

In electroluminescent (EL) devices, electrical energy is converted to visible light by a nonthermal process. The EL devices can be divided into two classes: (i) light-emitting diodes (LEDs), where the light is generated by electron–hole pairs, and (ii) high-energy EL devices where the high-energy electrons excite the emitting centers, dopant ions, to give luminescence. The latter EL devices can further be grouped in different types depending on the form of phosphors, powder vs thin film, and of the driving voltage, AC vs DC. Two of these types, namely, AC thin film electroluminescent (ACTFEL) devices and AC-driven powder EL devices are commercially available (Godlewski and Leskelä, 1994). Due to its good performance (operation brightness, viewing angle, ruggedness, temperature range) ACTFEL is the most important EL device type and has gained interest in demanding applications. In fact ACTFEL is the most reliable and longest lasting flat panel display technology on the market (King, 1996a; Kobayashi and Tanaka, 1996). Monochrome, yellow-emitting ACFTEL devices based on ZnS:Mn are commercially available in sizes ranging up to full size for working stations, the most common size being, however, the half-page. The big challenge in the ACTFEL technology is to produce multicolor and full-color devices. The lack of efficient blue phosphor has delayed the manufacturing of full-color devices, but multicolor displays have been available for some years. AC powder EL has found use in lighting applications, for example as a back light in liquid crystal displays (LCDs). The focus of this chapter is in materials needed in ACTFEL devices.

Numerous review articles have been published on TFEL devices and materials used therein, see, for example, Ono (1992, 1995, 1997), Godlewski and Leskelä (1994), Kobayashi and Tanaka (1996), Leskelä (1998), and Rack and Holloway (1998). In these reviews, the focus has usually been on the luminescent materials, and during the latest years, particularly on possible candidates for blue phosphors. The review articles coming from the industry emphasize the device technological issues (King, 1996b). In this chapter the current state of the art of materials needed in TFEL devices is highlighted. All materials — substrates, conductors, dielectrics, and phosphors — are included. Although the most space is given for phosphors, dielectrics and conductors are discussed as well because except in the book of Ono (1995) and the review by Rack and Holloway (1998), no presentation on all materials has been made. In the discussions on the phosphors, the most attention is given to the dominant phosphors (ZnS:Mn, ZnS:Tb, SrS:Ce) and the promising new blue phosphor SrS:Cu, while those rare-earth-doped ZnS, CaS, SrS, $CaGa_2S_4$, ZnF_2, and different oxide phosphors, which have not been as successful get less attention. In presenting the properties of the phosphors, EL luminance, efficiency, and emission color are described, while very little attention is paid to electrical properties such as space–charge, memory effects, and leading and trailing edge emission. The physics and mechanisms in EL devices are beyond the scope of this review and are covered in Chapter 2 of this book.

II. Device Structures and Characteristics

An ACTFEL device consists of a metal–insulator–semiconductor–insulator–metal (MISIM) structure deposited on a substrate which, in a traditional case, is transparent glass (Fig. 1). The central layer is the semiconducting thin film phosphor, which is sandwiched between two dielectric layers. The dielectrics have two functions: they protect the luminescent layer and limit the current going into it, and they store charge that enhances the internal electric field and increases the luminous efficiency of the device (King, 1996a). The electrodes complete the capacitive structure, and the electrode on the glass substrate must be transparent in the conventional device. The electrodes are patterned perpendicular to each other to form a matrix of picture elements (pixels).

Another ACTFEL structure is the so-called inverted structure, where the places of the transparent and metal electrodes have been changed and where the viewing is not made through the substrate (Haaranen et al., 1992) (Fig. 1). The inverted structure allows the use of opaque substrates, as is the case, for example, in the thick film dielectric hybrid EL devices (Wu, 1996).

The multicolor and full-color devices can be fabricated by both conventional and inverted structures. For phosphor layers four different approaches have been used in multicolor and full-color devices, namely (i) the stacked unpatterned phosphor system, which requires extra dielectric and transparent conducting layers (Barrow et al., 1986); (ii) patterned phosphors, where the blue, green, and red phosphor films are deposited and patterned separately (Barrow et al., 1988); (iii) "color by white" broadband phosphors, where the primary colors needed are filtered by patterned color filters on the top of the thin film stack (Tanaka et al., 1988b; Haaranen et al., 1992); and (iv) a hybrid structure that is based on two substrates, one

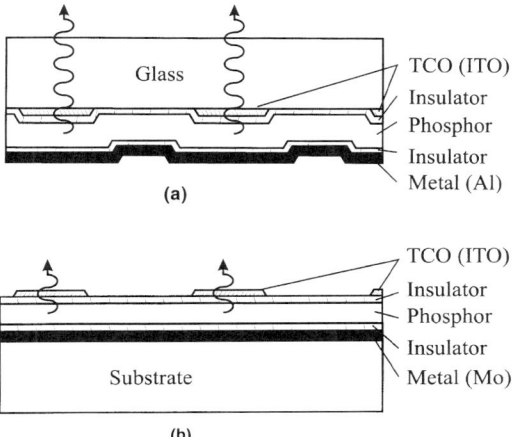

FIG. 1. Cross section of ACTFEL devices with (a) conventional and (b) inverted structures.

containing the blue phosphor and the other the green and red phosphors (Barrow et al., 1993). The different approaches have variations produced by different research groups. At the moment, the color by white approach using ZnS:Mn and SrS:Ce phosphors seems to be the most attractive (Törnqvist, 1997), although a better color gamut can be obtained with the dual substrate device using filtered ZnS:Mn, ZnS:Tb and $CaGa_2S_4$:Ce phosphors (Barrow et al., 1993).

There is a growing interest in small (0.5- to 2-in.), high-resolution (1000 lines per inch) displays for head mounted display applications. In EL this can be realized only with active matrix technology. The active matrix circuitry is fabricated on silicon-on-insulator (SOI) wafers, which enables the preparation of a high-voltage transistor in a small enough area (King, 1996a). All films deposited on the drive circuitry are continuous, which also means conformality requirements for the film processing technology. The first active matrix electroluminescent (AMEL) devices were based on green EL of ZnS:Tb, but white phosphor SrS:Ce–ZnS:Mn also appeared soon after (Khormaei et al., 1994; Aguilera and Aitchison, 1996; King, 1996a). In SrS:Ce–ZnS:Mn at 60 Hz, the yellow emission of ZnS:Mn dominates. In AMEL, the active matrix addressing enables the phosphor to be illuminated by a high frequency (5 kHz), which markedly increases the relative emission in the blue part of the emission spectrum, giving better white color chromaticity.

Important characteristics of the EL devices are luminance (L–V), luminous efficiency (η–V), current density (I–V), and transferred charge (ΔQ–V) vs voltage curves (Fig. 2). All curves show a threshold voltage (V_{th}) below which the insulator and phosphor layers in a series form the capacitance of the device. Above V_{th}, the phosphor layer looses its insulating properties and electrons in the interface states are injected into the phosphor layer. Simultaneously, luminescence is observed. The average field in the phosphor layer is of the order of 10^8 V/m, therefore setting high requirements for the film quality. The luminance and transferred charge increase rapidly above the threshold voltage but saturate later. The luminous efficiency is at its maximum just above V_{th}. The L–V curve should be steep to guarantee a small modulation voltage of the device from its off to on stage.

The luminescence is proportional to the transferred charge density in the phosphor layer. The excitation probability of a luminescent center is further dependent on the number of luminescent centers, the excitation cross section and mean free path of the hot electrons, and the values depending on the phosphor materials and their crystallinity. Thus the luminescent efficiency varies in different phosphor materials, but is preferably >1 lm/W. The rise time in EL devices is of the order of microseconds, while the decay time varies from nanoseconds to milliseconds, depending on the dopant ion–matrix combinations.

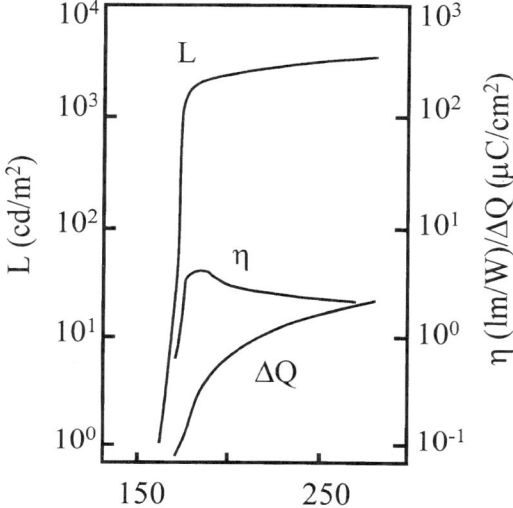

FIG. 2. Typical L–V, η–V, and ΔQ–V curves of an EL device.

The stability of the EL device is very important. The threshold voltage should not change, the L–V curve should not soften, and the luminance level should not decrease within a time of thousands of hours. Changes are normal but usually they occur in the beginning of the operation and therefore a high-frequency preaging of 50–200 h is made for the devices.

III. Film Deposition Techniques

Several techniques have been used in depositing luminescent material, insulator, and electrode layers for the TFEL devices. These techniques include various modifications of sputtering and evaporation, chemical vapor deposition (CVD), and atomic layer epitaxy (ALE), which all have their pros and cons in terms of film properties, yield, investment, operation costs, etc. When optimized, all these techniques can produce phosphor films with rather similar EL properties as has most widely been exemplified with ZnS:Mn and SrS:Ce. Sputtering and evaporation techniques are usually the most flexible and straightforward for codoping with different elements, whereas CVD and ALE have advantages in deposition of high-quality pinhole-free and conformal insulators. To simplify the production, techniques capable of depositing several films in one continuous process should be preferred.

Numerous comprehensive descriptions of the preceding deposition techniques in general can be found from text books and review articles, and their use in TFEL display manufacturing has been reviewed, for instance, by Leskelä (1989), Saunders and Vecht (1989), Leskelä and Niinistö (1990, 1992b), and Ono (1990, 1992, 1995). Therefore they are not dealt with in this context.

In preparation of the novel hybrid EL displays some of the thin film processes (i.e., deposition of metal electrode and dielectric layers) are replaced by thick film screen printing followed by high temperature (900°C) sintering and sol-gel planarization of the dielectric surface (Wu, 1996). The advantages of screen printing are simplicity and low cost, but it has limitations concerning the display resolution.

For the following discussions on the thin film materials, it is worth emphasizing here the important effects that the deposition methods often have on the properties of the films. In other words, the film properties are a result of two factors: the nominal film composition and the deposition method. Properties affected by the deposition method include the true film composition (stoichiometry of main components, impurities and their distribution), crystallinity (phase, orientation, crystallite size), morphology (grain size, density, surface roughness), stress, and adhesion. These all have important effects on TFEL display performance and must be recognized while comparing films and devices made with different methods, or even with the same method but with different setups or process parameters.

Taking ZnS:Mn as an example, the deposition method and/or temperature determine whether the crystalline phase is cubic or hexagonal, which have somewhat different emission peak wavelengths of 585 nm (orange-yellow) and 580 nm (yellow), respectively. The same factors also affect the thickness of the so-called dead layer; that is, the starting layer with such a poor crystallinity that no luminescence takes place. Yet another important, though less thoroughly understood property affected by the deposition method is the aging characteristics; that is, the shifts in the L–V curves during operation. Depending on the deposition method or precursors used, the L–V curves may shift toward either lower or higher voltages, and some softening (change of slope) may also occur.

IV. TFEL Phosphors

1. REQUIREMENTS FOR THE PHOSPHORS

According to Ono (1992, 1995) the requirements for ACTFEL phosphor matrix material are: (i) they must have a large bandgap to emit visible light from the doped luminescent centers, (ii) the material has to withstand high electric fields (10^8 V/m) without electric breakdown, (iii) the material must behave like an insulator below the threshold voltage for luminescence, (iv) the host material can be doped with an activator that gives bright emission at a desired wavelength under excitation, (v) the material must be able to tolerate annealing temperatures up to 600°C, and (vi) there must be a method to deposit the phosphor in thin film form. The luminescent centers must satisfy the following requirements: (i) they must be properly incorporated into the host lattice and emit the desired light, (ii) the cross section for impact excitation must be large, and (iii) they must be stable in the high electric fields used. The materials that fulfill these requirements are II–VI compounds (ZnS, ZnSe, CaS, SrS), some ternary sulfides ($CaGa_2S_4$, $SrGa_2S_4$), and some oxides doped with different elements or molecules. As noted, all the common thin film deposition techniques can be used in fabrication of phosphor layers for ACTFEL devices.

Table I lists structures and some physical properties of frequently studied phosphor host materials. In Table II, Shannon's (1976) effective ionic radii of the elements used as host materials and possible dopants and codopants are listed for comparison. Only a few rare-earth ions are mentioned in the table since the f elements have similar radii, and they follow the contraction rule with the increase of the atomic number. Note that the ionic radii are estimated for coordination number of 6 and they are not necessarily equal to the values found in every material. In the case of ZnS, the coordination number of Zn is 4.

Yellow-emitting ZnS:Mn is by far the most studied and used phosphor material in ACTFEL devices. Monochrome yellow–black panels based on this phosphor were commercialized in the early 1980s and a wide variety of products are now available. The paper by Kahng (1968) on $ZnS:REF_3$ (RE:.rare earth elements) phosphors started the EL color phosphor development that has made a rapid progress during the 1990s (Leskelä and Niinistö 1992a; Ono, 1995). For full-color panels, blue, green, and red phosphors are required. The best red ACTFEL phosphor today is filtered ZnS:Mn (Tuenge, 1992). Green-emitting ZnS:Tb has been successful and high efficiency and brightness is obtained with different luminescent centers, for example, ZnS:TbOF (Okamoto et al., 1988). The blue material has been a problem and the weakness of the materials available has delayed the fabrication of full-color devices. Current ACTFEL materials and their EL performances are collected in Table III.

For devices the luminance (in candela per square meter), efficiency (in lumens per watt), color chromaticity, and aging are the most crucial properties of the phosphors. The 1936 CIE (Commission Internationale de l'Eclairage) color coordinates of the primary CRT phosphors and the

TABLE I

STRUCTURE AND PHYSICAL PROPERTIES OF PHOSPHOR HOST MATERIALS

	ZnS	CaS	SrS	BaS	CaGa$_2$S$_4$	SrGa$_2$S$_4$
Structure	Cubic, zinc blende	Hex., wurtzite	Cubic, rock salt	Cubic, rock salt	Cubic, rock salt	Orth.	Orth.
Space group	F$\bar{4}$3m	P6$_3$mc	Fm3m	Fm3m	Fm3m	Fddd	Fddd
Lattice constant (Å)	$a = 5.41$	$a = 3.81$ $c = 6.26$	$a = 5.70$	$a = 6.02$	$a = 6.39$	$a = 20.09$ $b = 20.09$ $c = 12.11$ $z = 32$	$a = 20.84$ $b = 20.49$ $c = 12.21$ $z = 32$
Coordination number	4	4	6	6	6	Ca(8) Ga(4)	Sr(8) Ga(4)
Energy gap (eV)	3.7	3.8	4.4	4.3	3.8	4.2	4.4
Dielectric constant	8.3	8.1	9.3	9.4	11.3	15	14
Melting point (°C)	1020[a]	1700	>2000[b]	>2000	1200	1150	1200
Thermal expansion coefficient at RT (10^{-6}/K)	6.6	5.0	17	14	34		
Refractive index	2.36	2.36–2.38	2.14	2.11	2.16		

[a]Phase transition point.
[b]Decompose.

TABLE II

Effective Ionic Radii (Å) for Selected Ions with Coordination Number 6 (Shannon, 1976)

Zn^{2+}	0.74	S^{2-}	1.84	Mn^{2+}	0.83
Mg^{2+}	0.72	Se^{2-}	1.98	Cu^{+}	0.73
Ca^{2+}	1.00	O^{2-}	1.40	Ag^{+}	1.15
Sr^{2+}	1.18	F^{-}	1.33	Ce^{3+}	1.01
Ba^{2+}	1.35	Cl^{-}	1.81	Eu^{3+}	0.95
Ga^{3+}	0.62	Na^{+}	1.02	Eu^{2+}	1.17
Y^{3+}	0.90	K^{+}	1.38	Tb^{3+}	0.92
				Pb^{2+}	1.19

current dominant and emerging color TFEL phosphors are compared in Fig. 3. The luminance requirements, according to European Broadcasting Union (EBU) standard and the American National Standards Institute–Human Factors Society (ANSI–HFS) 100-1988 standard, are listed in Table IV, together with required subpixel luminances for two sets of fill factors.

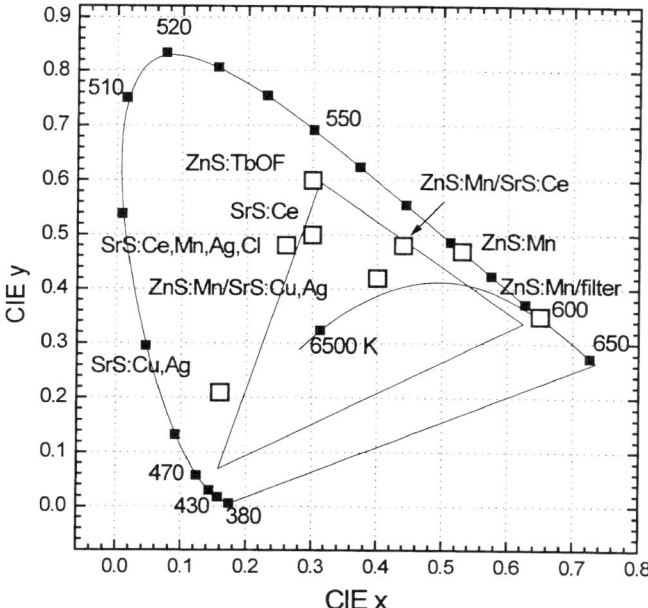

FIG. 3. CIE color coordinates of the CRT phosphors (solid squares) and the currently dominant TFEL phosphors.

TABLE III
CURRENT ACTFEL MATERIALS AND THEIR EL PERFORMANCE

EL phosphors	Luminance $L40$ (cd/m^2) at 60 Hz	Efficiency (lm/W)	CIE$_x$	CIE$_y$	Emission color	Reference
ZnS:Mn	300	5	0.53	0.47	Yellow	Ono (1995)
Zn$_{x-1}$Mg$_x$S:Mn	150		0.52	0.47	Green-yellow	Mikami et al. (1996)
ZnS:Mn–CdSe	70	0.8	0.65	0.35	Red	Tuenge et al. (1991)
ZnS:Mn–green filter	80		0.45	0.55	Green	Ono (1995)
SrS:Ce	100	0.8–1.6	0.30	0.50	Bluish-green	Leskelä (1998), King (1996b)
SrS:Ce–filter	10		0.13	0.18	Blue	Leskelä (1998)
SrS:Ce,Mn,Ag,Cl	170	1.5	0.26	0.48	Green-blue	Oberacker et al. (1997)
SrS:Ce,Mn,Ag,Cl–filter	19		0.10	0.15	Blue	Oberacker et al. (1997)
SrS:Cu	34	0.24	0.17	0.27	Blue	Sun and Chen (1997)
SrS:Cu,Ga,Ag	35	0.24	0.16	0.21	Blue	Sun (1998a)
SrS:Pb	~2		0.26	0.33	Blue-green	Nykänen (1992)
CaS:Pb	80		0.15	0.15	Blue	Yun (1999)
SrGa$_2$S$_4$:Ce	5	0.02	0.15	0.10	Blue	King (1996b)
CaGa$_2$S$_4$:Ce	10	0.04	0.15	0.19	Blue	King (1996b), Leskelä (1998)
ZnS:TbOF	100–120	1.3–1.7	0.30	0.60	Green	Leskelä (1998), Yoshino (1992)
ZnS:Tb	70	0.6	0.30	0.60	Green	Leskelä (1998)
ZnS:Sm,Cl	12	0.08	0.64	0.35	Red	Ono (1995)
ZnS:Tm,F	0.2	<0.01	0.11	0.09	Blue	Ono (1995)
CaS:Ce	10	0.1	0.27	0.52	Green	Ono (1995)
CaSSe:Eu	25	0.25	0.66	0.33	Red	Leskelä (1998)
CaS:Eu	12	0.05	0.68	0.31	Deep red	Leskelä (1998)
ZnGa$_2$O$_4$:Mn	10	0.2	0.08	0.68	Green	Rack and Holloway (1998)
ZnSi$_{0.5}$Ge$_{0.5}$O$_4$:Mn	~30	~0.2	0.27	0.67	Green	Kitai (1998), Minami (1998)
ZnF$_2$:Gd	~10				UV	Miura et al. (1991)
ZnS:Mn–SrS:Ce	470	1.5	0.44	0.48	Yellow-white	King (1996b)
ZnS:Mn–SrS:Cu	160	0.62	0.40	0.42	White	Sun (1998b)
SrS:Ce–CaS:Eu	17		0.35	0.36	White	Ono (1995)
SrS:Ce,Eu	32	0.4	0.41	0.39	White	Ono (1995)

TABLE IV

LUMINANCE REQUIREMENTS (AFTER TÖRNQVIST, 1992a)

Areal luminance (cd/m² at 60 Hz)			Required subpixel luminance (cd/m² at 60 Hz)					
Color	EBU	ANSI–HFS	Fill factor (%)	EBU	ANSI–HFS	Fill factor (%)	EBU	ANSI–HFS
Red	10.5	9.3	22	48	42	16	67	58
Green	20.6	23.0	22	93	104	22	93	104
Blue	3.9	2.7	22	18	12	28	14	9.6
White	35.0	35.0	66			66		

EBU: European Broadcasting Union Standard (R:G:B: = 30:59:11).
ANSI–HFS: American National Standard Institute–Human Factor Society 100-1988 Standard.
(R:G:B: = 27:65:8).

2. DOMINANT MATERIALS

a. ZnS:Mn

ZnS:Mn is among the earliest EL phosphor discovered (Destriau, 1937) and it is the most successful TFEL phosphor known today. There was an enormous amount of literature dealing with the EL of ZnS:Mn before the 1970s, mainly focusing on the AC and DC powders. During that period, TFEL ZnS:Mn had little progress largely due to the lack of good insulating and transparent conducting layers. The first announcement of a high-brightness and long-life ZnS:Mn TFEL device appeared from Inoguchi *et al.* (1974), who used the double insulating layer (MISIM) structure. The bright and efficient yellow emission from ZnS:Mn and its greatly improved stability has enabled its commercial applications since the early 1980s (Takeda *et al.*, 1983; Duncker, 1983). The monochrome ACTFEL ZnS:Mn display has been dominating the market, but filtered ZnS:Mn is also used as red and green components for multicolor and full-color displays. Today's ZnS:Mn TFEL device has achieved a brightness of 300–500 cd/m² with luminance efficiency of 3-6 lm/W at 60 Hz (Table III) and the operating life extends well beyond 30,000 h. Although a large amount of research has been devoted to ZnS:Mn as an EL phosphor, many aspects still remain to be solved regarding its physics, EL performance, and device reliability. In particular, realizing high-contrast and better gray-scale display have been challenging, and higher brightness and better stability are still sought after.

Almost all thin film deposition techniques have been used to deposit ZnS:Mn thin films. The techniques employed by the major commercial

ACTFEL display manufacturers are ALE (Planar International Ltd.), thermal evaporation (Planar America), and electron beam evaporation (Sharp Corporation). In PVD methods, high-purity ZnS:Mn single-source or multiple-source materials are used. In ALE, zinc and manganese chlorides or metalorganic compounds and H_2S are used as precursors (Soininen et al., 1998). Typical metalorganic precursors are diethylzinc (DEZ) and $Mn(thd)_3$ (thd = 2,2,6,6-tetramethyl-3,5-heptanedione). Despite the differences in their deposition techniques, the displays produced by these three manufacturers have comparable performances. All of them are able to produce in large volume at least 9-in. diagonal monochrome and multicolor displays based on ZnS:Mn with VGA format.

Several other techniques also demonstrate potential in producing high-quality ZnS:Mn thin films for EL devices. In metal–organic chemical vapor deposition (MOCVD), dimethylzinc (DMZ) and H_2S are the most commonly used source materials for high-quality ZnS thin films. Doping of manganese can be achieved using several cyclopentadienyl-type manganese compounds (Wright et al., 1982; Hirabayashi and Kozawakuchi, 1986; Kina et al., 1996; Su et al., 1997). Another chemical vapor deposition (CVD) technique, developed at Sharp Corporation (Mikami et al., 1991), uses H_2 and HCl as carrier gases to deliver Zn, S, and Mn from ZnS powder and Mn metal. Multisource deposition in which the ZnS:Mn film is reactively evaporated using Zn, S, and Mn elements has been used as well (Nire et al., 1994). It was thought that sputtering is not a favorable technique for the ZnS:Mn deposition due to the small crystallite size and nonstoichiometry of the product (Matsuoka et al., 1988; Ono, 1995). Nevertheless, Lewis et al. (1998) showed that the EL properties can be enhanced by fluxing for example KCl in their ZnS:Mn rf sputtering process. In particular, EL brightness and efficiency may exceed the state-of-the-art ALE or thermal evaporated ZnS:Mn thin films. The molecular beam epitaxy (MBE) technique is well known for its capability of depositing high-quality epitaxial ZnS films (Schön et al., 1997), but the high cost of the equipment is not appreciated in large-scale production. An interesting technique, successive ionic layer adsorption and reaction (SILAR), has been successfully used for ZnS:Mn film deposition at room temperature without vacuum. This technique, based on aqueous solutions with a principle similar to ALE, requires little investment and films can even be deposited on polymer substrates. Unfortunately the deposition rate is low and no substantial EL result has been reported so far (Lindroos et al., 1995).

ZnS has two main structures: cubic zinc blend and hexagonal wurtzite, which is stable above the phase transition temperature of 1020–1150°C (Fig. 4). In thin films, both structure modifications can be obtained at relatively low temperatures (<500°C). Most of PVD ZnS:Mn films pro-

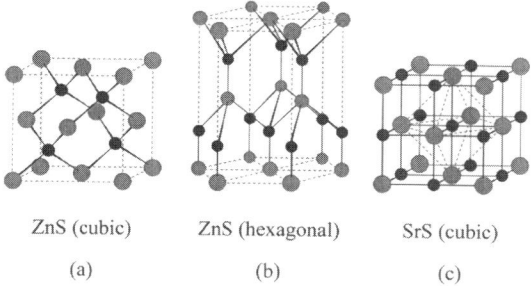

FIG. 4. Structures of (a) cubic zinc blend, (b) hexagonal wurtzite, and (c) rock salt.

duced at low deposition temperatures [room temperature (RT) to 200°C)] have the cubic phase but multisource deposited films may have mixed cubic and hexagonal phases at a substrate temperature of 350°C (Nire et al., 1994). The hexagonal phase becomes dominant in ALE ZnS:Mn at growth temperatures of 350–500°C (Soininen et al., 1998). A similar tendency is also seen in other CVD processes (Mikami et al., 1991; Kina et al., 1996). A number of works have reported that Mn doping in ZnS tends to result in the hexagonal phase (Mikami et al., 1991; Kreissl and Gehlhoff, 1984). Extended x-ray absorption fine structure (EXAFS) (Charreire et al., 1993) and electron-paramagnetic-resonance (EPR) (Kreissl, 1986; Chen et al., 1998) studies confirmed that Mn replaces Zn in the host lattice even when the Mn concentration is high. On the other hand, Yeom et al. (1996) reported that Mn occupies cubic Zn site in the hexagonal ZnS phase. This result seems inconsistent with the spectral observation. Nevertheless, the same oxidation state and similar ionic radii of Mn^{2+} (0.83 Å) and Zn^{2+} (0.74 Å) is advantageous in many respects, for example, there is no need of introducing charge compensator and it allows a high concentration of luminescence centers while maintaining good crystallinity in an efficient phosphor.

The wide bandgap of ZnS (cubic: 3.7 eV; hexagonal: 3.8 eV) enables high-energy electrons to excite the luminescent centers within the bandgap. The luminescence of ZnS:Mn originates from the $^4T_1-^6A_1$ transition within the 3d orbital of a Mn^{2+} ion. The $3d^5$ configuration of Mn is splitted into several energetic states by the tetrahedral crystal field. The influence of the crystal field is also reflected from the emission peak position; that is, in the cubic ZnS:Mn the emission maximum locates at about 585 nm, whereas a blue shift to about 580 nm is observed in the hexagonal ZnS:Mn. Typical EL spectra of ZnS:Mn thin films with cubic and hexagonal structures are shown in Fig. 5. The decay of an isolated Mn^{2+} ion emission in a ZnS host is long, in the range of milliseconds, due to the forbidden transitions (Benalloul et al., 1990).

Often blue photoluminescence emission can be observed in ZnS:Mn thin films together with the yellow Mn^{2+} emission (Fig. 5). It is well known that this blue emission is associated with the intrinsic defects in ZnS, although the reported peak location of the emission band varies somewhat from 420 to 460 nm and the broadband can extend to longer wavelengths seen as bluish-white emission (Tanaka et al., 1976a; Mikami et al., 1991). The blue emission is the result of the recombination of a self-activator center such as a Zn vacancy, with a shallow donor of various kinds depending on the different film deposition techniques. In the ALE ZnS films, blue PL having a short decay can be observed when using a $ZnCl_2$ precursor. Films made from Zn acetates do not show this blue emission, which gives a indication that the emission originates from chlorine impurities and has a donor–acceptor pair (DAP) nature. In Mn- and Tb-doped films this blue PL can

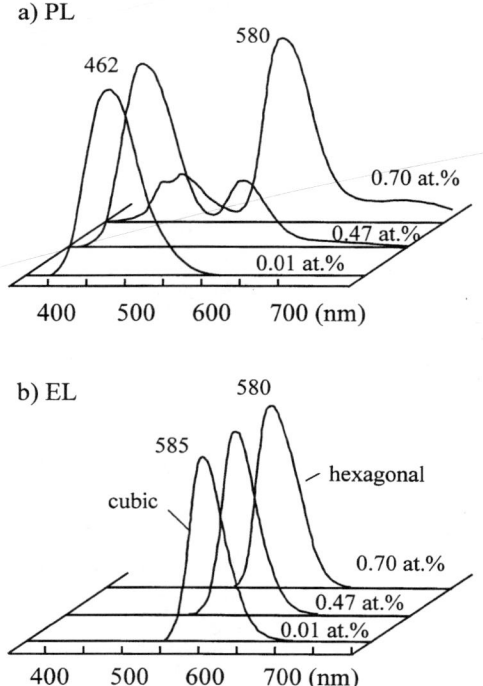

FIG. 5. PL and EL of cubic and hexagonal ZnS:Mn thin films. (From Mikami et al., 1991, with permission from Elsevier Science.)

be seen due to energy transfer to the dopant only in spectra recorded 2 μs after the laser pulse. In EL, both undoped and doped ALE ZnS films show no blue emission (Swiatek *et al.*, 1989).

The blue emission is also seldom reported in EL in films prepared with other deposition techniques. Tanaka *et al.* (1976a) did not obtain the blue emission in their evaporated films using a time-resolved measurement with a time interval of 5 μs. They used this as evidence for the impact excitation mechanism in ZnS:Mn electroluminescence. On the other hand, Okamoto and Miura (1986) showed that the broad host emission band at 460 nm can be observed with a time resolution of less than 2 μs and the blue emission is more efficient in Mn- or Tb-doped ZnS than in an undoped host. They further proposed that impact excitation may not be the only efficient excitation process, but radiative recombination of bound excitons at Mn or TbF centers followed by energy transfer to the dopant ions can be also efficient. Douglas *et al.* (1993) and Ang *et al.* (1995) reported that there does exist a "blue flash" in both ALE and evaporated ZnS:Mn, and they took this as the evidence of the presence of band-to-band impact ionization in the ZnS:Mn EL process as well.

The mechanism behind electroluminescence of ZnS:Mn thin films is a subject that has been intensively studied. The ease of producing ZnS:Mn thin films with good EL performance by various deposition techniques has made the ZnS:Mn TFEL device an outstanding example to study the basic physics of electroluminescence in general. Extensive reviews dealing with these issues can be found in the literature (Mach and Müller, 1982; Müller, 1984; Müller and Mach, 1988; Bringuier, 1994; Wager and Keir, 1997). It is not the purpose of this chapter to detail every aspects of electroluminescence; however, in the following text regarding the EL properties of the ZnS:Mn TFEL devices, topics such as memory effects, aging, and asymmetry are briefly discussed with more emphasis on the chemical aspects.

The luminance and efficiency of ZnS:Mn TFEL devices are affected by the manganese concentration. In films made with various deposition techniques, the optimum Mn concentration for good luminance and efficiency has been 0.5 ~ 2 mol% (Sasakura *et al.*, 1981; Törnqvist, 1983; Mauch *et al.*, 1988; Mikami *et al.*, 1991; Nire *et al.*, 1994; Kina *et al.*, 1996). A calculation shows that the critical concentration is around 0.8 mol % for randomly doped ZnS:Mn (Katiyar and Kitai, 1992). But even an optimum Mn concentration of 3 mol % has been reported (Schön *et al.*, 1997). However, a high Mn concentration leads to concentration quenching and shortening of the decay time. In the early experiments on both single-crystal and thin films, it was suggested that increasing the Mn concentration causes the formation of Mn pairs and clusters that have shorter decay times as compared with the isolated Mn^{2+} center (McClure, 1963; Vlasenko *et al.*, 1975). Meanwhile, a high Mn concentration increases the probability of energy migration among excited and nonexcited Mn ions and subsequently an energy loss to various traps in forms of red, infrared, or nonradiative centers takes place. Later studies found that the characteristic decay time shortening with increased EL luminance may be related to the cross-relaxation among two near-neighboring excited Mn^{2+} ions (Müller and Mach, 1988; Benalloul et al., 1990). Another interesting observation is that in a high electric excitation field, the L–V curve becomes steeper at higher Mn concentrations (Törnqvist, 1983). The changes of electrical properties of ZnS:Mn thin films under a high excitation field can be considered a result of free electron generation at high Mn concentrations, consequently increasing the nonradiative energy transfer.

It is well known that increasing the film thickness increases the EL luminance and efficiency. Maintaining a thick enough ZnS:Mn layer in a filtered multicolor or full-color EL display with inverted structure is particularly important because of a loss factor of more than 50% in such structures. In an ALE ZnS:Mn, the luminance and luminous efficiency increase as a function of film thickness but only after a dead layer of about 35 nm (Törnqvist et al., 1983). The thickness of the dead layer is much greater (>100 nm) in films made by other deposition techniques (Sasakura et al., 1981; Mikami et al., 1992; Nire et al., 1994). A transmission electron micoscopy (TEM) study of the cross sections of ALE and several PVD films provided interesting details of the microstructure (Fig. 6). At the initial growth stage, the electron-beam-evaporated film has much finer grains as compared with the ALE-grown film. Films made by sputtering have even smaller grains as compared with the previous two techniques (Theis et al., 1983). Therefore, one can correlate the thicker dead layer in PVD films to the poorer crystallinity at the beginning of the film growth. On the other hand, a comparison of ZnS films having a Mn-doped thin probe layer at different positions (i.e., from initial to the last stage of film growth) shows that the EL efficiency of the probe layer at the dead layer region is not much different from that at other parts of the film (Benoit et al., 1993). This leads to the conclusion that the electrons responsible for impact excitation must be much less accelerated in the dead layer regardless of the film morphology. The different thicknesses of the dead layers associated with different deposition techniques can be explained by the much finer gains in PVD ZnS:Mn films causing stronger deacceleration of excitation electrons at grain boundaries (Sakama et al., 1993) as compared with those in the ALE films.

Hysteresis of L–V curve, or the "memory effect," was discovered when the first stable TFEL device was built by the Sharp group (Takeda et al., 1975). Howard et al. (1982) and later Neyts et al. (1994) presented some

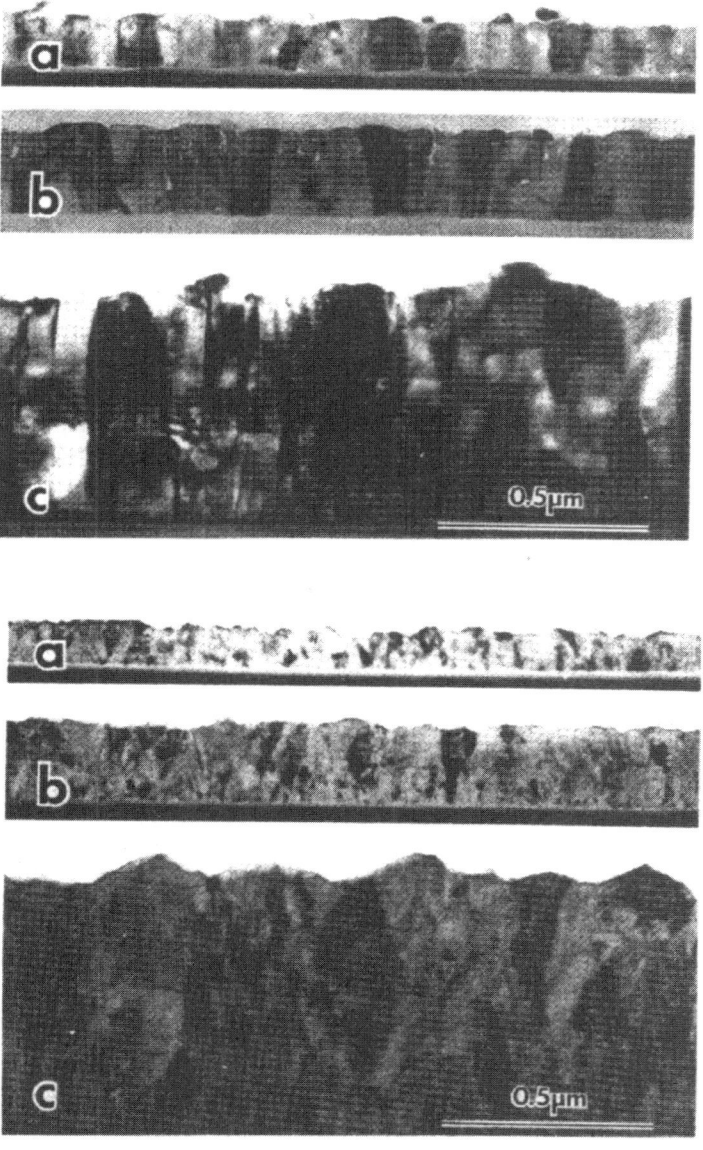

FIG. 6. Cross section TEM images of ALE grown (upper) and electron-beam-evaporated (lower) ZnS:Mn with different film thicknesses. (From Theis *et al.*, 1983, with permission from Elsevier Science).

theoretical models for the hysteresis. However, the memory effect is not observed in all ZnS:Mn films, and it is shown to be related to the Mn concentration, film thickness, deposition techniques, and insulating layers (Howard, 1981; Törnqvist and Korpela, 1982; Taniguchi et al., 1984). The hysteresis of ZnS:Mn thin films could be a valuable feature in decreasing the power consumption of the EL devices (Törnqvist, 1985), however, due to the difficulty of maintaining the hysteresis stability, the interest toward such devices seems to have declined.

Another important characteristic of a ZnS:Mn TFEL device is its aging behavior. Two distinctive changes in the L–V curve are known. As early as when the first successful ZnS:Mn TFEL device was reported, the L–V curve was found to shift to higher voltages during the initial operation and to stabilize after certain operating time (Inoguchi et al., 1974). Such a threshold voltage shift, often called as "positive shift" (P-shift) is most commonly observed in PVD ZnS:Mn films (Khormaei et al., 1989; Davidson et al., 1992). In contrast to this, ALE-prepared ZnS:Mn was reported to have a "negative shift" (N-shift) and to be accompanied by a softening of the L–V curve (Törnqvist, 1982); that is the slope of the L–V curve decreases (Fig. 7). Mikami et al. (1992) observed in their CVD ZnS:Mn that the P-shift was associated with lower deposition temperature ($<500°C$) and that an increased deposition temperature ($>500°C$) resulted in an N-shift. For an

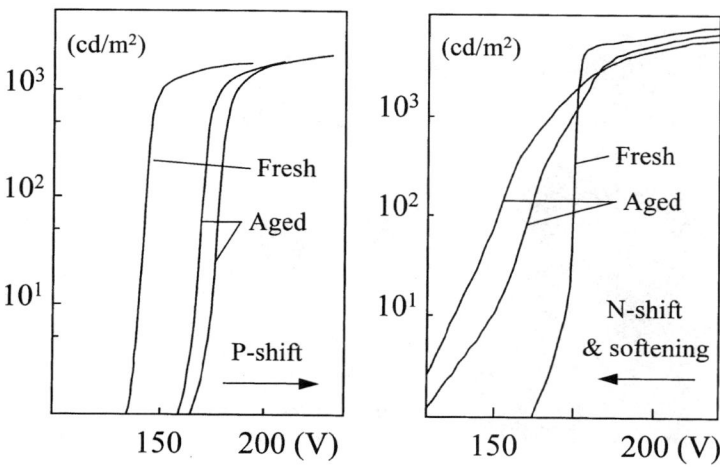

FIG. 7. Aging characteristics of ZnS:Mn TFEL devices: (left) P-shift in a CVD device and (right) N-shift and softening in an ALE device (after Mikami et al., 1992, and Soinen et al., 1998).

optimized film grown at 500°C, no shift was observed over an accelerated aging of 1000 h.

Although aging is observed in ZnS:Mn TFEL devices produced by almost all deposition techniques, its origin and chemical nature are poorly understood. Aging characteristics are closely associated with space charges of various kinds, which can be observed during the electrical characterization of the device. Howard (1981) proposed that the P-shift is associated with deep electron traps or positive space charges that accumulate near the ZnS–insulator interfaces in the ZnS layer. Khormaei et al. (1991) proposed that these deep level traps or space charges are most likely caused by intrinsic sulfur vacancies and atomic rearrangements at interfaces via thermalization. According to Mikami et al. (1992), the N-shift, in fact, consists of a P-shift but is accompanied with a much stronger softening than the P-shift ones.

Different suggestions have been offered to explain the distinct features in N-shift-type devices. Besides the positive space charge, which may originate from zinc vacancies, there may be shallow donor states present in the phosphor film (Mikami et al., 1992; Abu-Dayah et al., 1993a, 1993b, 1994; Kononets et al., 1996). A popular explanation for the nature of these donors is impurity ions in the ZnS, typically Cl^-, since chlorine-containing precursors were often used for the ZnS:Mn film deposition, for example, in ALE and CVD processes. It has been found, however, that despite replacing the chlorine-containing precursor with chlorine-free ones, the N-shift aging behavior of ALE ZnS:Mn TFEL devices still persists (Soenen et al., 1997). Some authors have thus offered O^{2-} ion as the source for shallow donors (Vlasenko et al., 1997). It is doubtful that naming a couple of impurity ions alone can explain the whole story of aging since the impurity contents, for example, oxygen in ALE ZnS thin film, is not necessarily higher as compared to that in evaporated ZnS films (Li et al., 1998a). Unfortunately, insufficient experimental data is available for a direct comparison of impurity contents in ZnS:Mn thin films prepared by different deposition techniques. Interestingly, an improvement of luminance, efficiency, and aging behaviors in sputtered ZnS:Mn can be achieved by a fluxing technique using chlorides (Lewis et al., 1998), despite the fact that traditionally chlorine has been regarded as an undesirable impurity ion that influences the EL performance in, for example, ALE and CVD ZnS:Mn films.

Note that aging behavior is a stabilization process that is intrinsic for the whole EL device and does not correspond to the degradation of the ZnS:Mn thin film during operation. One of the degradation mechanisms that should be considered is the presence of high-energy free electrons, which possibly catalyzes interface reactions during device operation (Müller and Mach, 1988). Several other chemical aspects regarding degradation have also been proposed. Lee et al. (1995) reported the reduction of the active Mn^{2+}

concentration after longtime operation. Wang *et al.* (1997) found that the brightness degradation in their MOCVD ZnS:Mn is associated with deep electron traps that originate from the reaction of Mn^{2+} ions with water molecules.

In addition to the aging with time, the L–V curve of ZnS:Mn is known to behave asymmetrically with equally applied voltage at different polarities. Most likely, asymmetric EL is related to the difference of the internal electric field between the two phosphor–insulator interfaces. In the excellent work done by Mikami *et al.* (1992), an asymmetric ZnS:Mn EL device was found closely associated to its aging characteristic as well; that is, only an N-shift-type device exhibits the strong asymmetry, while the P-shift-type device is much more symmetric even though the device structures are the same. When applying asymmetric voltage, the L–V curve of a symmetric device becomes asymmetric and exhibits characteristic N-shift-type aging behavior. It is thus suggested that internal asymmetric polarization during the device operation is responsible for the L–V curve softening and asymmetry of the ZnS:Mn TFEL devices.

Aging and asymmetry characteristics are challenging issues that are currently under discussion. The instability of the L–V curve limits, for example, the preparation of monochrome EL displays for gray-scale applications. Current technology uses either pulse amplitude or pulse width modulation for gray-scale driving by controlling the steep region of the L–V curve. If an EL device can achieve its application luminance within 30 V of the threshold voltage, then for 16 gray levels, a voltage variation of 1 or 2 V is required. Therefore an unstable aging behavior may affect the gray-scale performance considerably. The film thickness affects the threshold voltage and luminance as well. To fall in the voltage variation within 2 V, a less than 2% film thickness variation is required.

In the perspective of realizing full-color TFEL displays, the superior EL performance of the ZnS:Mn phosphor has shown its success as green and red components by using color filters (Haaranen *et al.*, 1992). The luminances of the filtered green and red using only ZnS:Mn have reached 80 and 75 cd/m² at 60 Hz, respectively (Ono, 1995). Yet there are still possibilities for further improvement. Mikami *et al.* (1996) have shown that there is a blue shift of the luminescence when adding Mg into evaporated ZnS:Mn to form a $Zn_{1-x}Mg_xS:Mn$ ($x \approx 0.1–0.3$) solid solution. Due to the increased bandgap and modified crystal field of Mn, the emission peak shifts from 580–585 to 575 nm, thus enhancing the green component by a factor of about 20%. Interestingly, Nakanishi *et al.* (1997) have found a red shift in a $Zn_{1-x}Mg_xS:Mn$ film due to the oxygen incorporation using the same deposition technique. The continuous improvements in the EL performance of ZnS:Mn have shown that the ZnS:Mn phosphor will remain the most important material for monochrome, multicolor, and full-color TFEL displays as long as there are no major breakthroughs in other phosphor systems.

b. ZnS:Tb

The unique properties of rare-earth (RE) ions, enabling light emission over the whole visible range, has been of a great importance in phosphor applications. Naturally, they are also potential candidates for multicolor and full-color TFEL applications. The EL studies on RE ions using ZnS as the host material started in late 1960s at Bell Laboratories. Following Kahng's (1968) LUMOCEN (luminescent molecular centers) ZnS:REF$_3$ TFEL devices, Chase et al. (1969) and Hamakawa et al. (1988) conducted extensive EL studies on RE fluorides incorporated in ZnS thin films. Up to now, almost all RE elements have been studied as dopants in EL phosphors using ZnS thin film as the host except La and Lu, which have filled electron configuration, Gd with emission energy of about 4 eV that is too high to be used in ZnS, and of course the radioactive Pm.

Among the ZnS:RE phosphors, only ZnS:Tb has been used for practical TFEL applications. Green-light-emitting ZnS:Tb is the second brightest monochrome phosphor after ZnS:Mn, which can be used in practical EL displays without filtering. Current ZnS:TbOF devices have achieved luminance of 100 cd/m^2 at 60 Hz driving frequency (Yoshino et al., 1992), which fullfills the minimum requirement for the green component in a full-color display, and an efficiency of 1.7 lm/W (2.6 mW) can be achieved under normal operating conditions.

It is known that the emission of Tb^{3+} in ZnS originates from the ^5D_J–^7F_J transitions within the 4f orbitals. Similar to the ZnS:Mn, the selection rules for the intrashell transitions are relaxed in such a site symmetry. Since the 4f orbital is shielded by the 5d and 6s orbitals, the crystal field has a fairly small influence on the position of the sharp line transitions in the emission spectrum. As shown in Fig. 11b, the dominant transition is $^5D_4$–$^7F_5$, located at about 550 nm. The color coordinates of the green light almost matches that of the CRT green phosphor (Fig. 3).

In the literature, the most controversy seems to be in the Tb^{3+} concentration in the ZnS host, since the ion size and charge mismatch results in small solubility. Kahng (1968) realized early that it is nessecery to incorporate a third element (i.e., ZnS:TbF$_3$), to increase the activator concentration, and hence the luminance of his evaporated TFEL device. Followed by such a guideline, a variety of phosphor compositions were studied with several deposition techniques. Kobayashi et al. (1984) showed that codoping of F$^-$,

Br^-, and O^{2-} are favorable for electron-beam-evaporated ZnS:Tb^{3+} but N^{3-} and H^- are not, and the highest optimum Tb concentration is achieved by terbium fluoride complex centers. Ogura *et al.* (1986) found in their sputtered ZnS:Tb,F that F is lost away on annealing at 600°C and the best brightness is correlated with the F:Tb ratio close to 1. Mita *et al.* (1988) confirmed that electron-beam-evaporated ZnS:Tb,F contains a higher F:Tb ratio but has a lower EL performance as compared with that of the sputtered one after annealing at 600°C. On the other hand, Hirabayashi *et al.* (1988) reported that the EL luminance and efficiency of their MOCVD ZnS:Tb,F_x are independent of the F:Tb ratio in the range of 1.4–2.6. In the work done by Mikami et al (1991), it was concluded that the F:Tb ratio is dependent on the Tb concentration; that is, the ratio is close to 1 at 0.5 at. % Tb doping, but a TbF_3 center forms in a high Tb concentration.

Nevertheless, a number of papers reported the optimum concentration of Tb–F complex centers in ZnS to vary from 1.4 mol % up to even 7 mol % (Kobayashi *et al.*, 1985b; Mikami *et al.*, 1987; Tuenge *et al.*, 1981). The Tb concentrations in ZnS EL phosphors are so high that no blue transitions from the $^5D_3$ level have ever been observed in the spectra. Later works by Okamoto *et al.* (1988, 1989a) showed that the luminance of the ZnS:Tb,F is sensitive to the oxygen impurity, and a bright ZnS phosphor system based on TbOF complex centers was developed with a rather high activator concentration (4.0 mol %).

High luminance ZnS:Tb thin films have also been prepared by ALE without any codoping (Härkönen *et al.*, 1990; Kong *et al.*, 1994). However, the ALE ZnS:Tb has a layered structure with TbS_x layers separated by thicker ZnS layers. Only about 1 mol % of Tb^{3+} can be incorporated into the ALE ZnS without reducing the crystal quality and the EL luminance. Charreire *et al.* (1992) found that the Tb^{3+} centers in ALE ZnS:Tb films are actually in the form of isolated $(TbO)^+$ clusters similar to Tb_2O_5S, which apparently explains the high luminance (100 cd/m^2), which is comparable to those of ZnS:TbOF TFEL phosphors.

Studies of the decay of the ZnS:Tb,F_x system showed that an increase of Tb concentration or the F:Tb ratio (x) results in an increased decay time (Kobayashi *et al.*, 1985b; Benalloul *et al.*, 1985). This is in strict contrast with the ZnS:Mn phosphor, in which the decay time shortens with increasing Mn concentration as the result of concentration quenching. It was thus suggested that the presence of codopant centers is an essential condition for high luminance and efficiency in Tb-doped ZnS TFEL phosphors. It is difficult to speculate where the codopants are situated or whether codopants fill the defect centers in the host matrix, since most of mentioned phosphor systems contain so high a Tb concentration that there is a tendency to form new Tb-containing phases.

According to Kahng's (1968) proposal, the formation of the so-called lumocens, or complex centers, increases the solubility of large, trivalent terbium activators in the ZnS matrix, reduces the nonradiative energy losses through the host lattice, and importantly, increases the impact excitation cross section. Furthermore, unsymmetric crystal field around Tb^{3+} center is further distorted by electronegative ions such as O^{2-} and F^-, and therefore the selection rule is more relaxed for the $4f$ intrashell transitions (Godlewski and Leskelä, 1994). As demonstrated by Kato *et al.* (1994, 1997), several different complex centers may coexist in ZnS:RE and it is reasonable to assume that these complex centers may have differences in the EL processes.

Despite many similarities, the electroluminescence characteristics of ZnS:Tb devices may have differences compared to ZnS:Mn. One of the debates concerns the EL excitation mechanism. Krupka (1972) proposed that direct impact excitation is the dominant mechanism in ZnS:Tb,F and that the impact ionization of electron–hole pairs in the host followed by energy transfer to Tb^{3+} centers is an inefficient process. Such an interpretation, similar to that of ZnS:Mn EL devices, has gained wide support from many authors with their experimental data (Kobayashi *et al.*, 1973; Tanaka *et al.*, 1976a; Mikami *et al.*, 1988). On the other hand, Marrello *et al.* (1981) compared their Mn-, TbF_3-, and Ag-doped ZnS probe layer EL devices and concluded that there is significant difference between ZnS:Mn and ZnS:TbF_3 with respect of the excitation mechanism; that is, the direct impact excitation of ZnS:TbF_3 may not be the primary process, but energy transfer plays an important role.

Okamoto and Miura. (1986) found that the well-known host blue emission by PL excitation can also be observed in their EL ZnS:Tb,F devices, and similar results were obtained in ZnS:Mn EL devices as well. As mentioned previously, they offered yet another model: hot-electron impact generates electron–hole pairs at the Tb–F complex center, and the Tb is excited by the energy transfer via electron–hole pair recombination in ZnS. Consistent with such a model, Okamoto *et al.* (1989) further believed that the ZnS:TbOF system is efficient because of the increased excitation efficiency due to the increased lattice distortion and enlarged impact cross section of the ZnS host around Tb^{3+} resulting from the incorporation of oxygen.

Miura *et al.* (1997, 1994) conducted a series of works using RE-ions-doped $Zn_{1-x}Cd_xS$, $Zn_{1-x}Mg_xS$, and $ZnS_{1-x}Se_x$ solid solutions as hosts that have either narrower or wider bandgaps than ZnS. They concluded that EL excitation of a ZnS:TbOF involves energy transfer back from Tb^{3+} centers to the host; however, their conclusion is based on the direct impact excitation model. Another interesting conclusion from Miura *et al.* (1997) is that cross-relaxation process between the $^5D_3$–$^5D_4$ excited states to $^7F_6$–$^7F_0$ ground states takes place efficiently (Fig. 8).

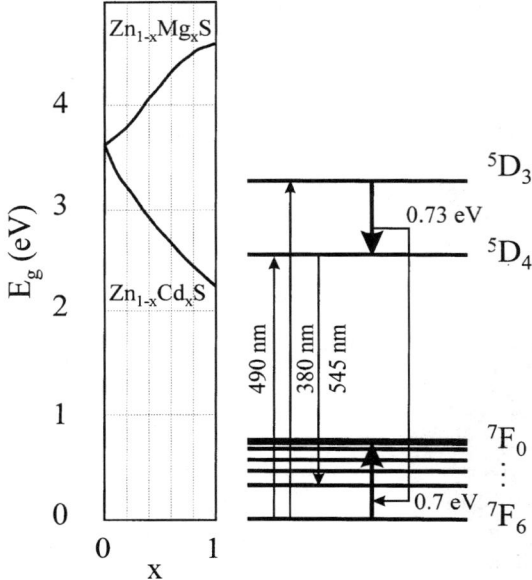

FIG. 8. Bandgap energies for $Zn_{1-x}Cd_xS$ and $Zn_{1-x}Mg_xS$ and the energy levels of Tb^{3+} ion and the possible cross-relaxation process (after Miura et al., 1997).

The stability of ZnS:Tb-based TFEL devices has been studied by several groups. Tuenge (1989) reported about 8% luminance decrease after 3000 h of operation at 1 kHz, which is fairly acceptable for display application. Khomchenko et al. (1990) found that the degradation is mainly caused by segregation of Tb centers, especially in highly doped ZnS:Tb. In the case of the ZnS:TbOF, Okamoto et al. (1989a) found that the F ion is responsible for L–V curve softening.

Despite the valuable feature of sharp line transitions and good chromaticity, ZnS:Tb is rarely used in TFEL display production. Manufacturers still use filtered ZnS:Mn phosphor for multicolor and full-color displays largely due to its comparable brightness and well-established processes. To surpass the filtered ZnS:Mn, major improvements on the brightness and EL performance of ZnS:Tb phosphor are still needed.

c. *SrS:Ce*

As mentioned before, monochrome yellow ZnS:Mn ACTFEL displays were commercialized first, followed by the development of green and red colors. Simultaneous development in processing and drive electronics made multicolor displays possible. The first larger multicolor, yellow–green–red EL panel was announced in 1991 (Laakso *et al.*, 1991), and the first 10-in. full-color TFEL device was presented in 1993 by Barrow *et al.* However, the lack of good blue phosphor has delayed the development and full commercialization of the full-color displays.

Strontium sulfide doped with cerium has been the most studied blue phosphor. Its emission is, however, blue-green having a maximum at 480 nm and a shoulder at 535 nm (color coordinates $x = 0.3$, $y = 0.5$) (Barrow *et al.*, 1984). In conventional SrS:Ce EL devices, the efficiency level has been 0.4–0.5 lm/W, independent of the applied fabrication technique. It is possible to obtain pure blue EL emission by using a blue filter, but 80–90% of the original luminance is lost. Even with this loss, the luminance level is much higher than that from ZnS:Tm.

Efforts have been made to improve the SrS:Ce brightness, color purity, and stability by codoping, minimizing defects and oxidation, modifying the lattice, filling the possible vacancies, using flux agents, and annealing. Films of SrS:Ce have been deposited by several techniques, including electron beam evaporation (Tanaka *et al.*, 1985), reactive evaporation (Tohda *et al.*, 1991; Mauch et al 1992), sputtering (Ohnishi and Okuda, 1989), multisource deposition (Tanda *et al.*, 1989), atomic layer epitaxy (Leppänen *et al.*, 1991), chemical vapor deposition (Samuels *et al.*, 1996), and molecular beam epitaxy (Tong *et al.*, 1997a). The performances of the devices obtained by the various deposition and processing techniques are very similar and it can not be stated that one method is the best, but all the methods have their own advantages and disadvantages.

In evaporation of SrS:Ce films cerium is introduced from a separate source and halides are natural choices as the precursors. Both CeF_3 (Barrow *et al.*, 1984) and $CeCl_3$ (Mauch *et al.*, 1992) have been used. The $CeCl_3$ molecule is very moisture-sensitive and its inert handling requires care. The CeF_3 molecule is more inert than the chloride and may therefore result in better EL aging properties (Poelman *et al.*, 1993). The doping with CeF_3 results in a blue shift (20 nm) in EL when compared to $CeCl_3$-doped samples (Oberacker and Schock, 1995a). In samples prepared or annealed at high temperatures, the Ce:halide ratio is close to 1:1, indicating the decomposition of the dopant molecule. Similar to the case of ZnS:Tb,F, it is also believed that in SrS the ratio 1:1 is close to optimum (Tanda *et al.*, 1989). A similar blue shift was also detected in SrS:Ce devices when codoped with LiF, while LiCl codoping did not change the spectrum (Oberacker and Schock, 1996a). In practical film deposition, $CeCl_3$ is often used as a dopant because of the lower temperature required in the evaporation and the good EL properties obtained (Hüttl *et al.*, 1996b).

The experiments on halogen codoping of films made by ALE have shown that chlorine codoping ($SiCl_4$ as a precursor) results in good initial brightness values, but the devices age rapidly (Soininen et al., 1993), while those codoped with fluorine were more stable (Nykänen et al., 1995). The emission in both chlorine- and fluorine-codoped samples was quite greenish, however. To act as a charge compensator the halogenide ion must locate in an interstitial site. The concentration of the halogenide ions must be low because the excess easily forms a strontium halide phase, which is fatal for EL (Ylilammi and Ranta-aho, 1992). The F ion implantation studies on ALE-deposited SrS:Ce films have shown that the copoding may cause a slight blue shift in the emission spectrum, and that the codoping does not decrease the luminance if the sample is annealed after the implantation (Li et al., 1998b).

Because of the oxidation state mismatch between Ce^{3+} and Sr^{2+}, SrS:Ce films have been codoped with K^+ (Tanaka et al., 1986), $CuBr_2$ (Mach et al., 1990), or Na^+ (Soininen et al., 1996), with K^+ being the most often used. The experiments with powders have shown that, based on EPR measurements, only a small amount of Ce (0.03 at. %) can be incorporated into the SrS lattice without alkaline codoping (Fouassier and Garcia, 1996). PL emission maximum is reached at a Ce level of 0.03 mol %, but when codoped with Na both PL and EL saturate at a Ce concentration of 0.2 mol % (Hüttl et al., 1996a). The emission of Ce shifts to longer wavelengths with increasing Ce concentration: from 481 nm (0.1 mol %) to 492 nm (1 mol %), and as a consequence of the reabsorption of the emitted light, the relative intensity of the short-wavelength component of the emission shows a pronounced decrease with respect to the long-wavelength component. The blue component in a sample with 1 mol % Ce is reduced by a factor of 1.5 with respect to the green component. Unfortunately, in thin films, the red shift is even more pronounced than in powders, which shows that in films some additional defects together with the concentration effect cause the red shift (Hüttl et al., 1996c).

In ALE deposition codoping with sodium has not shown clear positive effects on the EL of SrS:Ce devices (Soininen et al., 1996). If the sodium concentration is small, there are no negative effects, but with high sodium concentrations and especially if the sodium is locating at the phosphor-insulator interface, the EL brightness is decreased (Li et al., 1998a). Ion implantation of ALE SrS:Ce films with Na and K showed that with potassium it is possible to get a small blue shift in the spectrum without a loss in the brightness. Sodium implantation did not improve the properties of ALE SrS:Ce films (Li et al., 1998b).

Sulfur vacancies have been considered to drastically reduce the emission of SrS:Ce, therefore S has been coevaporated during phosphor layer deposition to guarantee stoichiometric SrS (Kobayashi et al., 1985a; Onisawa et al., 1988). The consumption of S can be reduced and better results can be obtained at lower growth temperature by cracking the sulfur molecules before they strike the growth surface (Okamoto et al., 1990b). Addition of a sulfur flux decreases the oxygen content and increases the density, crystallinity, and luminance of SrS:Ce films (Onisawa et al., 1988). The addition of sulfur seems to be deposition-dependent: in PVD it has been reported to be beneficial but not in ALE and CVD. The use of a H_2 atmosphere during evaporation of SrS:Ce films has been reported to improve significantly the luminance via reducing the oxygen content and affecting the orientation of the crystallites (Gao et al., 1992). In reactive evaporation deposition of SrS:Ce a cracker for sulfur is routinely used (for example, Hüttl et al., 1996b) but elementary sulfur was also obtained when ZnS or Ga_2S_3 were used as sulfur precursors (Velthaus et al., 1995; Inoue et al., 1997). A dramatic increase in the EL efficiency (1.5 lm/W), luminance ($L_{40} = 1700\,cd/m^2$, 1 kHz), and stability has been reported for ZnS as a sulfur precursor. However, these improvements were later attributed to the incorporation of Zn at Sr vacancies (Mauch et al., 1995). Ga has been reported to have a positive effect on the SrS:Ce grains obviously via a flux formation and this effect may also be active when Ga_2S_3 is used as a sulfur precursor (Naman et al., 1996).

Leppänen et al. (1991) tried to enhance the blue emission in SrS:Ce by changing the environment of Ce^{3+} ions by replacing Sr with Ca or Ba in the SrS lattice, but the emission maximum shifts to longer wavelengths. It seems quite unusual that replacing Sr by both smaller and larger atoms would cause red shift. The crystal field and molecular orbital calculation of O'Brien et al. (1998) confirmed the increased red shift of Ce emission from CaS to MgS as compared with that in the SrS lattice. In the case of BaS, the red shift was attributed to the stronger covalence as compared to SrS and, as a consequence, the energy of the lowest excitation state is lowered. Saanila et al. (1998) also verified the red shift in ALE-deposited BaS:Ce films. About the same behavior was observed in $Sr_{1-x}Zn_xS:Ce,F$ samples (Takahashi et al., 1990). When replacing sulfur by selenium a shift in emission color can be seen, and in SrSe:Ce the PL emission maximum is at 470 nm and the shoulder at 527 nm (Yamashita, 1987). The shift of 10 nm is important with respect to the color coordinates and the shift has also been observed in EL measurements (Oseto et al., 1989; Poelman et al., 1992). Besides the spectral shift, the addition of Se has been reported to cause an increase in EL brightness, to improve stoichiometry, and to reduce oxygen level and anion vacancies (Poelman et al., 1992; Vercaemst et al., 1995; Yang et al., 1995). Tentative results from ALE-deposited $SrS_{1-x}Se_x:Ce$ showed that the Se incorporation decreased the EL brightness. This may be due to the method used, where Se was introduced as element. It may be possible that selenium has not completely reduced to selenide (Ihanus et al., 1997,

1999). However, the benefits of the Se addition have not been verified in the best SrS:Ce films ($L > 100\,\mathrm{cd/m^2}$), thus leaving open the real advantage of selenium.

Good crystallinity of the phosphor layer has a positive effect on the EL luminance. The crystallinity and luminescent properties can be improved by raising the deposition temperature (Onisawa et al., 1991a). A higher deposition temperature together with sulfur coevaporation also affects the preferential orientation (texture) of the evaporated SrS film. The use of sulfur coevaporation favors the (200) texture while devices prepared without sulfur coevaporation and treated with H_2S after deposition showed the (111) reflection as the most pronounced. However, the influence of texture on the EL properties of SrS:Ce is unclear since both (111) and (200) oriented films deposited by ALE give similar luminance. On the other hand, it is difficult to study the effect of the orientation since when going from one orientation to the other, the preparation conditions change so much that other properties of the film also simultaneously change. The FWHM of an XRD reflection from a SrS film with a good crystallinity is of the order of 0.2°.

Postdeposition annealing of SrS:Ce,K in an Ar-S atmosphere at 630°C doubles the luminance when compared to as-deposited samples (Tanaka et al., 1991). The positive effect of annealing is attributed to the improvement of crystallinity. An advantage of post-deposition annealing is that the temperature during deposition can be lower (Okamoto et al., 1993). The annealing can even restore the luminance level of SrS:Ce devices degraded during the operation (Leppänen et al., 1991).

The crystallinity of SrS:Ce films has been improved by growing the film on crystalline ZnS films (Onisawa et al., 1991a), for example as sandwiches of SrS:Ce between ZnS layers. Besides enhancing the crystallinity and orientation, the ZnS layers may protect the SrS:Ce film and improve the acceleration of electrons. In ZnS–SrS:Ce multilayer devices, an enhanced trailing edge emission (TEE) was detected (Mauch et al., 1992) and because the TEE was found to be dependent on the SrS:Ce film thickness (Tanaka et al., 1993) the multilayer structures were studied as possible high-efficiency devices. In fact, stable, highly luminescent devices (efficiency close to 1 lm/W; filtered blue 14 cd/m², 60 Hz) have been obtained by using multilayer (ZnS–SrS:Ce)$_n$ structures (Velthaus et al., 1992, 1994; Kobayashi, 1994). In developing multicolor EL devices, it was realized that these "protecting" ZnS layers can also be doped with Mn, resulting in a white phosphor (Ohmi et al., 1992b; Leppänen et al., 1993).

Alkaline earth sulfides are hygroscopic and moisture in the evaporation source has been reported to decrease the luminance of SrS:Ce TFEL device, but on the other hand, a controlled oxidation of the SrS surface to $SrSO_3$ may improve the EL properties (Onisawa et al., 1991b). When well crystalline and stoichiometric, SrS films are surprisingly stable against air atmosphere. The experiments with a high-temperature XRD showed that the first signs of oxidation in an ALE-made (111) oriented SrS film appear at 500°C (Madarasz et al., 1996). On the other hand, a comparison between ALE- and evaporation-made SrS:Ce samples showed that the oxygen content in the ALE-made samples is clearly lower than in those made by evaporation (Li et al., 1996). The EL performance was the opposite, evaporated samples were better than ALE samples. It seems that where the oxygen is located is more important than its total concentration. Maybe oxygen at the grain boundaries is not as harmful as that inside the bulk.

Vacancies at the Sr sites have been believed to be an important reason for the weak luminescence properties of SrS:Ce films when compared to powders that have been made at a high temperature and therefore are more perfect crystals. Especially, degradation of the devices has been proposed to be caused by the vacancies (Takahashi et al., 1992). Strontium vacancies are easily formed in PVD-deposited films where the sulfur deficiency is corrected with treatments in a S atmosphere or when an excess of S is used in evaporation. As mentioned earlier, the positive effects of Zn incorporation into SrS:Ce were detected in connection to the studies of new sulfur precursors (Mauch et al., 1995). Addition of Zn up to 5 mol % improves the luminescence properties of SrS:Ce (Lee et al., 1996) and improvements in crystal quality (FWHM = 0.15–0.18°) and aging of the EL devices have also been detected (Ohmi et al., 1997). In this respect, manganese has been found to be even better codopant than zinc (efficiency of 1.6 lm/W, decay time close to that measured for powder, 27 ns) (Hüttl et al., 1996b). Recently Velthaus et al. (1997) reported the codoping of SrS:Ce with silver, which is believed to act as a charge compensator. The Ag-codoped devices are the best reported in the literature so far (efficiency close to 2 lm/W, luminance of 142 cd/m^2 at 60 Hz, $x = 0.26$, $y = 0.47$). The role of silver is unclear and its participation in the luminescence process, unlike in the case of SrS:Cu, has not been reported.

Films of SrS:Ce have been codoped with other RE ions to achieve white EL which when combined with color filters can result in full-color displays. The codopants studied were Eu^{2+}, Pr^{3+}, Nd^{3+}, Sm^{3+}, and Ho^{3+} (Abe et al., 1989; Okamoto et al., 1990a). Strong interaction (energy transfer) has been observed between Eu^{2+} and Ce^{3+} ions, while interactions with the other ions were much weaker. Reasonably strong white emission has been obtained in double-doped SrS:Ce,K,Eu films—better than in a stacked SrS:Ce,K–CaS:Eu multilayer structure (Tanaka et al., 1988b) but worse than in the SrS:Ce–ZnS:Mn devices. It is worth mentioning that the addition of Pr has been used to improve the color balance in SrS:Ce,Pr–ZnS:Mn in hybrid EL devices without loosing the high luminance ($L_{50} > 300 \text{ cd/m}^2$, 60 Hz) (Wu, 1996).

In conclusion, a continuous improvement has taken place during the 1990s in the efficiency and stability of the SrS:Ce thin films. Although the properties of the thin films are far behind those of the powders, the level reached by several film deposition techniques is high enough for applications in "color by white" concept using the SrS:Ce–ZnS:Mn phosphor stack. (Table III). The weak blue emission is a problem in normal EL devices, but in the AMEL devices blue emission in the SrS:Ce–ZnS:Mn phosphor is significantly improved because of the use of high frequencies (5 kHz) (King, 1996c). A better understanding of the material properties of SrS:Ce is still needed and will be the key issue in further improvement of the blue emission. Reaching a conclusion from the numerous papers written on SrS:Ce TFEL devices is difficult because of the very difficult conditions used for the preparation. The reported "significant improvements" made by codoping and other treatments are obtained in comparison to samples that have not been the state-of-the-art references.

3. EMERGING MATERIALS

a. SrS:Cu, SrS:Ag,Cu

Sun *et al.* (1997a) reported a new and bright blue TFEL phosphor based on SrS:Cu. This denoted substantial progress in the development of a new blue phosphor because the progress in SrS:Ce, although constant, is limited by the dominating greenish color always present in its emission. The EL emission of SrS:Cu reported by Sun *et al.* (1997a) is a broadband between 390 and 620 nm with a peak at 470 nm (Fig. 9). Because a great deal of the emission is at the deep blue, the color coordinates are real blue $x = 0.17$ and $y = 0.27$, and the brightness (34 cd/m^2, 60 Hz) and efficiency (0.24 lm/W) are higher than those of thiogallates and the aging properties are good. Optimum Cu concentration is about 0.2–0.3 at. % (Sun, 1998a).

Copper is known as a dopant in CRT phosphors based on ZnS, and PL properties of Cu$^+$ in alkaline-earth sulfides were studied already long ago (Lehmann, 1970). Powder DCEL devices showing green emission were reported by Vecht *et al.* (1981) and TFEL properties of SrS:Cu,I were reported in 1985 (Kane *et al.*). Copper has also been studied as a charge compensator in connection with the SrS:Ce thin films and a high efficiency has been reported in a sputter-deposited SrS:CeF$_3$,CuBr$_2$ device (Mach *et al.*, 1990). Also a blue shift in Ce emission as well as an increase in grain size were detected as influences of copper, as mentioned earlier (Tong *et al.*, 1997b). The nature of the emission in Cu-doped alkaline-earth sulfides was considered as of the donor–acceptor type, but the results of Yamashita *et*

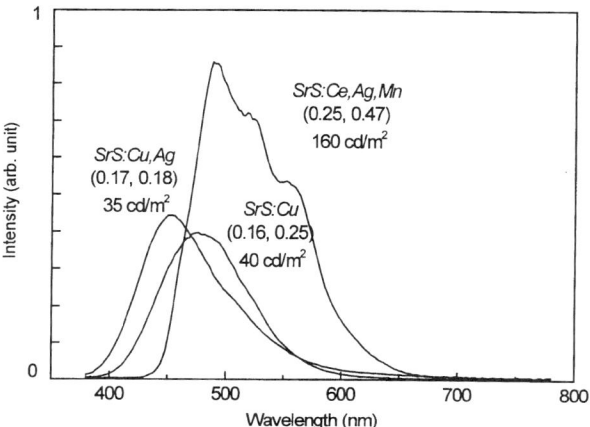

FIG. 9. EL emission spectra of SrS:Ce,Ag,Mn, SrS:Cu, and SrS:Ag,Cu (Troppenz et al., 1998).

al. (1991, 1993) and Sun et al. (1997a) indicate that the emission originates from Cu^+ emission centers. This conclusion is in agreement of the well-understood isolate Cu^+ emission in alkali halide hosts having the same crystal structure as SrS.

SrS:Cu thin films are being studied by several groups and films have been deposited by different methods. The PL at room temperature shows either two bands at 460 and 520 nm or a coalesced broadband peaked at 470–490 nm. When the temperature is lowered, the emission bands narrows and only one band at about 520 nm is observed (Park et al., 1997; Ohmi et al., 1998; Li et al., 1998c). The EL emission is essentially the same including the low temperature behavior (Troppenz et al., 1998). The PL excitation spectrum of SrS:Cu shows at least two excitation band at 283 and 315, which can be attributed to the $^1A_{1g}-^1T_{2g}$ and $^1A_{1g}-^3E_g$ transitions of isolated Cu^+ centers (Fig. 10). The decay time of the isolated Cu^+ center is of the order of 80–100 µs at 80 K but becomes much faster at room temperature (Yamashita et al., 1993a; Li et al., 1999b). Completely green EL emission from SrS:Cu is possible (Li et al., 1998c, 1999b) as was also in the case of DCEL devices (Vecht et al., 1981). The origin of the green emission can be caused by Cu pairs.

A disadvantage in SrS:Cu devices is the need for high-temperature annealing at $T > 650°C$ (preferably 750–810°C) for good EL emission in films made by sputtering, evaporation, or ALE at low substrate temperatures (Sun et al., 1997; Menkara et al., 1998; Soininen et al., 1998). On the

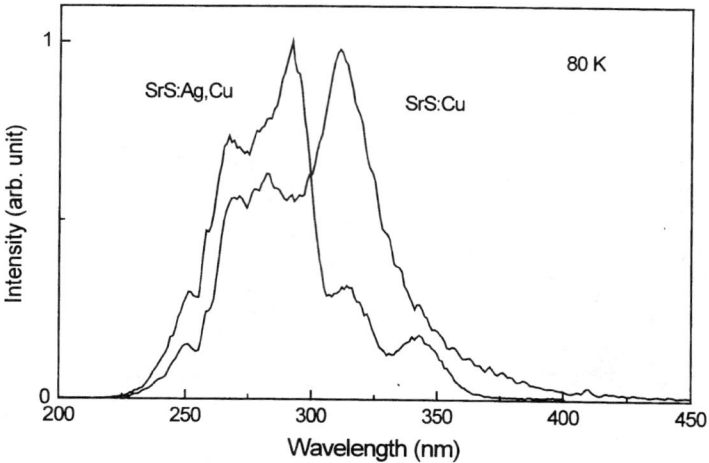

FIG. 10. Excitation spectra of SrS:Cu and SrS:Cu,Ag at 80 K.

other hand, the post-depositional annealing may cause some sulfur vacancies in the matrix as determined by charge deep level transient spectroscopy (Menkara et al., 1998). The vacancies may enhance the charge injection to SrS:Cu and thereby improve strongly the EL performance. The annealing can be avoided if a substrate temperature of 600°C is employed in the deposition (Velthaus et al., 1998).

Sun et al. (1997a) suggested in their first paper that CaS:Cu or SrS:Ag,Cu may be even better blue phosphors than SrS:Cu. The doubly doped SrS:Ag,Cu devices show a strong deep blue emission in both PL and EL: $x = 0.17–0.18$, $y = 0.13–0.26$ depending on the Ag concentration (Sun, 1998a). The luminance of SrS:Cu,Ag devices is even better than that of SrS:Cu devices with equivalent efficiency (Table III). Silver is very mobile in the deposition conditions and possibly also during the operation of the EL device, however, stable devices have been obtained.

The luminescence of singly doped SrS:Ag has been studied in detail. The PL shows only weak emissions at 430 nm and almost no EL at room temperature. The emissions can be enhanced by cooling, but then a new PL band appears in the UV region (Troppenz et al., 1998). The PL excitation spectrum of SrS:Ag shows the Ag band at 281 and 295 nm, and similar excitation bands are also found in SrS:Ag,Cu (Park et al., 1998). It is likely that the blue emission is mainly contributed by Ag-related emission centers; however, such emission is not observed in SrS:Ce,Ag (Velthaus et al., 1997; Li et al., 1999a) in which the Ag probably acts entirely as a charge compensator. The luminescence mechanism of SrS:Ag,Cu is not fully understood, and energy transfer between Ag and Cu ions has been under discussion. Decay studies show that at least Cu^+ emits rather independently in SrS:Ag,Cu. In that case, the improvement observed in Ag emission when codoped with Cu remains unclear. The role of possible cation pairs forming in the films must be studied in the future.

The SrS:Cu–ZnS:Mn multilayer structure and the possible need to enhance green emission have been studied for white EL. The EL emission of SrS:Cu, similar to that of SrS:Ce films, show a strong frequency dependence and in the preliminary experiments, no benefit was achieved by adding ZnS:Tb or SrS:Ce green emitters.

The studies of Cu and Ag,Cu doping show that it is still possible to find new materials for TFEL. They also show how complicated the interaction between the emitting ions may be. More detailed studies of EL mechanisms are needed to gain an understanding of the EL process and simultaneously new EL phosphors.

4. Materials of Minor Importance

This section summarizes those phosphor materials that have been examined in the past but were found to be not efficient enough for practical applications.

a. Rare-Earth-Doped Zinc Sulfides

Most of the REs incorporate in ZnS as trivalent ions. The sharp line $f-f$ transitions are parity forbidden and the luminescence is inefficient. One of the major concerns is the solubility of RE ions in the ZnS matrix due to the large ionic radii of REs (1.01–0.86 Å) and the charge mismatch of RE^{3+} as compared with Zn^{2+} (0.74 Å). Therefore, most of the literature reporting ZnS:RE contains codopants of various kinds such as F^-, O^{2-}, Cl^-, Br^-, Li^+, Na^+, and P^{5+} ions or combinations of them. These codopants together with REs form complex centers that are believed to be important for enhancing the EL luminance and efficiency. Even in the ALE-deposited ZnS:Tb and ZnS:Tm thin films without intentional codoping, at least O associated with the RE activators can be found (Charreire et al., 1993).

Extensive reviews of the photo- and electroluminescence and the mechanism of RE in II–VI compounds have been made, for example, by Boyn (1988), Leskelä and Niinistö (1992a), and Godlewski and Leskelä (1994). Figure 11 shows the energy levels of RE^{3+} ions, and for a better

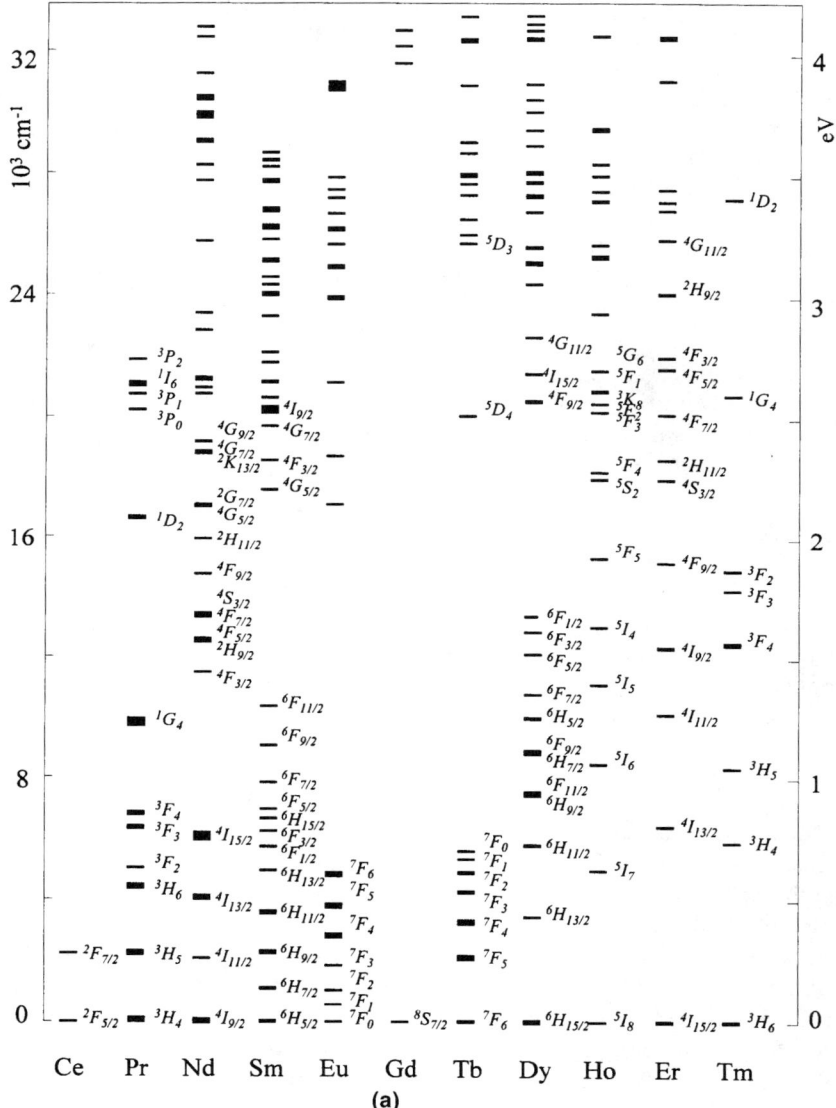

FIG. 11. (a) Energy levels of the RE^{3+} ions and (b) the EL spectra of ZnS:RE TFEL devices (after Boyn, 1988, and Hamakava et al., 1988).

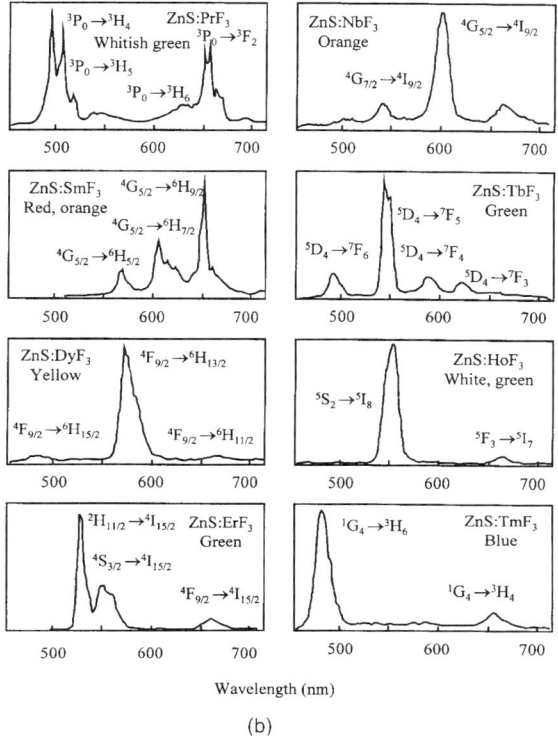

(b)

FIG. 11. *Continued*

comparison, the typical electroluminescence spectra of ZnS:REs are presented as well.

Some RE ions, such as Ce^{3+} and Eu^{2+}, have broadband emission spectra due to the allowed $d-f$ transitions. In the case of ZnS:Ce, the $5d-4f$ fully allowed transition results in broadband blue-green emission similar to that of the well-known SrS:Ce (Tammenmaa et al., 1986; Sohn et al., 1993). However, the reported efficiency of TFEL ZnS:CeF$_3$ is less than 0.1% (Tanaka, 1984), which is much lower than the estimated 12% in ZnS:Ce,Li powder (Kawai and Hoshina, 1981). Reports on the EL of ZnS:Eu^{2+} are sparse. The Eu^{2+}-doped ZnS shows a broadband blue in EL due to allowed $4f-4f5d$ transition, but the emission extends to green, and red emission from Eu^{3+} is difficult to avoid (Chase et al., 1969).

Among the RE-activated ZnS, the red-emitting ZnS:Sm has the highest luminance value next to ZnS:Tb, with a chromaticity close to that of the standard CRT red phosphor. Typically, three emission lines locate at about 568, 604, and 651 nm. Relative intensities of the dominant transitions influence the CIE color coordinates and luminance in red. Early studies by Chase *et al.* (1969) showed that the emission of their thermal evaporated ZnS:Sm,F is reddish-orange. Tohda *et al.* (1986) used P as a codopant and the chromaticity was improved in their sputtered ZnS:Sm,P thin films, but the luminance was poor. In 1987, Hirabayashi *et al.*, reported ZnS:Sm,Cl using the MOCVD technique. They obtained a luminance of $1000 \, cd/m^2$ at 5 kHz and good CIE color coordinates $x = 0.64$ and $y = 0.35$. This is regarded to be so far the best EL record for ZnS:Sm; however, the luminance is much less than that of the filtered ZnS:Mn.

ZnS:Tm is known to be a good CRT phosphor having good color coordinates and therefore it was natural to study $ZnS:TmF_3$ as a first blue EL phosphor, but its luminance was quite poor ($0.2 \, cd/m^2$; 0.01 lm/W) (Kobayashi *et al.*, 1985b; Ono, 1995). In CL, this material shows blue emission, but IR emission dominates in EL. Tanaka *et al.* (1992) studied SrS:Tm as well as multilayer structures of $(ZnS:Tm-ZnS)_n$, $(SrS:Tm-ZnS)_n$, and $(ZnS:Tm-SrS)_n$, where n is the number of the repeat units. The SrS matrix and the multilayer structures only slightly increased the blue EL emission from Tm^{3+} in comparison to ZnS:Tm. The spectroscopy of Tm^{3+} in ZnS is reasonably well known (Müller, 1995) but explanations for its behavior are missing. Our EXAFS studies on rare-earth-doped ZnS thin films have shown that the rare-earth dopants form oxysulfide centers, $ZnS:Tm_2O_2S$ (Charreire *et al.*, 1993, 1994) and we expect that the oxygen-containing, nonideal environment of Tm is the reason for the relaxation as IR emission. Charreire *et al.* (1996) tried to fabricate very pure ZnS:Tm films by ALE and prove the role of oxygen in the EL properties, but the films were not pure enough to show any significant enhancement in the blue emission.

ZnS:Ho and ZnS:Pr are two attractive white phosphors. An emission spectrum of ZnS:Ho contains multiple line transitions in blue (465 and 489 nm), green (550 nm), and red (657 nm). The wavelength locations match the CRT phosphors. However, it is important to maintain a balanced luminescence intensity at each emission wavelength to obtain white light. Contrary to most of the ZnS:RE studied, codopants such as F should be avoided in ZnS:Ho. In electron-beam-evaporated ZnS:Ho, Zhong *et al.* (1988) found that the blue component is suppressed when codoped with F and the EL of $ZnS:HoF_3$ is mostly green. On the other hand, white light ($x = 0.34$, $y = 0.41$) can be achieved by doping with metallic Ho. The filtered ZnS:Ho has luminance of 7, 160, and $22 \, cd/m^2$ in blue, green, and red, respectively, under 1-kHz drive (Hao *et al.*, 1994). The other white phosphor, ZnS:Pr, has two groups of transition lines forming two bands around

490 and 660 nm. Unfortunately, it can not be used in the filtered RGB EL applications due to having only two bands in emission.

Although a considerable amount of research has been focused on the study of both applications and fundamentals, most of the ZnS:RE TFEL phosphors are still too dim to be used in practice. One of the drawback is the low solubility of large-size, charge-mismatched RE ions in the ZnS host. This problem is magnified as the cross sections are small in the ZnS:RE EL phosphors. As shown by Kobayashi et al. (1985b), the optimum concentrations of Sm (0.14 mol %) and Tm (0.43 mol %) are even lower than that of Tb (1.4 mol %) in ZnS films. Another intrinsic factor influencing the EL is the cross-relaxation (Kobayashi et al., 1985b). Cross-relaxation can be common in RE-doped phosphors due to the closely distributed energy levels from both excited and ground states. Such an energy process is facilitated when the dopant concentration is high, consequently, the probability of nonradiative energy loss is increased. In the case of ZnS:Ho, it may be possible that the blue line transitions are reduced through cross-relaxation processes to a lower energy (Zhong et al., 1988). It is known that all applications of Sm as dopant ions have been hindered by the cross-relaxation process.

The preceding discussions show that the intrinsic efficiency of ZnS:RE is lower as compared to ZnS:Mn. Low excitation efficiency, possibly due, for example, to insufficient hot electrons (Mueller-Mach et al., 1996b) and a small cross section for impact excitation (Jiang et al., 1994) contribute to the low luminance level as well. Nevertheless, these limits have not prevented further improvements of ZnS:RE in TFEL applications. As shown by Mueller-Mach et al. (1996a), many RE-doped ZnS phosphors are capable of being operated under aggressive driving condition. In particular, the luminance of ZnS:Ho can almost match that of ZnS:Mn at a very high driving frequency.

b. *Thiogallates*

Rare-earth-doped thiogallates have been studied as CRT phosphors since the early 1970s (Peters and Baglio, 1972) and the luminescent properties have been reported for different rare-earth-doped samples (Garcia et al., 1982; Davolos et al., 1989). Ce^{3+}-doped alkaline earth thiogallates were introduced by Tuenge (1992) and Barrow et al. (1993) as a new family of blue EL phosphors (see Fig. 12). A deep blue emission was measured from the thiogallate films and the emission band maximum varied from 445 to 459 nm depending on the alkaline earth host ion (Barrow et al., 1993; Benalloul et al., 1993). The EL luminance of $CaGa_2S_4$:Ce was reasonable

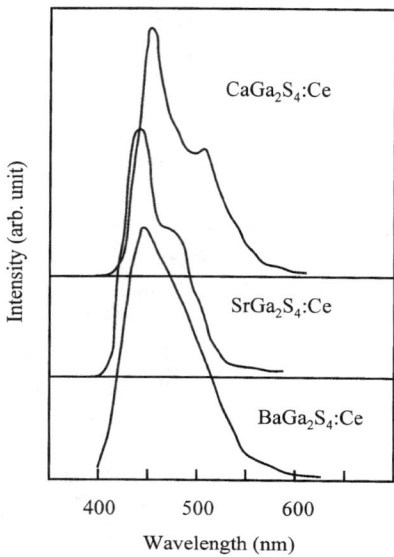

FIG. 12. EL spectra of Ce-doped thiogallates. (From Sun et al., 1994, with permission of The Electrochemical Society, Inc.)

(10 cd/m^2, $\eta = 0.03$ lm/W, $x = 0.14$, $y = 0.19$), while SrGa$_2$S$_4$:Ce has lower luminance but better color purity ($x = 0.14$, $y = 0.10$). The eye response for the deep blue emission of the Sr compound is low, and the best thiogallate phosphors were achieved by mixing the Ca and Sr host cations. The aging properties of MGa$_2$S$_4$:Ce were better than those of SrS:Ce in the early 1990s. At this moment, filtered SrS:Ce,Ag has a better blue EL (19 cd/m^2, $x = 0.10$, $y = 0.15$) (Oberacker et al., 1997) than Ce-doped thiogallates. The bandgaps of the thiogallates are between 4.2 and 4.4 eV, which make them comparable to alkaline earth sulfides, but the dielectric constants (14–15) are higher than that of SrS (8–9) (Kobayashi and Tanaka, 1996) and insulators with high dielectric constants are required in devices based on thiogallates. The band structure of the thiogallates is still unknown.

The first thiogallate films were prepared by sputtering and postdepositional annealing. The requirement for EL is good crystallinity, and to achieve that, the films must be annealed at temperatures above 700°C, which prohibits the use of soda lime glass substrates. Later the films have been prepared by thermal evaporation and MBE, using elements or sulfides as source materials (Oberacker et al., 1995b; Inoue et al., 1995). Oberacker et al. (1995b) also used LiF or NaF as fluxing agent to improve the crystallization of the films. A comparison of the thiogallate powders and thin films shows that, as in the case of SrS:Ce, the crystal quality of thin films is much poorer resulting in nonradiative transitions to killer centers, and thus loss in EL efficiency. The decay time curves of Ce^{3+} and Eu^{2+} in strontium thiogallate powders are exponential with $\tau = 24$ and 480 ns, respectively. In sputter-deposited $SrGa_2S_4$:Ce films, the decay time is strongly nonexponential but the EL properties can be improved by Na codoping (Eichenauer et al., 1996) or fluorine addition (Sun et al., 1994). Oxygen codoping up to 4 at. % has been reported to drastically improve the luminous efficiency, possibly because of filling the sulfur vacancies and thereby eliminating the nonradiative transition paths at point defects or grain boundaries (Sun et al., 1997b). Oxygen also causes a blue shift in the Ce emission by affecting the crystal field of Ce (Rack et al., 1995).

Thiogallates can be used without filtering as blue phosphors in full-color EL devices, and a 10-in. VGA prototype based on dual substrates and patterned phosphors (ZnS:Tb for green and filtered ZnS:Mn for red) was reported by Planar America in 1993 (Barrow et al., 1993). An improved version that can be considered as a commercial full-color EL display was introduced in 1994 (Barrow et al., 1994). However, the high temperatures required for the preparation of the thiogallate thin films, and the low EL efficiency, which requires an expensive dual-substrate systems, show that the thiogallates are not the solution for the blue phosphor problem in EL displays. SrS:Cu, although it also requires high-temperature annealing, has better performance and is a more promising blue material than the thiogallates. Liu et al. (1998) examined a variety of RE-ion-doped thiogallates thin films on less expensive ceramic substrates made by Westaim. Their results are encouraging.

c. *Rare-Earth-Doped Alkaline Earth Sulfides*

Compared to the abundant literature dealing with SrS:Ce, little work has been done on other RE-doped alkaline earth sulfides. One of the EL phosphors studied in relative detail is CaS:Eu, europium being divalent.

Europium has both divalent and trivalent stable oxidation states. Divalent Eu ions are commonly regarded as the activator centers in CaS due to the observed broadband emission centered at 650 nm owing to the allowed $4f^65d-4f^7$ transition. The broadband red emission is dominant even when europium is introduced as Eu^{3+}. Time-resolved spectroscopy reveals that the Eu^{3+} sharp line emission may coexist, depending on the phosphor preparation condition, especially when the dopant concentration is high or oxygen is present in the phosphor (Yamashita et al., 1993b; Abe et al., 1990; Xian et al., 1989). These two emission centers can be easily distinguished since the decay of Eu^{2+} center is faster than red emission from the $^5D_0$ level of the Eu^{3+} center. Reported decay values of CaS:Eu^{2+}

powder are of the order of 0.7 µs (Pham-Thi et al., 1992) and 0.7–1.3 µs Yamashita et al., 1993b, 1995). A comparison of ALE-grown CaS:Eu and ZnS:Eu shows that doping of Eu^{2+} in CaS is straightforward due to the same charge and similar ionic radius of Eu^{2+} as compared with the host cation (Leskelä et al., 1988), while europium at percent level concentrations in ZnS host remains primarily as trivalent ion (Tammenmaa et al., 1986).

EL luminance of CaS:Eu^{2+} thin films is much lower than the filtered ZnS:Mn. Several methods have been used to increase the luminance and efficiency in physically deposited films, such as codoping F, Cu, and Br (Ohnishi et al., 1988), incorporating excessive S (Kobayashi et al., 1985a; Ohnishi et al., 1988), adding H_2 in the sputtering process (Aozasa et al., 1990), using high substrate temperatures (Tanaka et al., 1986), and increasing deposition rate (Onisawa et al., 1989). Partial replacement of the CaS host with $CaS_{1-x}Se_x$ increases the luminance and efficiency, but the emission spectrum undergoes a blue shift due to the crystal field influence (Yoshioka et al., 1989; Poelman et al., 1995). A blue shift also occurs when CaS is replaced with SrS, but the EL performance has not exceeded that of CaS:Eu^{2+} (Tanaka et al., 1986). Up to the present time, a luminance of about 12 cd/m^2 has been achieved by electron-beam-evaporated CaS:Eu^{2+} when driven at 60 Hz (Tanaka et al., 1986; Leskelä et al., 1988). Surprisingly little work has been done using CVD techniques. We have prepared CaS:Eu^{2+} by ALE deposition. About 2 cd/m^2 was achieved at 60 Hz without optimizing the deposition conditions (Leskelä et al., 1988).

Oxygen has been considered to be influential over the EL performance and the stability of CaS:Eu^{2+} thin films since the host material is hygroscopic and chemically less stable (Abe et al., 1990). Degradation can occur when the phosphor is sandwiched between the insulating oxide layers. Replacing the oxide layers by nitrides has been seen to improve the stability (Yoshida et al., 1986). A white light TFEL device was reported using stacked SrS:Ce^{3+}–CaS:Eu^{2+} layers (Ono et al., 1989).

Another often-studied phosphor is white-emitting Pr-doped SrS. It has relatively narrow bands at the bluish-green and red regions ($x = 0.38$, $y = 0.40$). The EL brightness (30 cd/m^2, 60 Hz) and efficiency (0.1 lm/W), although five times higher than those in ZnS:Pr, are too low for practical applications (Tanaka et al., 1988b; Kong et al., 1995). Codoping with Ce improves the EL brightness of SrS:Pr and the emission color remains white. Ce brings its own emission and no energy transfer occurs between the Ce and Pr ions (Abe et al., 1988). The high threshold voltage in SrS:Pr,Ce devices has been lowered by thin tantalum oxide interlayers (Horng et al., 1997).

Okamoto, Nakazawa, and Tsuchiya (1989b, 1995) studied the EL of the most RE in the SrS matrix, and found that the EL performance and emission color may be different for SrS:RE and ZnS:RE. One of the factors that should be considered is the difference in crystal structures. SrS and CaS are crystals of the rock salt type, where the cation has octahedral neighboring sulfides as compared to the tetrahedrally coordinated ZnS structure (Fig. 4). Furthermore, ionic radii of REs are much closer to Ca^{2+} and Sr^{2+} than to Zn^{2+}. The difference in the crystal field results in differences in the transition probabilities in the excited RE centers. Thus, the energies and relative intensities of the sharp line emissions observed are different in RE-doped alkaline earth sulfides as compared to the ZnS:RE, and may therefore result in the differences in luminescence color. Another difference may be the excitation mechanism involved in these two phosphor types, keeping in mind that CaS and SrS have indirect band structures as compared with the direct band structure in ZnS. Experiments show that impact ionization is dominant in $SrS:Ce^{3+}$ and $CaS:Eu^{3+}$ (Tanaka et al., 1988a; Karpinski et al., 1993), and quite possibly, such an excitation process is also significant in some other RE-doped alkaline earth sulfides (Okamoto et al., 1989b, 1995). In contrast, as discussed previously, the impact ionization process is less clear in, for example, ZnS:Tb, and there are still questions as to whether or not the impact excitation is the dominant one.

Similar to the ZnS:RE phosphors, the EL performances of most of the RE- doped alkaline earth sulfides are far from being used in practical TFEL devices, except the well-known SrS:Ce, despite the fact that ionic radii of RE are more favorable in CaS and SrS compared with ZnS. Sohn and Hamakawa (1992) attributed the higher EL luminance of SrS:Ce over that of ZnS:Ce to the differences in crystal fields and band structures of the hosts. Besides the SrS:Ce case, a number of other alkaline earth sulfide phosphors such as CaS:Eu and SrS:Pr show better EL performance than the ZnS counterparts. On the other hand, ALE-grown ZnS:Tb (Härkkonen et al., 1990) had higher luminance as compared with CaS:Tb (Leskelä et al., 1988). So far ZnS:Ho has also shown to have better EL performance than SrS:Ho (Mueller and Mach et al., 1997).

d. *Lead-Doped Phosphors*

Lead-doped CaS has a deep blue EL when the proper lead concentration is used (Nykänen et al., 1992). However, the luminance of the CaS:Pb device is low (2.5 cd/m^2, 300 Hz). Isolated Pb^{2+} ions show UV emission in CaS and the blue emission originates from lead pairs. Lead centers in SrS also emit in the UV region when isolated and in green as pairs (Yamasihita et al., 1984). Both UV and green or bluish white EL (17 cd/m^2, 300 Hz) have

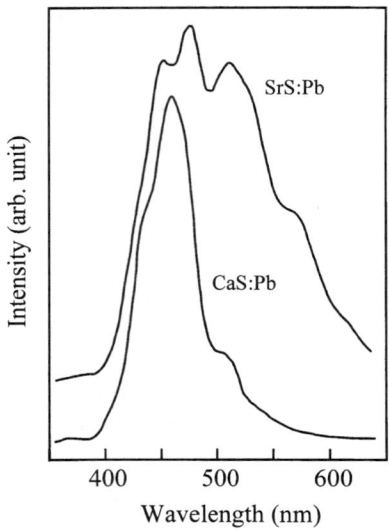

FIG. 13. EL spectra of CaS:Pb and SrS:Pb TFEL devices (after Nykänan et al., 1992).

been reported for SrS:Pb films (Nykänen et al., 1992; Oberacker et al., 1996b). Figure 13 illustrates the EL spectra of CaS:Pb and SrS:Pb.

Lead ions have been studied as a sensitizer for Mn and Ce in SrS (Matsuyama et al., 1989; Soininen et al., 1992; Oberacker et al., 1996b). Energy transfer occurs between lead and manganese and EL emission is improved, but the green hue of the SrS:Mn emission shifts to bluish-white when codoped with Pb. In SrS:Ce,Pb, the EL properties decrease when compared with the pure SrS:Ce, probably due to the strong field quenching caused by the lead ions. However, Yun et al.'s (1999) restudy of ALE CaS:Pb shows that it is possible to obtain high luminance (80 cd/m^2 at 60 Hz) with pure blue color ($x = 0.15$, $y = 0.15$). The great improvement on the EL performance shows that lead-doped phosphors may become good blue EL materials once the excellent results are verified.

e. Oxides and Oxysulfides

Electroluminescence of silicate, borate, phosphate, and tungstate powders were reported in the 1960s (Jones, 1964). Recently, there is growing interest in developing TFEL devices based on zinc silicate and zinc or alkaline earth gallates doped by transition metal or RE ions. These phosphors are

TABLE V

EL Properties of Oxide Phosphors (After Minami, 1998)

Phosphor	V_{th}	E_g	L_{max} (10^3/60 Hz)	η_{max}	Color	CIE_x	CIE_y
$Zn_2Si_{0.73}Ge_{0.23}O_4$:Mn	170		4220/809	0.75	G	0.27	0.66
$Zn_2Si_{0.7}Ge_{0.3}O_4$:Mn	110		1751/206	2.53	G	0.27	0.67
Zn_2SiO_4:Mn	160	5.4	3020/230	0.78	G	0.25	0.70
$CaGa_2O_4$:Mn	150		2790/592	0.25	Y	0.48	0.52
$ZnGa_2O_4$:Mn	135	4.4	758/235	1.2	G	0.08	0.68
Zn_2GeO_4:Mn	55		341/39	0.08	G	0.33	0.65
$ZnAl_2O_4$:Mn	130		21	0.01	G	0.15	0.71
Ga_2O_3:Mn	110	4.7	1018/227	1.7	G	0.12	0.68
CaO:Mn	390	5.5	55	0.02	R	0.60	0.39
GeO_2:Mn	180	3.9	7	0.02	R	0.52	0.41
$ZnGa_2O_4$:Cr	250	8	196/9	0.02	R	0.58	0.40
Ga_2O_3:Cr	175	4.4	375/34	0.04	R	0.65	0.29
Zn_2SiO_4:Tl	280	4.7	16	0.02	B	0.14	0.12
CaO:Pb	470	5.4	6	—	B	0.17	0.11
$CaGa_2O_4$:Eu	225	6.5	215/19	0.03	R	0.69	0.33
$ZnGa_2O_4$:Eu	130		144	0.14	R	0.57	0.39
Y_2O_3:Eu	175		62	0.01	R	0.58	0.40
Ga_2O_3:Eu	210	4.4	153	0.02	R	0.59	0.39
Gd_2O_3:Eu	235	4.7	5	—	R	0.52	0.35
$CaGa_2O_4$:Dy	215	5.3	30	0.02	Y	0.48	0.52
Y_2SiO_4:Mn	235		13	0.05	B	0.18	0.14
$Sr_2P_2O_7$:Mn	145		6	0.01	W	0.33	0.27

chemically very stable but often require high processing temperatures. Consequently, the structures of these TFEL devices are different compared with conventional TFEL display structures. An oxide TFEL device usually has a single insulating layer (mostly $BaTiO_3$) underneath the oxide phosphor sandwiched between the top transparent and back metallic electrodes. The insulating layer can be a thin film or the substrate. In the former case, a substrate (polished $BaTiO_3$, Si, or ceramic) is then required.

Minami et al. (1990, 1995, 1997a, 1997c) introduced a series of works on a variety of oxide thin film systems including for example, green Zn_2SiO_4:Mn and $ZnGa_2O_4$:Mn, red $ZnGa_2O_4$:Cr, multiple-color $ZnGa_2O_4$:RE, and multicomponent or codoped oxides (Table V). These thin films are typically deposited by rf magnetron sputtering on thick $BaTiO_3$ ceramic sheets, but solution coating without expensive vacuum technology has also been exploited (Minami et al., 1997b). Bright multicolor EL can be obtained from these phosphors, but high-temperature annealing at about 1000°C is required.

Xiao et al. (1997b) demonstrated that the EL performance can be improved when Si is partially substituted by Ge in a $ZnSi_{0.5}Ge_{0.5}O_4$:Mn (optimized composition) system with a reduced annealing temperature, and the $BaTiO_3$ substrate is replaced with Si or an alumina ceramic made by Westaim. In such a TFEL device, Mn emission undergoes a red shift from 528 to 535 nm with increasing Ge content in $ZnSi_xGe_yO_4$:Mn. When annealed near 700°C, a maximum luminance of more than 430 cd/m^2 at 60 Hz with efficiency of 0.8 lm/W was achieved. The color purity is even better than the green ZnS:TbOF phosphor. However, to achieve high luminance the oxide EL phosphors typically operate at a higher voltage since their L-V curves are much less steep compared to those of the sulfide phosphors. The annealing temperature required for $ZnSi_xGe_yO_4$:Mn is still too high for TFEL device using a glass substrate.

Kitai and his coworkers (Xiao and Kitai 1997a; Kitai et al., 1997) reported that thin films of RE-doped Ga_2O_3 and alkaline earth gallates exhibit EL at a processing temperature as low as 200°C. Annealing may improve the EL brightness, but high-temperature annealing often quenches the EL. Good EL luminance can be obtained when the phosphors are still in an amorphous state. An as-deposited (250–350°C) Eu activated alkaline earth gallate phosphor can achieve a maximum luminance of about 330 cd/m^2 at 60 Hz. However, the highly insulating properties and the device structure determine that the threshold voltage (>200 V) is higher as compared to the sulfide phosphors (~ 150 V), and the L-V slope is much less steep. These RE-doped alkaline earth gallates TFEL devices have about 7% decrease in luminance after about 400 h driven at 400 Hz.

Rare-earth oxides and oxysulfides, such as Y_2O_3 and Y_2O_2S, are efficient photoluminescence and cathodoluminescence phosphor materials. Tanaka et al. (1976b), Benoit et al. (1981), and Shanker et al. (1992) demonstrated that EL is possible in powder RE-doped RE oxides and oxysulfides. Unlike the RE-doped II–VI sulfides, the trivalent activators are readily soluble in RE_2O_3 and RE_2O_2S and these host materials are chemically very stable. These features are advantageous, considering that emission from different RE ions covers the whole visible range. On the other hand, these host materials are also electrically highly insulating, which may hinder the EL process. To overcome this drawback, Suyama et al. (1982) utilized a $(Y_2O_3:Eu–ZnS)_n$ thin film multilayer structure. They pointed out that maintaining a sufficient thickness of the ZnS layers is important for obtaining high EL luminance. No EL was obtained in single-layer Y_2O_3:Eu, but about 140 cd/m^2 at 5 kHz was obtained from a $(Y_2O_3:Eu–ZnS)_n$ multilayered TFEL device. Later works on Y_2O_3S:Eu (Sowa et al., 1992, 1993) and Y_2O_3S:Tb (Ohmi et al., 1992a) also confirmed the requirement of a ZnS layer to obtain efficient EL.

The EL mechanisms of such multilayered TFEL devices are, however, unclear. It was proposed that the ZnS layers provide hot carriers acceleration toward the insulating phosphor layer where impact excitation takes place. On the other hand, Sowa et al. (1992) and Ohmi et al. (1992) observed EL with a device structure of Y_2O_3–ZnS–Y_2O_3:Eu and Y_2O_2S:Tb–ZnS, respectively, regardless of driving polarities. Probably a more complicated mechanism is involved. In the work done by Kitai et al. (1997), fairly bright EL emission was obtained in Ga_2O_3:Eu without any ZnS layer, that is, with a device structure of ITO–Ga_2O_3:Eu–insulator–metal electrode.

In summary, the oxide phosphors generally exhibit better chemical stability compared with the sulfide phosphors and have long operating times. Despite some drawbacks, such as high processing temperature or high dielectric constant, the wide variety of host materials and activators and the rapid progress make oxide phosphors attractive as potential candidates for efficient, high-brightness TFEL devices.

f. Metal Fluoride Phosphors

Phosphors of metal halides have long been subjects of extensive studies, but reports on TFEL applications are sparse. Zinc fluoride and calcium fluoride have been investigated as host materials for TFEL applications. These fluorides are chemically stable and have large bandgap (ZnF_2: 7 eV; CaF_2: 12 eV), which enables the transmission of UV light.

ZnF_2:Mn thin film has an orange EL emission (Morton and Williams, 1979). ZnF_2 doped with Tb, Gd, and Tm emits in green, ultraviolet, and bluish white, respectively (Bernard et al., 1982; Miura et al., 1991, 1992). Blue EL of Eu^{2+} doped CaF_2 has been also reported (Miura et al., 1992; Ranta-aho and Ylilammi 1993; Chatterjee et al., 1997). Cho et al. (1997) demonstrated a possibility of achieving full-color TFEL device utilizing UV-emitting ZnF_2:Gd EL thin film hybrid with RGB PL phosphors (e.g., blue $BaMgAl_{14}O_{23}$:Eu^{2+}, green ZnS:Cu,Al, and red YVO_4:Eu^{3+}. As illustrated in Fig. 14, the EL emission of ZnF_2:Gd is at 311.5 nm, which acts as an excitation source for the PL phosphors. It is thus crucial to obtain a high intensity of UV emission, which was found to be dependent on the choice of Gd precursors in their CVD ZnF_2:Gd thin films. Luminances of 50, 30, and 10 cd/m^2 can be achieved for red, green, and blue, respectively, when the ZnF_2:Gd EL phosphor is driven at 5 kHz.

Similar to the oxide phosphors, the large bandgaps of the fluorides make the operating voltage high and the L–V curves are less steep as compared to the sulfide phosphors. Ranta-aho and Ylilammi (1993) tried to add ZnS as buffer layers but the effects are yet unclear. On the other hand, fluoride

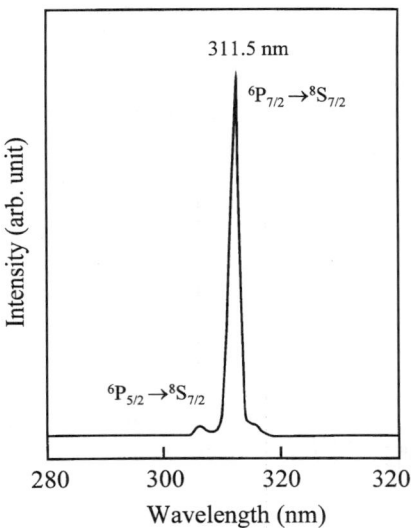

FIG. 14. EL emission of ZnF_2:Gd. (From Miura *et al.*, 1991, with permission from the *Japanese Journal of Applied Physics*.)

films can be deposited at relatively low temperature. Subsequent annealing requires only 200–500°C. Since the ionic radii of RE are close to that of Ca^{2+}, epitaxyl CaF_2:Eu^{2+} grown on Si by MBE could incorporate as much as 4 at. % Eu^{2+} without significant degradation of the surface morphology (Fang *et al.*, 1995). Up to 31.2 wt% Er^{3+} has been doped into CaF_2 film (Adachi *et al.*, 1996). So far, the EL performances of fluorides phosphors are poor and they are not as promising as the oxide phosphors.

5. COLOR BY WHITE

"Color by white," proposed by Tanaka and his coworkers in the 1980s, has become the most appropriate approach for realizing the full-color TFEL applications. The use of a white phosphor in TFEL devices is attractive because it could act as both monochromic white material and multicolor or full-color display could be realized by filtering white light through color filters. White EL has been achieved by three different ways: (i) by a single dopant (ZnS:Ho, SrS:Pr), (ii) by double doping (SrS:Pr,Ce or SrS:Ce,Eu), and (iii) by stacked layers (ZnS:Mn–SrS:Ce, SrS:Ce–CaS:Eu, or ZnS:Mn–SrS:Ag,Cu).

In the just mentioned first approach, ZnS:Ho has three groups of sharp line emissions at the RGB regions, and thus emits good white light ($x = 0.34$, $y = 0.41$). Although attractive, it gives only about 30 cd/m² luminance at 60 Hz. Luminance of SrS:Pr also reaches about the same level and has comparable color gamut as ZnS:Ho, but this phosphor has only two group of emission lines at the bluish-green and red regions and it can not be filtered with blue and green filters. Codoping with a broadband emitter such as Ce in SrS:Pr extends the phosphor application to full color. The other doubly doped SrS:Ce,Eu has broadband emission covering almost the whole visible region from about 470 to 700 nm due to the emission of both Ce and Eu. There is energy transfer from Ce to Eu but that does not quench the Ce emission dramatically. Color coordinats are good ($x = 0.41$, $y = 0.39$) and the luminance also reaches 30 cd/m². Stacked SrS:Ce–CaS:Eu has luminance of only about 20 cd/m². All of these phosphors have just one-third or less of the required luminance for practical display applications.

Stacked phosphor structure is somewhat more complicated in fabrication; however, it allows use of the best phosphor of choice. A prototype device with ZnS:Mn–SrS:Ce stacked layers shows the highest luminance of 470 cd/m² at 60 Hz. The luminance is suitable for display applications and it is expected to be further enhanced owing to recently published results from codoped SrS:Ce,Ag and SrS:Ce,Pr. An unsatisfactory color gamut due to the missing pure blue is the greatest and the last challenge for commercialization of the full-color TFEL displays. As mentioned previously, the white color purity is better in AMEL devices due to the high operation frequency, which results in a more balanced white emission spectrum.

To realize the ultimate color-by-white, true blue phosphor must be used. The emerging blue materials SrS:Cu and SrS:Ag,Cu are promising. The first studies on ZnS:Mn/SrS:Cu samples already reached a luminance of 160 cd/m² at 60 Hz with color coordinates of $x = 0.40$ and $y = 0.42$, which is sufficient for display applications. The commercialization of the full-color TFEL devices is expected in the foreseeable future.

None of the mentioned color-by-white approaches, however, have utilized the important advantage of TFEL, that is, the films in the DSD structure are thin and thus transparent. Therefore TFEL display that can be seen-through is one of the most attractive feature. A monochrome see-through TFEL display using ZnS:Mn has been already brought to the market. For multicolor and full-color devices, filtered structures can not be employed but phosphors having sharp line monochromic emission and pure RGB colors must be used. Rare earths provide promising features. Currently, however, only ZnS:Tb has reached the required luminance with a good green color, but blue and red phosphors are missing. Looking back over the long path of searching the full-color TFEL phosphors, more effort is needed to realize good white see-through TFEL devices.

V. Other Materials

1. SUBSTRATES

Glass is the most common substrate material used in the TFEL devices. The choice of the glass type is affected by a number of physical and chemical properties (Moffatt 1994, 1996; Dumbaugh and Bocko, 1990). The substrate must be thermally and chemically stable under the conditions used to manufacture the TFEL devices. An approximation for the maximum temperature at which glass can be used is given by its strain point, which is still well below the temperature where viscous flow and deformation begins within the time regime used in manufacturing. Nevertheless, temperatures exceeding the strain point and approaching the annealing point have also been used, but require properly designed temperature ramps. To avoid thermal stresses, the thermal expansion coefficient of glass should match that of the thin film materials. Yet another important thermal property is compaction, or thermal shrinkage, which occurs due to viscous relaxation of glass at temperatures that are even hundreds of degrees below the strain point. The higher the resolution of the display, the more important is the compaction. In general, for a given processing temperature, compaction decreases with an increasing strain point. Compaction may be compensated in the device design, and it may be decreased by preshrinking the glass, but this adds cost and risk of damaging the surface.

The various etching and cleaning procedures expose the substrate to aggressive chemicals, either acidic or alkaline. These solutions may cause preferential dissolving of the glass (e.g., of the nonsilica part into acidic and of the silica part into alkaline solutions), and thereby affect the surface properties and visual appearance of the glass. The alkali metal content of glass is also very important because alkali metals, especially sodium, migrate easily from the glass into the TFEL structure. Therefore near-zero alkali glasses having typical alkali contents of 0.1–0.2 wt% (Dumbaugh and Bocko, 1990) are preferred. Alternatively, a diffusion barrier layer (e.g., amorphous Al_2O_3 or SiO_2) can be added to block the sodium diffusion. Other important properties of the glass substrate include smooth, clean, and damage-free surfaces, high and scatter-free transmittance, rigidity, and cost.

The most common glass used in the TFEL devices is barium aluminoborosilicate which is best known as Corning 7059 glass. It is a near-zero alkali glass with a strain point of 593°C, and is thus suitable for most of the TFEL manufacturing processes. If higher temperature processes are required, special high-temperature glasses must be used, for example, alkaline-earth boroaluminosilicate glass (Corning 1733, strain point 640°C) or alkaline-earth aluminosilicate (Corning 1737, strain point 666°C) (Moffatt, 1994). A low-cost alternative for the above-near-zero alkali glasses is soda lime silicate glass covered with an ion barrier to prevent the sodium outdiffusion. The processing temperatures of this glass are, however, limited to about 500°C.

As the price of glass increases with increasing maximum processing temperature, ceramic substrates are a potential choice for those TFEL display structures that require high-temperature processing. Naturally, opaque ceramic substrates can be used only in the inverted TFEL structures. Because ceramic sheets are usually much rougher than glass, planarization is required before thin film deposition. In the novel hybrid TFEL device of Westaim, alumina substrate is used (Wu, 1996). An interesting approach to preparing TFEL displays on multilayer ceramic-on-metal circuit boards has been suggested (Sreeram *et al.*, 1997). In these devices, the driver circuits will be located on the rear of the substrate and connected to the display electrodes by conductive vias through the substrate, thus enabling smaller and simpler display packaging. Finally, note that thin and high-permittivity ceramic sheets may at the same time serve both as substrates and as lower insulator layers (Miyata *et al.*, 1991; Horng *et al.*, 1997) (Fig. 15).

In AMEL devices, the EL films are deposited on single-crystal silicon-on-insulator wafers to which the required driving circuitry has already been integrated (Aguilera and Aitchison, 1996; King, 1996c; Khormaei *et al.*, 1994). SOI wafers are needed to ensure effective isolation of the high-illumination voltage from the low-voltage circuitry. For interconnects, refractory metals must be used to make the circuitry compatible with the subsequent high-temperature processes in the EL film stack preparation. The large difference in thermal expansion coefficients between Si and EL films has required some process modifications, such as lower deposition temperatures and thinner films.

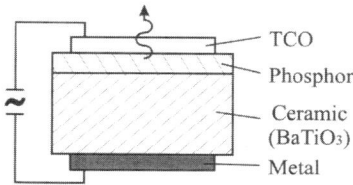

FIG. 15. Structure of a TFEL device with a ceramic substrate that serves at the same time as an insulator (after Miyata *et al.*, 1991).

2. Transparent Electrodes

Polycrystalline indium-tin-oxide (ITO) films are by far the most extensively used transparent electrodes in TFEL displays. ITO is a degenerate n-type wide-bandgap semiconductor having a cubic structure of In_2O_3 in which 5–10% of In^{3+} ions are replaced by Sn^{4+} ions (Jarzebski, 1982, 1982; Chopra et al., 1983). The free electrons originate from both the substitutional Sn^{4+} ions and oxygen vacancies. The oxygen vacancy formation is a reversible process affected mainly by temperature and partial pressure of oxygen, thereby making the electric properties of uncovered ITO sensitive to its thermal treatment conditions, which must be taken into account while planning the deposition process of the next film. Otherwise, the thermal and chemical stability of ITO is sufficient for normal TFEL display structures and processes. However, if high-temperature annealings are used, interdiffusion between ITO and insulators, especially Ta_2O_5, may take place, degrading the insulator properties (Satoh et al., 1997; Tiku and Rustomji, 1989). ITO films are also easy to pattern with normal lithographic techniques. Optimized ITO films have resistivities around $1 \times 10^{-4}\,\Omega\,cm$, so 300- to 500-nm-thick films are typically used to obtain sheet resistances below $5\,\Omega/\square$. Such films have about 90% transmittances in the visible region. In AMEL devices, the transparent electrode film is continuous and may thus have somewhat higher sheet resistance (i.e., be thinner than the patterned electrodes used in the other devices).

Ohmic losses in the ITO electrodes constitute a significant fraction of the overall power consumption in TFEL displays. In addition, the RC time constant of the ITO column electrodes set upper limits to the frame frequency and the number of lines in the display. A clever way of decreasing the resistance of the ITO lines is to add narrower metal lines, so-called bus bars, on top of them (Törnqvist et al., 1991; Törnqvist 1992b; Haaranen et al., 1992; Nire et al., 1992). In the inverted structure, the bus bars can be made thick enough to ensure an order of magnitude reduction in the electrode resistance. Though part of the EL emission is blocked out by the bus bars, the overall transmission remains essentially constant because thinner and therefore less absorbing ITO films can be used.

Because of the added complexity in topography, bus bars on top of ITO can be used only in the inverted structures. In addition, extra patterning steps are needed. An attractive and simple alternative might be to add the metal bus bars to the sidewalls of the ITO lines by electrodeposition right after ITO etching before removing the photoresist (Liu et al., 1991, 1994). In this configuration, there is no direct light blocking, and with carefully planned undercutting during the ITO etching the topography may be preserved.

Alternatives to ITO would be interesting for two reasons: to decrease sheet resistances without sacrificing the transmittance and to avoid the rather high costs of indium. However, only a few true alternatives exist. The one with the most potential is zinc oxide doped with Al or F, which was demonstrated to exceed ITO in optical and electrical performance (Gordon, 1996). ZnO:Al and ZnO:F are also easier to etch than ITO and their raw materials are cheaper. Nonetheless, so far they have been used only in some experimental TFEL devices. Tin dioxide films doped with fluorine are extensively used in energy-conserving windows and photovoltaics, for example, and therefore on-line deposition systems have been installed to float glass lines offering an economic way for producing SnO_2:F covered soda lime silicate glass (Gerhardinger and McCurdy, 1996). The sodium outdiffusion from the glass is prevented by an appropriate layer, usually of SiO_2, deposited in the same on-line process. However, sheet resistances of $5\,\Omega/\square$ may be difficult to reach with acceptable transmittance unless bus bars are added. In addition, wet etching of SnO_2 is complicated.

3. METAL ELECTRODES

Aluminum is the dominantly used metal electrode in the traditional TFEL structures, whereas molybdenum and tungsten are the most common choices for the inverted structures. Important characteristics of metal electrodes include low resistivity, good adhesion to the adjacent films, no migration at high electric fields, patternability, and low cost. In addition, in the inverted structure where the metal is deposited before the insulators and luminescent material, the metal is required to have good thermal stability and thermal expansion coefficient close to that of glass substrate, that is about $4 \times 10^{-6}/°C$ for high-temperature glasses and $9 \times 10^{-6}/°C$ for soda lime silicate glass. Therefore Al, having a low melting point and large expansion coefficient, can not be used in the inverted structures but Mo and W meet these requirements well (Table VI). Their resistivities are, however, twice the resistivity of Al (Table VI), and correspondingly thicker films are needed.

The optical reflectance (R) of the metal electrode has a twofold effect. High reflectance, like that of Al, ensures high luminance but results in a poor contrast by reflecting the ambient light. Therefore contrast enhancement with circular polarizers or neutral density gray filters is usually needed even though this decreases luminance (Törnqvist, 1992b). Low reflectance, in turn, means that a significant fraction $(1 - R)$ of both the emitted and ambient light is absorbed in the reflection, thus decreasing the luminance but improving the contrast. In the inverted multicolor structures, absorbing

TABLE VI

Physical Properties of the Most Common TFEL Display Metal Electrode Materials
(CRC Handbook of Chemistry and Physics, CRC Press, Boca Raton, FL)

Metal	Resistivity ($\mu\Omega$ cm)	Reflectance[a] (%)	Melting point (°C)	Thermal expansion coefficient (10^{-6}/°C)
Al	2.71	91–92	660	23.1
Mo	5.47	54–59	2623	4.8
W	5.39	46–52	3422	4.5

[a]Within a wavelength range 400–700 nm.

Mo or W rear electrodes, together with the front color filters and black back layers, result in an excellent contrast. Therefore, no additional contrast enhancement circular polarizers or filters are needed, which partially counterbalances the luminance losses caused by the absorbing electrode (Haaranen et al., 1992). In fact, high contrast is often preferred over high luminance. For example, the new superhigh contrast (SHC) display introduced by Sharp incorporates a new light-absorbing layer in the thin film structure for reducing reflections from the back electrode (Mikami et al., 1996). Besides, using low-reflectance metals, contrast enhancing layers can be realized also with two-layer interference black coatings made of Al_2O_3 and semitransparent Inconel, for example (Li et al., 1994). Finally, if distortions in the emission spectra are to be avoided, the reflectance should be rather constant within the wavelength range of interest (i.e., the metals should be noncolored).

4. Insulators

The role of insulator films is to limit current transport across the TFEL device. Since high electric fields of about 1–2 MV/cm are typically used, high dielectric strength is required from the insulator materials. In addition, the films must be pinhole-free over the entire display area. Therefore, the deposition of insulator layers is a key step in preparation of reliable TFEL devices.

The insulator layers should also have large capacitance because this increases luminance and facilitates low operating voltages. Below the threshold voltage, TFEL device may be approximated by two ideal capacitors connected in a series; one (C_P) representing the phosphor layer with a thickness d_P and relative permittivity ε_P, and the other one (C_I) combining the two insulator layers (C_{I1} and C_{I2}) into a one effective capacitor

(Howard, 1977; Ono, 1995):

$$C_P = \varepsilon_0 \varepsilon_P / d_P \tag{1}$$

$$C_I = C_{I1} C_{I2} / (C_{I1} + C_{I2}) = \varepsilon_0 \varepsilon_I / d_I \tag{2}$$

where capacitances are per the unit area, and ε_I and d_I are the relative permittivity and thickness of the effective insulator layer.

The voltage applied to the TFEL device (V_a) is divided across the phosphor (V_P) and insulator (V_I) layers in a proportion

$$V_a = V_I + V_P = E_I d_I + E_P d_P \tag{3}$$

where E_I and E_P are the electric fields within the insulator and phosphor layers. From electrostatic continuity one obtains

$$\varepsilon_0 \varepsilon_I E_I = \varepsilon_0 \varepsilon_P E_P \tag{4}$$

which together with the Eq. (3) gives

$$V_P = E_P d_P = [\varepsilon_I d_P / (\varepsilon_I d_P + \varepsilon_P d_I)] V_a \tag{5}$$

The purely capacitive behavior of the phosphor layer persists up to a certain threshold field $E_{p.th}$, above which current begins to flow through the phosphor layer. Rearrangement of Eq. (5) gives the corresponding threshold voltage V_{th} applied to the device:

$$V_{th} = [1 + (\varepsilon_P d_I / \varepsilon_I d_P)] E_{P.th} d_P \tag{6}$$

Since EL is excited by the electrons flowing through the phosphor layer, V_{th} is often called also as a turn-on voltage. Clearly, to have low V_{th}, and thereby low operating voltage, the insulator material should have high permittivity and it should be made as thin as possible without causing a risk of its breakdown. Uniform luminescence characteristics, in turn, require that d_I and d_P must be constant within about 5% over the entire display.

In principle, V_{th} could be decreased also by decreasing d_P [Eq. (6)] but this would decrease also the luminance, L:

$$L = (4/\pi) \eta f (\varepsilon_0 \varepsilon_I / d_I)(V_a - V_{a.th}) E_{P.th} d_P \tag{7}$$

where η is luminous efficiency and f is the drive frequency. In fact, also this equation points out the importance of the high insulator capacitance $\varepsilon_0 \varepsilon_I / d_I$.

Above V_{th} the current flow through the phosphor layer clamps the phosphor field to $E_{P.th}$ with a consequent increase of V_I [from Eq. (3)]

$$V_I = V_a - E_{P.th} d_P \qquad (8)$$

The maximum applied voltage $V_{a.max}$ may be taken a sum of V_{th} and modulation voltage V_M. When inserted to Eq. (8) and using Eq. (6) for V_{th}, one obtains the lower limit to the thickness d_I of an insulator having a breakdown field E_{BD}:

$$\begin{aligned} V_{I,\max} &= (V_{th} + V_M) - E_{P.th} d_P \\ &= V_M + (\varepsilon_P d_I / \varepsilon_I d_P) E_{P.th} d_P < E_{BD} d_I \end{aligned} \qquad (9)$$

In practice, the device reliability is guaranteed by using about 200-nm-thick insulator films, corresponding to $d_I = 400$ nm for identical top and bottom insulators, which are thick enough to give sufficient safety margins and eliminate pinholes and other weak points often existing in thinner layers.

A minimum requirement for the breakdown field E_{BD} of a given insulator material with ε_I, or more generally, for the product $\varepsilon_0 \varepsilon_I E_{BD}$, is obtained from Eq. (4):

$$\varepsilon_0 \varepsilon_I E_{BD} > \varepsilon_0 \varepsilon_P E_{P.th} \qquad (10)$$

This statement contains the trivial requirement that out of the two capacitors in series, the first one to break must be the phosphor layer. The above is, however, only a minimum requirement. From the device efficiency point of view it is desirable that $\varepsilon_0 \varepsilon_I E_{BD} \geqslant 3\varepsilon_0 \varepsilon_P E_{P.th}$ (Howard, 1977). For ZnS ($\varepsilon_P = 8.3$, $E_{P.th} = 1.5$ MV/cm), $\varepsilon_0 \varepsilon_P E_{P.th} = 1.1 \,\mu\text{C/cm}^2$, to be compared with the values listed in Table VII for insulators.

The product $\varepsilon_0 \varepsilon_r E_{BD}$ indicates the maximum charge density that can be stored to a capacitor made of a given dielectric material. Therefore it is a useful figure of merit for evaluating insulator films (Table VII). Its importance is further emphasized by the observation that, in general, dielectrics with high permittivity have low electric field strength and vice versa (Table VII). It must be noted, however, that the breakdown fields, and thereby also the $\varepsilon_0 \varepsilon_r E_{BD}$ values, are statistical quantities and depend on the measurement method and the definition of breakdown, that is, whether the breakdown field corresponds to the destructive breakdown or certain leakage current density, like 1 or 100 $\mu\text{A/cm}^2$. Also the electrode material may have an effect on the I–V and breakdown characteristics of the insulator, the important parameters being the thermal stability of the electrode–insulator interface

TABLE VII
DIELECTRIC PROPERTIES OF POTENTIAL INSULATOR LAYER MATERIALS

Material	ε_r	E_{BD} (MV/cm)	$\varepsilon_0\varepsilon_r E_{BD}$ (μC/cm^2)	Breakdown mode[a]	Reference
SiO_2	4–5	6–11	2–5	SH	Ono (1995), Ohwaki (1987)
SiN_x	4	>9	3	SH	Fukao et al. (1988)
SiO_xN_y	6–7	7–8	4–4.5	SH	Ono (1995), Tiku and Rustomji (1989)
SiO_xN_y/Ta_2O_5	10–12	6	6	SH	Ono (1995), Tiku and Rustomji (1989)
SiO_2/TiO_2	6–9	2–4	2	—	Nakayama (1992)
Al_2O_3	8–9	5–8	3.5–6	SH	Ono (1995), Tiku (1989), Fujita (1983)
Si_3N_4	8	6–8	4–6	SH	Ono (1995), Fujita (1983)
SiAlON	8	8–9	5–6	SH	Ono (1995), Tiku and Smith (1984)
Al_2O_3/TiO_2	9–18	5–6.7	5.5–10	SH	Skarp (1984), Sun and Khormaei (1992)
Al_2O_3/Ta_2O_5	10–20	4–6.6	2–6	SH	Tiku (1989), Kattelus (1993), Kukli (1996a)
Y_2O_3	12	3–5	3–5	SH	Ono (1995), Fujita (1983)
HfO_2	13–16	1–1.5	1–2	P	Kattelus (1993), Kukli (1996b)
$BaTiO_3$[b]	14–20	3	3–5	SH	Ono (1995), Venghaus (1982)
Sm_2O_3	15	2–4	3–5	SH	Ono (1995)
$Ta_xSn_yO_z$	17–23	3–5	5–7	—	Fujikawa (1997)
Ta_2O_5/HfO_2	19–23	2.5–5	6–9	SH	Kattelus (1993), Kukli (1996b)
ZrO_2	20	1	2	—	Kukli (1996a)
Ta_2O_5/TiO_2	20	7	12	—	Ono (1995)
$BaTa_2O_6$[b]	22–23	3.5–5	7–10	SH	Ono (1995), Sun (1992), Fujita (1983)
Ta_2O_5	23–25	1.5–3	3–7	SH/P	Ono (1995), Yoshida (1989), Tiku (1984), Tiku (1989), Fujita (1983), Ohwaki (1987)
Ta_2O_5/ZrO_2	25–28	2–2.5	4–6	—	Kukli (1996a)
$(Nb_{1-x}Ta_x)_2O_5/ZrO_2$	31–33	3	8	—	Kukli (1997)
$PbNb_2O_6$	41	1.5	5	SH	Ono (1995)
TiO_2	60	0.2	1	P	Ono (1995)
$Sr(Zr,Ti)O_3$	100	3	26	P	Ono (1995)
$SrTiO_3$	140	1.5–2	19–25	P	Ono (1995), Fujita (1983)
$PbTiO_3$	150–190	0.5	7–9	P	Ono (1995), Fujita (1983), Okamoto (1981), Fukao (1988)

Best results of each material are usually shown. The notation AB/CD represents stacked insulator with a variable number of layers. Note that E_{BD} values reported depend on its definition. [a]SH = self-healing, P = propagating. [b]Amorphous.

and the metal work function (Okibayashi *et al.*, 1991; Matsuhashi and Nishikawa, 1994; Sun and Chen, 1997; Nire *et al.*, 1992). Because of differences in film qualities, the dielectric properties are naturally dependent also on the deposition method. Furthermore, because some oxides, such as Ta_2O_5 and TiO_2, may easily become oxygen deficient, their dielectric properties are sensitive to thermal treatment conditions.

The dielectric breakdown can take place by two distinct mechanisms (Fujita *et al.*, 1983). The preferred one is the self-healing mechanism where the breakdown site becomes electrically isolated from its surrounding because of local removal of the insulator and the overlying electrode. This mechanism ensures that the few defects that inevitably always exist in the large-area insulator films do not result in a device failure. The other breakdown mechanism is propagating, which grows in size without restriction and eventually destroys the device. The type of breakdown mode is material dependent (Table VII), but inconsistencies may exist because of different deposition methods and electrode materials.

Yet another important electrical property related to the insulator films, though affected also by the phosphor material and by the deposition methods of these films, is the distribution of electron trap states at the insulator–phosphor interfaces. Tunneling of electrons from these states above a threshold field $E_{P.th}$ is the initiating step in the current flow through the phosphor. When accelerated in a high field in the phosphor layer, the electrons gain enough energy to impact excite the luminescent centers. The deepness of the interface states determines at which field the tunneling begins, while their density determines the number of electrons available. If the interface states are too shallow, tunneling occurs at such low fields that the electron acceleration is insufficient for luminescent center excitation. Tunneling out of deep states, in turn, requires high electric fields (i.e., high operating voltages). In other words, $E_{P.th}$ is not solely dependent on the phosphor material, but also on the phosphor–insulator interface properties. For example, $E_{P.th}$ of ZnS:Mn has been found to vary from 1.41 to 1.58 MV/cm, depending on the insulator material and the highest luminous efficiency correlated with the highest $E_{P.th}$ (Tiku and Smith, 1984; Sun and Khormaei, 1992). Interface state distributions are, however, difficult to predict since they are dependent on many factors, and their detailed experimental studies have also been rather limited (Kobayashi *et al.*, 1992; Petre *et al.*, 1995). Changes in the interface states are common in the beginning of the device operation causing aging (i.e., shifts in the L–V curves), but after a certain time they become stabilized. Therefore, the displays are usually preaged at the factory with a high-frequency driving. In the worst case, changes in the interface states lead to the degradation of the device (Chubachi *et al.*, 1992).

To summarize, important properties of insulator films include: (i) high electric field strength, (ii) high permittivity, (iii) pinhole-free structure, (iv) self-healing breakdown mode, (v) appropriate and stable interface state distribution, and (vi) good thickness uniformity. In addition, the insulators must have (vii) good conformality and (viii) adhesion to the adjacent electrode and phosphor layers, be (ix) stress-free and (x) stable during the upper electrode patterning, and (xi) have no detrimental interface reactions. Preferably, the insulator films should be amorphous because polycrystalline films lead to rough interfaces and they contain grain boundaries through which electrons can flow and ions migrate.

As noted, dielectrics with high permittivity would be ideal insulator materials but they suffer from low breakdown fields and from propagating breakdown mechanisms (Table VII). Therefore, dielectrics with moderate permittivity but high E_{BD} have usually been preferred. However, no single insulator material has gained a dominating position, as can be noticed, for example, by comparing the recent reports of the commercial manufacturers. Sharp has announced to use SiN_x–SiO_2 (Mikami *et al.*, 1996; Okibayashi *et al.*, 1991) or SiN_x–Al_2O_3 (Okibayashi *et al.*, 1991) while Planar America (previous Planar Systems) has reported SiON and $BaTa_2O_6$ (Sun, 1990; Sun and Khormaei, 1992) and Al_2O_3–TiO_2 (ATO) (Sun and Khormaei, 1992; Aguilera and Aitchison, 1996), the latter of which is predominantly used also by Planar International (Leppänen *et al.*, 1993). Only in the hybrid EL displays of Westaim ferroelectric lower insulators are used, but these are thick films made by screen printing (Wu, 1996).

The insulator films used in commercial and laboratory TFEL devices have often been composite structures rather than homogeneous single layers. Bi-layer, tri-layer, as well as multilayer insulator structures have been used. In the composite structures, advantageous characteristics of two or more materials are combined in realizing insulators that are at the same time reliable and efficient. Such combinations include, for example, high permittivity in one layer and high breakdown field in another one, or one material with good electrical properties and the other with stable interfaces and good adhesion to the phosphor and electrode layers. If the stacked insulator consists of only a few relatively thick layers, its $\varepsilon_0\varepsilon_r E_{BD}$ value is the same as that of the highest $\varepsilon_0\varepsilon_r E_{BD}$ of its components (Ohwaki *et al.*, 1987), but if the constituent layers are numerous and very thin, their properties may differ from the thicker layers and consequently $\varepsilon_0\varepsilon_r E_{BD}$ of the composite may be much higher than the $\varepsilon_0\varepsilon_r E_{BD}$ values of the components (Kukli *et al.*, 1996a, 1996b, 1997).

Structures based on Ta_2O_5 serve as an illustrative example of the composite insulators. Ta_2O_5 has relatively high permittivity of about 25 but, when deposited by sputtering, its adhesion to ITO is poor and the breakdown mode is propagating (Yoshida et al., 1989; Tiku and Rustomji, 1989). In addition, Ta_2O_5 has been observed to react with ZnS:Mn during high-temperature annealing. These problems can be avoided by adding 20 to 40-nm-thick SiO_2 on both sides of Ta_2O_5, that is, by constructing a three-layer structure SiO_2–Ta_2O_5–SiO_2 (Yoshida et al., 1989). Also Si_3N_4 interlayers have been used between ZnS:Mn and Ta_2O_5 to stabilize the interface against degradation (Lee et al., 1991). However, in another study Ta_2O_5 layers between ZnS:Mn and Si_3N_4 were found to improve adhesion and breakdown characteristics (Tiku and Smith, 1984). Obviously, the film deposition methods have their own contributions to the interface characteristics.

In addition, to improve reliability, multilayer insulator structures have also been constructed to modify the charge injection characteristics of the insulator–phosphor interface. One approach has been to place a layer of low-resistivity dielectric material between the phosphor and better resistivity insulator (Ohwaki et al., 1990; Mita et al., 1987). In this configuration, the low-resistivity insulator serves as an electron injecting and trapping layer, while the high-resistivity layer limits the current through the device. On the other hand, if a thin-high resistivity buffer layer is added between electron injecting insulator and phosphor layers, the electrons become accelerated while tunneling through the barrier, thus increasing the luminous efficiency (Fukao et al., 1989). Other ways examined for increasing the interface state density include an addition of an extra interface layer, for example, of Al (Britton et al., 1992) or Ge (Kobayashi et al., 1982), and doping of the phosphor layer close to the interface (Rack and Holloway, 1998).

Thin (about 10 nm) insulator layers have also been incorporated into the phosphor layers resulting in increased luminances and efficiencies (Horng et al., 1997; Thomas and Cranton, 1993; Cranton et al., 1997). These improvements have been attributed to an energy gain of electrons tunneling through the thin insulator layers, and redistribution of the phosphor field. In addition, in the case of Ta_2O_5 barriers in SrS:Pr,Ce the threshold voltage decreased and transferred charge density increased (Horng et al., 1997). However, just the opposite was observed when Y_2O_3 barrier layers were incorporated into ZnS:Mn (Thomas and Cranton, 1993; Cranton et al., 1997). In the latter case, the importance of hot electron tunneling was evidenced by observing a decrease of luminance when the barrier layer thickness was increased from 10 to 40 nm (Cranton et al., 1997). A possible negative effect of the barrier layers in the phosphor is a reduced crystallinity of the phosphor.

In principle, multilayer insulators can also be used in optimizing the optical efficiency of the TFEL device by making use of light interference in the multilayer structure. In practice, however, the computation of the outcoupling of light generated within the TFEL device is a difficult optical engineering problem. On the other hand, relying on a simpler optical engineering task of maximizing the transmittance through the glass–ITO–insulator–ZnS:Mn–insulator structure, the TFEL device efficiency was successfully enhanced by replacing single SiON insulators with insulators consisting of four sublayers with optimized refractive indexes and thicknesses (Ryu *et al.*, 1994). The generality of this approach remains to be verified, however.

VI. Conclusions

The development of the TFEL phosphors can be summarized as follows: the 1970s were the time of ZnS:Mn, the 1980s were the time of commercialization of yellow-emitting ZnS:Mn device and development of rare-earth-doped ZnS phosphors, and the 1990s has been the time of SrS:Ce and the commercialization of multicolor devices. The phosphor materials have been studied very thoroughly for CRT and lighting applications and it is difficult to find any completely new luminescent material. The discovery of Ce-doped thiogallates and Cu-doped SrS for EL application shows, however, that "old phosphors" can be brought to EL applications and they can provide new, useful features. Mg doping in ZnS:Mn shifts the emission toward green and shows that old EL phosphors can be modified to a more useful form for multicolor applications.

A comparison of the powder and thin film phosphors, especially that made with SrS:Ce, shows that the key issue in developing good-performance ACTFEL devices is the quality of the thin films. The deposition technique, the trace element content originating from the deposition, codoping in the case when the dopant ion and host ion have a valence mismatch, the use of fluxing agents and the filling of vacancies are important factors to be considered in making good films. The characterization of the doped SrS films by ion-beam techniques showed that the situation is very complicated: the purity of the phosphor material is not the only important factor, but the impurity tolerance is dependent on whether the impurites are located in the bulk or in grain boundaries.

The insulator–phosphor interface is an important part of the whole ACTFEL device. The nature and energy levels of the traps in the interface determine the electrical behavior of the device: threshold voltage, effciency, and aging. The traps have a very important effect on the shape of the time-dependent emission curve with different polarities (leading edge and trailing edge emission, spikes, and bands). The interface is a result of the materials and deposition techniques used and its optimization is a demanding task.

The development of EL phosphor materials has made steady progress during 1990s and careful electrical characterization of devices made by different groups has deepened the understanding of the functioning of ACTFEL devices. The development in these two key areas make the commercialization of full-color devices possible.

References

Abe, Y., Onisawa, K., Tamura, K., Nakayama, T., Hanazono, M., and Ono, Y. A. (1988). Multi-color electroluminescent devices utilizing SrS:Pr,Ce phosphor layers and color filters. *Electroluminescence, Springer Proc. Phys. Vol. 38* (Shionoya, S., and Kobayashi, H., eds.), Springer-Verlag, Berlin, pp. 199–202.

Abe, Y., Onisawa, K., Tamura, K., Nakayama, T., Hanazono, M., and Ono, Y. A. (1989). White light emitting thin film electroluminescent cells with SrS:Pr,Ce active layer and their application to multicolor electroluminescent devices. *Jpn. J. Appl. Phys.* **28**, 1373–1377.

Abe, Y., Onisawa, K., Ono, Y. A., and Hanazono, M. (1990). Effects of oxygen in CaS:Eu active layers on emission properties of thin film electroluminescent cells. *Jpn. J. Appl. Phys.* **29**, 1495–1498.

Abu-Dayah, A., Kobayashi, S., and Wager, J. F. (1993a). Internal charge-phosphor field characteristics of alternating-current thin–film electroluminescent devices. *Appl. Phys. Lett.* **62**, 744–746.

Abu-Dayah, A., Wager, J. F., and Kobayashi, S. (1993b). Electrical characterization of atomic layer epitaxy ZnS:Mn alternating-current thin-film electroluminescent devices subject to various waveforms. *J. Appl. Phys.* **74**, 5575–5581.

Abu-Dayah, A., and Wager, J. F. (1994). Aging studies of atomic layer epitaxy ZnS:Mn alternating-current thin-film electroluminescent devices. *J. Appl. Phys.* **75**, 3593–3598.

Adachi, K., Yao, T., Taniuchi, T., Kasuya, A., Miles, R. H., Uda, S., and Fukua, T. (1996). Epitaxial growth of Er^{3+}-doped CaF_2 by molecular beam epitaxy. *Jpn. J. Appl. Phys.* **35**, L435–L437.

Aguilera, M., and Aitchison, B. (1996). Fabricating high resolution AMEL flat panel displays. *Solid State Technol.* Nov., 109–116.

Ang, W. M., Pennathur, S., Pham, L., Wager, J. F., Goodnick, S. M., and Douglas, A. A. (1995). Evidence for band–to–band impact ionization in evaporated ZnS:Mn alternating-current thin-film electroluminescent devices. *J. Appl. Phys.* **77**, 2719–2724.

Antson, J. (1982). Atomic layer epitaxy: present status. *SID 1982 Digest* **13**, 124–125.

Aozasa, M., Kato, K., Nakaama, T., and Ando K. (1990). CaS:Eu,F thin film electroluminescent devices prepared by RF sputtering with hydrogen-argon mixture gas. *Jpn. J. Appl. Phys.* **29**, 1997–2002.

Barrow, W. A., Coovert, R. E., and King, C. N. (1984). Strontium sulphide: the host for a new high-efficiency thin film EL blue phosphor. *SID 1984 Digest* **15**, 249–250.

Barrow, W. A., Coovert, R. E., King, C. N., and Ziuchkovski, M. J. (1988). Matrix-addressed full-color TFEL display. *SID 1988 Digest* **19**, 284–286.

Barrow, W. A., Tuenge, R. T., and Ziuchkovski, M. J. (1986). Multicolor TFEL display and exerciser. *SID 1986 Digest* **17**, 25–28.

Barrow, W. A., Coovert, R. C., Dickey, E., King, C. N., Laakso, C., Sun, S. S., Tuenge, R. T., Wentross, R., and Kane, J. (1993). A new class of blue TFEL phosphors with application to a VGA full-color display. *SID 1993 Digest* **24**, 761–764.

Barrow, W. A., Coovert, R. C., Dickey, E., Flegal, T., Fullman, M. *et al.* (1994). A high contrast, full color, 320.256 line TFEL display. *Conference Records of the 1994 Int. Display Research Conference*, New York, pp. 448–451.

Benalloul, P., Benoit, J., and Geoffroy, A. (1985). TbF_3 complex center in ZnS ACTFEL devices. *J. Cryst. Growth* **72**, 553–558.

Benalloul, P., Benoit, J., Mach, R. Müller, G. O., and Reinsperger, G. U. (1990). Decay of ZnS:Mn emission in thin films-revisited. *J. Cryst. Growth* **101**, 989–993.

Benalloul, P., Barthou, C., Benoit, J., Eichenauer, L., and Zeinert, A. (1993). IIA-III$_2$-S$_4$ ternary compounds: new host matrices for full color thin film electroluminescent devices. *Appl. Phys. Lett.* **63**, 1954–1956.

Benalloul, P., Barthou, C., Benoit, J. (1998). SrGa$_2$S$_4$:RE phosphors for full colour electroluminescent displays. *J. Alloy Comp.* **275–277, 709–715**.

Benoit, J., Benalloul, P., and Blanzat, B. (1981). Rare earth complex dopants in a.c. thin-film electroluminescent cell. *J. Lumin.* **23**, 175–190.

Benoit, J., Barthou, C., and Benalloul, P. (1993). Excitation efficiency in thin-film electroluminescent devices: probe layer measurements. *J. Appl. Phys.* **73**, 1435–1442.

Bernard, J. E., Martens, M. F., Morton, D. C., and Williams, F. E. (1982). *1982 Int. Display Research Conference, Conference Record*, Cherry Hill, New Jersey, p. 20.

Boyn, R. (1988). 4f–4f Luminescence of rare-earth centers in II-VI compounds. *Phys. Status Solidi B* **148**, 11–47.

Bringuier, E. (1994). Tentative anatomy of ZnS-type electroluminescence. *J. Appl. Phys.* **75**, 4291–4312.

Britton, J. D., McClure, J. C., and Singh, V. J. (1992). Modification of dielectric-phosphor interface in a.c. thin film electroluminescent display devices. *Proc. 6th Int. Workshop on Electroluminescence*, El Paso, TX, pp. 286–291.

Charreire, Y., Marbeuf, A., Tourillon, G., Leskelä, M., Niinistö, L., Nykänen, E., Soininen, P., and Tolonen, O. (1992). EXAFS study of terbium activated zinc sulfide thin films. *J. Electrochem. Soc.* **139**, 619–621.

Charreire, Y., Svoronos, D.-R., Ascone, I., Tolonen, O., Niinistö, L., and Leskelä, M. (1993). Extended x-ray absorption fine structure studies of luminescent centers in II-VI thin films. *J. Electrochem. Soc.* **140**, 2015–2019.

Charreire, Y., Tolonen-Kivimäki, O., Leskelä, M., Cortes, R., Nykänen, E., Soininen, P., and Niinistö, L. (1994). EXAFS study of thulium doped zinc sulfide thin films. *2nd International Conference on f-Elements, Abstracts*, Helsinki, p. 326.

Charreire, Y., Garpon, C., Boulon, G., Leskelä, M., Nykänen, E., Soininen, P., and Niinistö, L. (1996). Identification of luminescent rare earth clusters in ZnS thin films prepared by rf-sputtering and atomic layer epitaxy. *2nd Int. Conf. Science and Technology of Display Phosphors, Extended Abstracts*, San Diego, CA, pp. 123–126.

Chase, E. W., Hepplewhite, R. T., Krupka, D. C., and Kahng, D. (1969). Electroluminescence of ZnS lumocen devices containing rare-earth and transition-metal fluorides. *J. Appl. Phys.* **40**, 2512–2519.

Chatterjee, T., McCann, P. J., Fang, X. M., Remington, J., Johnson, M. B., and Michellon, C. (1997). Visible electroluminescence from Eu:CaF$_2$ layers grown by molecular beam epitaxy on p-Si (100). *Appl. Phys. Lett.* **71**, 3610–3612.

Chen, C., Husurianto, S., Lu, X., and Koretsky, M. D. (1998). The effect of processing conditions on crystal orientation and structure in ZnS:Mn thin films. *J. Electrochem. Soc.* **145**, 226–229.

Cho, Y. J., Hirakawa, T., Sakiyama, K., Okamoto, H. and Hamakawa, Y. (1997). ZnF$_2$:Gd thin film electroluminescent device. *Appl. Surf. Sci.* **113–114**, 705–708.

Chopra, K. L., Major, S., and Pandya, K. D. (1983). Transparent conductors—a status review. *Thin Solid Films* **102**, 1–46.

Chubachi, Y., Aoyama, K., and Koyama, S. (1992). Effects of insulating layer to the degradation mechanism of SrS:CeCl$_3$ electroluminescent devices. *Proc. 6th Int. Workshop on Electroluminescence*, El Paso, TX, pp. 111–116.

Cranton, W. M. Thomas C. B., and Stevens, R. (1997). The barrier layer effect in high-luminance TFEL devices. *SID 1997 Digest* **28**, 866–869.

Davidson, J. D., Wager, J. F., and Kobayashi, S. (1992). Aging studies of evaporated ZnS:Mn alternating-current thin-film electroluminescent devices. *J. Appl. Phys.* **71**, 4040–4048.

Davolos, M. R., Garcia, A., Fouassier, C., and Hagenmuller, P. (1989). Luminescence of Eu^{2+} in strontium and barium thiogallates. *J. Solid State Chem.* **83**, 316–323.

Destriau, G. (1937). Recherches Expérimentales sur les actions du champ Électrique sur les Sulfures Phosphorescentsi. *J. Chim. Phys.* **34**, 117–124.

Douglas, A. A., Wager, J. F., Morton, D. C., Koh, J. B., and Hogh, C. P. (1993). Evidence for space charge in atomic layer epitaxy ZnS:Mn alternating-current thin-film electroluminescent devices. *J. Appl. Phys.* **73**, 296–299.

Dumbaugh, W. H., and Bocko, P. L. (1990). Substrate glasses for flat-panel displays. *SID 1990 Digest* **21**, 70–72.

Duncker, J. (1983). Large information board using TFEL devices. *SID 1983 Digest* **14**, 42–43.

Eichenauer, L., Jarofke, B., Mertins, H.-C., Dreyhsig, J., Busse, W., Gumlich, H.-E., Benalloul, P., Barthou, C., Benoit, J., Fouassier, C., and Garcia, A. (1996). Optical characterization of europium and cerium in strontium thiogallate thin films and powders. *Phys. Status Solidi A* **153**, 515–527.

Fang, X. M., Chatterjee, T., McCann, P. J., Liu, W. K., Santos, M. B., Shan, W., and Song, J. J. (1995). Eu-doped CaF_2 grown on Si(100) substrates by molecular beam epitaxy. *Appl. Phys. Lett.* **67**, 1891–1893.

Fouassier, C., and Garcia, A. (1996). Luminescence of rare-earth doped alkaline-earth sulfide and alkaline–earth thiogallate powders. In *Inorganic and Organic Electroluminescence* (Mauch, R. H. and Gumlich, H.-E., eds.), Wissenschaftliche & Technik Verlag, Berlin, pp. 313–318.

Fujikawa, H., Noda, K., Tokito, S., and Taga, Y. (1997). Electrical properties of Ta-Sn-O films on indium tin oxide electrodes. *Appl. Surf. Sci.* **113–114**, 714–717.

Fujita, Y., Kuwata J., Nishikawa, M., Tohda, T., Matsuoka, T., Abe, A., and Nitta, T. (1983). Large scale AC thin-film electroluminescent display panel. *Japan Display '83* 76–79.

Fukao, R., Fujikawa, H., and Hamakawa, Y. (1988). High-brightness low-driving-voltage green color thin-film electroluminescent devices. *Appl. Surf. Sci.* **33–34**, 1229–1235.

Fukao, R., Fujikawa, H., and Hamakawa, Y. (1989). Improvement of luminous efficiency in ZnS:Tb,F thin-film electroluminescent devices using ferroelectric $PbTiO_3$ and silicon nitride as carrier accelerating buffer layers. *Jpn. J. Appl. Phys.* **28**, 2446–2449.

Gao, Q. Z., Mita, J., Tsuruoka, T., Kobayashi, M., and Kawamura, K. (1992). High luminance white EL devices using SrS:Ce,Eu,K films deposited in a H_2 atmosphere. *J. Cryst. Growth* **117**, 983–986.

Garcia, A., Fouassier, C., and Dougier, P. (1982). Photo- and cathodoluminescent properties of erbium-doped thiogallates. *J. Electrochem. Soc.* **129**, 2063–2069.

Gerhardinger, P. F., and McCurdy, R. J. (1996). Float line deposited transparent conductors— implications for the PV industry. *Mater. Res. Soc. Symp. Proc.* **426**, 399–410.

Godlewski, M., and Hommel, D. (1986). Eu^{2+} photocharge transfer processes in ZnS crystals determined by photo-ESR measurements. *Phys. Status Solidi A* **95**, 261–268.

Godlewski, M., and Leskelä, M. (1994). Excitation and recombination processes during electroluminescence of rare earth-activated materials. *Crit. Rev. Solid State Mater.* **19**, 199–239.

Gordon, R. G. (1996). Preparation and properties of transparent conductors. *Mater. Res. Soc. Symp. Proc.* **426**, 419–429.

Haaranen, J., Törnqvist, R., Koponen, J., Pitkänen, T., Surma-aho, M., Barrow, B., and Laakso, C. (1992). A 9-in. diagonal high–contrast multicolor TFEL display. *SID 1992 Digest* **23**, 348–351.

Hamakawa, Y., Fukao, R., and Fujikawa, H. (1988). Tunable color electroluminescent cells. *Optoelectronics-Devices Technol.* **3**, 31–46.

Hao, G., Zhong, G.-Z., and Li, C.-C. (1994). ACEL characteristics in ZnS:Ho thin film and full-color device. *Proc. Int. Workshop on EL*, Beijin, pp. 387–392.

Härkönen, G., Härkönen, K., and Törnqvist, R. (1990). Green-emitting thin film electroluminescent device grown by atomic layer epitaxy. *SID 1990 Digest* **21**, 232–235.

Hirabayashi, K., and Kozawaguchi, H. (1986). ZnS:Mn electroluminescent device prepared by metal-organic chemical vapor deposition. *Jpn. J. Appl. Phys.* **25**, 711–713.

Hirabayashi, K., Kozawaguchi, H., and Tsujiyama, B. (1987). Color electroluminescent devices prepared by metal organic chemical vapor deposition. *Jpn. J. Appl. Phys.* **26**, 1472–1476.

Hirabayashi, K., Kozawaguchi, H., and Tsujiyama, B. (1988). F/Tb ratio dependence of the photoluminecent and electroluminescent characteristics in MOCVD-prepared $ZnS:TbF_x$ green-emitting electroluminescent devices. *Jpn. J. Appl. Phys.* **27**, 587–591.

Horng, R.-H., Wuu, D.-S., and Kung, C.-Y. (1997). Characterization of thin-film electroluminescent devices with multiple Ta_2O_5 interlayers incorporated into SrS:Pr,Ce phosphor. *Jpn. J. Appl. Phys.* **36**, 7245–7249.

Howard, W. E. (1977). The importance of insulator properties in a thin-film electroluminescent device. *IEEE Trans. Electron Devices* **ED-24**, 903–908.

Howard, W. E. (1981). Memory in thin-film electroluminescet devices. *J. Lumin.* **23**, 155–173.

Howard, W. E, Sahni, O., and Alt, P. (1982). A simple model for the hysteretic behavior of ZnS:Mn thin film electroluminescent devices. *J. Appl. Phys.* **53**, 639–647.

Hüttl, B., Troppenz, U., Velthaus, K. O., Ronda, C. R., and Mauch, R. H. (1996a). Luminescence properties of $SrS:Ce^{3+}$. *J. Appl. Phys.* **78**, 7282–7288.

Hüttl, B., Velthaus, K.-O., Troppenz, U., Herrmann, R., and Mauch, R. H. (1996b). SrS:Ce,Mn,Cl — a novel efficient EL phosphor. *J. Cryst. Growth* **159**, 943–946.

Hüttl, B., Velthaus, K.-O., Troppenz, U., Kreissl, J., and Mauch, R. H. (1996c). Luminescence properties of SrS:Ce thin films. In *Inorganic and Organic Electroluminescence* (Mauch, R. H., and Gumlich, H.-E., eds.), Wissenschaftliche & Technik Verlag, Berlin, pp. 319–324.

Ihanus, J., Ritala, M., and Leskelä, M. (1997). *Electrochem. Soc. Proc.* **97–25**, 1423–1428.

Ihanus, J., Ritala, M., Leskelä, M., Nykänen, E., Niinistö, L., Lambers, E., and Holloway, P. H. (1999), to be published.

Inoguchi, T., Takeda, M., Kakahara, Y., Nakata, Y., and Yoshida, M. (1974). Stable high brightness thin-film electroluminescent panels. *SID 1974 Digest*, p. 84.

Inoue, Y., Tanaka, K., Okamoto, S. Kobayashi, K., and Fujimoto. I. (1995). Blue electroluminescent $SrGa_2S_4$:Ce thin films grown by molecular beam epitaxy. *Jpn. J. Appl. Phys.* **34**, L180–L181.

Inoue, Y., Tanaka, K., Okamoto, S., Kobayashi, K., and Takizawa, K. (1997). Ce-activated SrS thin film electroluminescent devices fabricated by multi source deposition using Ga_2S_3 precursor. *Jpn. J. Appl. Phys.* **36**, 4335–4338.

Jarzebski, Z. M. (1982). Preparation and properties of transparent conducting oxide films. *Phys. Status Solidi A* **71**, 13–41.

Jiang, X.-Y., Zhang Z.-L., Zhao, W.-M., Lui, Z.-G., and Xu, S.-H. (1994). A quantitative evaluation of the excitation mechanism of Tm^{3+} in a ZnS thin film. *J. Phys. Condens. Matter.* **6**, 3279–3290.

Jones, S. (1964). Electrolumiescence in oxide phosphors. *J. Electrochem. Soc.* **111**, 307–310.

Kahng, D. (1968). Electroluminescence of rare-earth and transition metal molecules in II–VI compounds via impact excitation. *Appl. Phys. Lett.* **13**, 210–212.

Kane, J., Harty, W., Ling, M., and Yocom, P. N. (1985). New electroluminescent phosphors based on strontium sulfide. *1985 Int. Display Research Conference, Conference Record*, pp. 163–166.

Karpinska, K., Swiatek, K., Godlewski, M., Niinistö, L., Leskelä, M. (1993). Rare-earth excitation mechanism in wide band gap II–VI compounds. *Acta Phys. Polonica A* **84**, 959–962.

Katiyar, M., and Kitai, A. H. (1992). Effect of organized doping on concentration quenching in ZnS:Mn. *J. Lumin.* **52**, 309–312.

Kato, A., Katayama, M., Mizutani, A., Hattori, Y., Ito, N., and Hattori, T. (1994). Satellite peak generation in the electroluminescence spectrum of ZnS:Sm grown by metalorganic chemical vapor deposition with Cl codoping. *J. Appl. Phys.* **76**, 3206–3208.

Kato, A., Katayama, M., Mizutani, A., Ito, N., and Hattori, T. (1997). ZnS:Tm grown by metalorganic chemical vapor deposition with Cl codoping. *J. Appl. Phys.* **81**, 445–450.

Kattelus, H., Ylilammi, M., Salmi, J., Ranta-aho, T., Nykänen, E., and Suni, I. (1993). Electrical properties of tantalum based composite oxide films. *Mater. Res. Soc. Symp. Proc.* **284**, 511–516.

Kawai, H., and Hoshina, T. (1981). Cathodoluminescent properties of ZnS:Ce,Li phosphor. *Jpn. J. Appl. Phys.* **20**, 1241–1247.

Khomchenko, V. S., Kononec, Ya. F., Vlasenko, N. A., Mach, R., Reinsperger, G. U., Selle, B., and Reetz, R. (1990). Thin film electroluminescence of ZnS:Tb^{3+}. *J. Crys. Growth* **101**, 994–998.

Khormaei, R., Wager, J. F., and King, C. N. (1989). Improved stability of ZnS:Mn ACTFEL devices. *SID 1989 Digest* **20**, 65–68.

Khormaei, R., King, C. N., Coovert, R. E., and Wager, J. F. (1991). Stabilization of ZnS:Mn ACTFEL devices through processing modifications. *SID 1991 Digest* **22**, 74–77.

Khormaei, R., Thayer, S., Ping, K., King, C., Dolny, G., Ipri, A., Hsueh, F.-L., Stewart, R., Keyser, T., Becker, G., Kagey, D., and Spitzer, M. (1994). High-resolution active-matrix electroluminescent display. *SID 1994 Digest* **25**, 137–139.

Kina, H., Yamada, Y., Maruta, Y., and Tamura, Y. (1996). ZnS:Mn thin-film electroluminescent devices prepared by metalorganic chemical vapor deposition. *J. Cryst. Growth* **169**, 33–39.

King, C. N. (1996a). Electroluminescent displays. *J. Vac. Sci. Technol. A* **14**, 1729–1735.

King, C. N. (1996b). Electroluminescence: an industry perspective. *J. SID* **4**, 153–156.

King, C. N. (1996c). Active Matrix EL. *Inorganic and Organic Electroluminescence* (Mauch, R. H., and Gumlich, H.-E., eds.), Wissenschaftliche & Technik Verlag, Berlin, pp. 375–380.

Kitai, A. H., Xiao, T., Liu, G. and Li, J. H. (1997). Doped amorphous and crystalline gallium oxides and alkaline-earth gallates as EL materials. *SID 97 Digest* **28**, 419–422.

Kobayashi, H. (1994). SrS-ZnS electroluminescent materials. *J. Cryst. Growth* **138**, 1010–1016.

Kobayashi, H., and Tanaka, S. (1996). The present and future prospects of electroluminescent phosphors. *J. SID* **4**, 157–163.

Kobayashi, H., Tanaka, S., Sasakura, H., and Hamakawa, Y. (1973). The electron injection mechanism of the electroluminescent ZnS:Tb^{3+} films. *Jpn. J. Appl. Phys.* **12**, 1854–1861.

Kobayashi, H., Tueta, R. J., and Menn, R. (1982). Thin film ZnS:Mn ac-electroluminescent device with a Ge layer. *IEEE Trans. Electron Devices* **ED-29**, 1626–1629.

Kobayashi, H., Tanaka, S., Kunou, T., Shiiki, M., and Sasakura, H. (1984). Varoius Tb-compound luminescent centers in ZnS thin-film electroluminescent devices. *Proc. SID* **25**, 187–192.

Kobayashi, H., Tanaka, S., Shanker, V., Shiiki, M., and Deguchi, H. (1985a). Evaporated CaS thin films for AC electroluminescence devices. *J. Cryst. Growth* **72**, 559–562.

Kobayashi, H., Tanaka, S., Shanker, V., Shiiki, M., Kunou, T., Mita, J., and Sasakura, H. (1985b). Multicolor electroluminescent ZnS thin films doped with rare earth fluorides. *Phys. Status Solidi A* **88**, 713–720.

Kobayashi, S., Wager, J. F., and Abu-Dayah, A. (1992). Distribution of trapped electrons at interface states in ACTFEL devices. *Proc. 6th Int. Workshop on Electroluminescence*, El Paso, TX, pp. 234–239.

Kong, W., Fogarty, J., and Solanki, R. (1994). Atomic layer epitaxy of ZnS:Tb thin film electroluminescent devices. *Appl. Phys. Lett.* **65**, 670–672.

Kong, W., Fogarty, J., Solanki, R., and Tuenge, R. T. (1995). White light emitting SrS:Pr electroluminescent devices fabricated via atomic layer epitaxy. *Appl. Phys. Lett.* **66**, 419–421.

Kononets, Ya. F., Törqvist, R., and Vlasenko, N. A. (1996). On physcal model of aging behaviour of atomic layer epitaxy ZnS:Mn AC TFEL devices. *Inorganic and Organic Electroluminesce* (Mauch, R. H., and Gumlich, H.-E., eds.), Wissenschaftliche & Technik Verlag, Berlin, pp. 259–262.

Kreissl, J. (1986). A quantitative EPR analysis of ZnS:Mn powders of different crystal structure. *Phys. Status Solidi A* **97**, 191.

Kreissl, J., and Gehlhoff, W. (1984). EPR investigations of ZnS:Mn and ZnSe:Mn. *Phys. Status Solidi A* **81**, 701–707.

Krupka, D. C. (1972). Hot-electron impact excitation of Tb^{3+} luminescence in $ZnS:Tb^{3+}$ thin films. *J. Appl. Phys.* **43**, 476–481.

Kukli, K., Ihanus, J., Ritala, M., and Leskelä, M. (1996a). Properties of Ta_2O_5 based dielectric nanolaminates deposited by atomic layer epitaxy. *J. Electrochem. Soc.* **144**, 300–306.

Kukli, K., Ihanus, J., Ritala, M., and Leskelä, M. (1996b). Tailoring the dielectric properties of HfO_2-Ta_2O_5 nanolaminates. *Appl. Phys. Lett.* **68**, 3737–3739.

Kukli, K., Ritala, M., and Leskelä, M. (1997). Properties of $(Nb_{1-x}Ta_x)_2O_5$ solid solutions and $(Nb_{1-x}Ta_x)_2O_5$-ZrO_2 nanolaminates grown by atomic layer epitaxy. *Nanostruct. Mater.* **8**, 785–793.

Laakso, C., Khormaei, R., King, C., Härkönen, G., Pakala, A., Pitkanen, T., Surma-aho, M., and Törnqvist, R. (1991). A 9 inch diagonal, compact, multicolor TFEL display. *Int. Display Research Conference, Conference Record*, pp. 43–44.

Lee, Y. H., Chung, I.-J., and Oh, M.-H. (1991). Possible degradation mechanism in ZnS:Mn alternating current thin-film electroluminescent display. *Appl. Phys. Lett.* **58**, 962–964.

Lee, Y. H., Kim, D. H., Ju, B. K., Song, M. H., Hahn, T. S., Choh, S. H., and Oh, M. H. (1995). Decrease of the number of the isolated emission center Mn^{2+} in an aged ZnS:Mn electroluminescent device. *J. Appl. Phys.* **78**, 4253–4257.

Lee, S. T., Kitagawa, M., Ichino, K., and Kobayashi, H. (1996). Preparation and photoluminescence characteristics of $Zn_xSr_{1-x}S$:Ce. *Appl. Surf. Sci.* **110–101**, 656–659.

Lehmann, W. (1970). Alkaline earth sulfide phosphors activated by copper, silver, and gold. *J. Electrochem. Soc.* **117**, 1389–1393.

Leppänen, M. Leskelä, M., Niinistö, L., Nykänen, E., Soininen, P., Tiitta, M. (1991). Blue emitting ACTFEL structure based on cerium-activated strontium sulfide grown by atomic layer epitaxy. *SID 1991 Digest* **22**, 282–284.

Leppänen, M., Härkönen, G., Pakkala, A., Soininen, E., and Törnqvist. R. (1993). Broadband double layer phosphor for an inverted filtered RGB electroluminescent display. *13th Int. Display Research Conference, Eurodisplay '93, Conference proceedings*, Strasbourg, pp. 229–232.

Leskelä, M. (1989). Electroluminescent materials grown by atomic layer epitaxy. *Electroluminescence, Springer Proc. Phys.*, Vol. 38 (Shionoya, S., and Kobayashi, H. eds.), Springer-Verlag, Berlin, pp. 204–209.

Leskelä, M. (1998). Rare earths in electroluminescent and field emission display phosphors. *J. Alloys Comp.* **275–277**, 702–708.

Leskelä, M., and Niinistö, L. (1990). Chemical aspects of the atomic layer epitaxy (ALE) process. In *Atomic Layer Epitaxy* (Suntola, T., and Simpson, M., eds.), Blackie and Son, Glasgow, pp. 1.

Leskelä, M., and Niinistö, L. (1992a). Applications of rare earths in full-colour EL displays. *Mater. Chem. Phys.* **31**, 7–11.

Leskelä, M., and Niinistö, L. (1992b). Thin films processed by ALE for applications in ACTFEL devices. *Proc. 6th Int. Workshop on Electroluminescence*, El Paso, TX, pp. 249–256.

Leskelä, M., Mäkelä, M., Niinistö, L., Nykänen, E., and Tammenmaa, M. (1988). Electroluminescent calcium sulphide thin films doped with Tb^{3+} and Eu^{2+}. *Chemtronics* **3**,

113–115.
Lewis, J., Waldrip, K. E., Davidson, M. R., Moorehead, D., Sun, S.-S., Holloway, P. (1998). Improved brightness and efficiency in electroluminescent thin film phosphors by fluxing. *4th Int. Conf. on the Science and Technology of Display Phosphors, Extended Abstracts*, Bend, OR, pp. 227–230.
Li, L., Dobrowolski, J. A., Sullivan, B. T., Simpson, R., and Bajcar, R. C. (1994). High-contrast TFEL displays with additional optical interference components. *SID Digest* **25**, 140–143.
Li, W.-M., Lappalainen, R., Jokinen, J., Ritala, M., Leskelä, M., Soininen, E., and Hüttl, B. (1996). Ion beam analysis of SrS:Ce thin films. In *Inorganic and Organic Electroluminescence* (Mauch, R. H., and Gumlich, H.-E., eds.), Wissenschaftliche & Technik Verlag, Berlin, pp. 157–160.
Li, W.-M., Ritala, M., Leskelä, M., Lappalainen, R., Jokinen, J., Soininen, E., Hüttl, B., Nykänen, E., and Niinistö, L. (1998a). Elemental characterization of electroluminescent SrS:Ce thin films. *J. Appl. Phys.* **84**, 1029–1035.
Li, W.-M., Ritala, M., Leskelä, M., Lappalainen, R., Karjalainen, M., Soininen, E., Barthou, C., Benalloul, P., Benoit, J., Nykänen, E., and Niinistö, L. (1998b). Ion implantation of SrS:Ce thin films. *J. SID*, in press.
Li, W.-M., Ritala, M., Leskelä, M., Lappalainen, R., Soininen, E., Niinistö, L., Barthou, C., Benalloul, P., and Benoit, J. (1998c). Codoping of ALE SrS:Ce and SrS:Cu thin films by ion implantation. *4th Int. Conf. on the Science and Technology of Display Phosphors, Extended Abstracts*, Bend, OR, pp. 259–262.
Li, W.-M., Ritala, M., Leskelä, M., Lappalainen, R., Soininen, E., Niinistö, L., Barthou, C., Benalloul, P., and Benoit, J: (1999a). Improved blue luminescence in Ag-codoped SrS:Ce thin films made by atomic layer epitaxy and ion implantation. *Appl. Phys. Lett.* in press.
Li, W.-M., Ritala, M., Leskelä, M., Niinistö, L., Soininen, E., Sun, S.-S., and Tong, W. S. (1999b). Photoluminescence and electroluminescence of SrS:Cu and SrS:Ag,Cu,Ga thin film electroluminescent devices. *J. Appl. Phys.*, in press.
Lindroos, S., Kanniainen, T., Leskelä, M., and Rauhala, E. (1995). Deposition of manganese-doped zinc sulfide thin films by the successive ionic layer adsorption and reaction (SILAR) method. *Thin Solid Films* **263**, 79–84.
Liu, J., Laverty, S. J., Maguire, P., McLaughlin, J., Molloy, J., and Anderson, J. (1991). Conductivity enhancement of transparent electrode by side-wall copper electroplating. *SID 1993 Digest* **24**, 554–557.
Liu, J. S, Laverty, S. J., Maguire, P., McLaughlin, J., and Molloy, J. (1994). The role of an electrolysis reduction in copper-electroplating on transparent semiconductor tin oxide. *J. Electrochem. Soc.* **141**, L38–L40.
Liu, G., Lobban, K., and Bailey, P. (1998). Investigation of rare-earth-doped thiogallates as potential phosphors for TDEL displays. *SID 1998 Digest* **29**, 648–651.
Mach, R., and Müller, G. O. (1982). Physical concepts of high-field, thin-film electroluminescence devices. *Phys. Status Solidi A* **69**, 11–66.
Mach, R., Müller, G. O., Schnuerer, E., Selle, B., and Ohnishi, H. (1990). Efficiency of alkaline earth sulfide electroluminescent devices. *Acta Polytechn. Scand., Ser. Appl. Phys. Ph.* **170**, 197–202.
Madarasz, J., Leskelä, T., Rautanen, J., and Niinistö, L. (1996). Oxidation of alkaline-earth-metal sulfide powders and thin films. *J. Mater. Chem.* **6**, 781–787.
Marrello, V., Samuelson, L., Onton, A., and Reuter, W. (1981). Probe layer measurements of electroluminescence excitation in ac thin-film devices. *J. Appl. Phys.* **52**, 3590–3599.
Matsuhashi, H., and Nishikawa, S. (1994). Optimum electrode materials for Ta_2O_5 capacitors for high- and low-temperature processes. *Jpn. J. Appl. Phys.* **33**, 1293–1297.
Matsuoka, T., Kuwata, J., Nishikawa, M., Fujita, Y., Tohda, T., and Abe, A. (1988). A study

of the crystallographic and luminescent characteristics of ZnS:Mn films prepared by an RF magnetron sputtering method for AC thin-film electroluminescent devices. *Jpn. J. Appl. Phys.* **27**, 592–596.

Matsuyama, I., Yamashita, N., and Nakamura, K. (1989). Photoluminescence of the SrS:Mn^{2+} phosphor and Pb^{2+} sensitized luminescence of SrS:Mn^{2+},Pb^{2+} phosphor. *J. Phys. Soc. Jpn.* **58**, 741–751.

Mauch, R. H., Menner, R., and Schock, H. W. (1988). Comparison of ZnS:Mn ACTFEL devices prepared by manganese diffusion and coevaporation. *J. Cryst. Growth* **86**, 885–889.

Mauch, R. H., Velthaus, K. O., Bilger, G., and Schock, H. W. (1992). High efficiency SrS,SrSe:$CeCl_3$ based thin film electroluminescent devices. *J. Cryst. Growth* **117**, 964–968.

Mauch, R. H., Velthaus, K. O., Hüttl, B., Troppenz, U., and Herrmann, R. (1995). Improved SrS:Ce,Cl TFEL devices by ZnS co-evaporation. *SID 1995 Digest* **26**, 720–723.

McClure, D. S. (1963). Optical spectra of exchange coupled Mn^{++} ion pairs in ZnS:Mn. *J. Chem. Phys.* **39**, 2850–2855.

Menkara, H. M., Park, W., Chaichimansour, M., Jones, T. C., Wagner, B. K., Summers, C. J., and Sun, S.-S. (1998). Evaporation and characterization of SrS:Cu,Ag electroluminescent devices. *4th Int. Conf. on the Science and Technology of Display Phosphors, Extended Abstracts*, Bend, OR, pp. 191–194.

Mikami, A., Ogura, T., Tanaka, K., Taniguchi, K., Yoshida, M., and Nakajima, S. (1987). Tb-F emission centers in ZnS:Tb,F thin film electroluminescent devices. *J. Appl. Phys.* **61**, 3028–3034.

Mikami, A., Ogura, T., Tanaka, K., Taniguchi, K., Yoshida, M., and Nakajima, S. (1988). Excitation process of the Tb emission center in a ZnS:Tb,F thin-film electroluminescent device. *J. Appl. Phys.* **64**, 3650–3657.

Mikami, A., Terada, K., Okibayashi, K., Tanaka, K., Yoshida, M., and Nakajima, S. (1991). Chemical vapor deposition of ZnS:Mn electroluminescent films in a low-pressure halogen transport system. *J. Cryst. Growth* **110**, 381–394.

Mikami, A., Terada, K., Okibayashi, K., Tanaka, K., Yoshita, M., and Nakajima, S. (1992). Aging characteristics of ZnS:Mn electroluminescet films grown by a chemical vapor deposition technique. *J. Appl. Phys.* **72**, 773–782.

Mikami, A., Yashima, I., and Kajikawa, F. (1996). New developments in ZnS type EL displays. In *Inorganic and Organic Electroluminescence* (Mauch, R. H., and Gumlich, H.-E., eds.), Wissenschaftliche & Technik Verlag, Berlin, pp. 370–374.

Minami, T. (1998). Oxide phosphor thin-film electroluminescent devices using thick insulating ceramic sheets. *4th Int. Conf. on the Science and Technology of Display Phosphors, Extended Abstracts*, Bend, OR, pp. 195–198.

Minami, T., Miyata, T., Takata, S., and Fukuda, I. (1990). High-luminance green Zn_2SiO_4:Mn thin-film electroluminescent devices using an insulating $BaTiO_3$ ceramic sheet. *Jpn. J. Appl. Phys.* **30**, L117–L119.

Minami, T., Kuroi, Y., Takata, S., and Miyata, T. (1995). Multicolor-emitting $ZnGa_2O_4$ phosphors for emitting layer of TFEL devices. *Asia Display '95*, pp. 821–824.

Minami, T., Kuroi, Y., Yamada, H., Kubota, Y., and Takata, S. (1997a). TFEL devices using $ZnGa_2O_4$ Phosphors co-doped with Mn and Cr. *Proc. SID*, pp. 350–353.

Minami, T., Sakagami, Y., and Miyata, T. (1997b). New type multicolor electroluminescent devices fabricated without vacuum process. *3rd Int. Conf. on the Science and technology of Display Phosphors, Extended Abstracts*, Huntington Beach, CA, p. 37.

Minami, T., Yamada, H., Kubota, Y., and Miyata, T. (1997c). New high-luminance TFEL devices using Mn-activated CaO-Ga_2O_3 phosphors. *Proc. SID*, pp. 354–357.

Mita, J., Koizumi, M., Kanno, H., Hayashi, T., Sekido, Y., Abiko, I., and Nihei, K. (1987).

ZnS:Mn thin film electroluminescence devices having doubly-stacked insulating layers. *Jpn. J. Appl. Phys.* **26**, L541–L543.

Mita, J., Koizumi, M., Kanno., H. Hayashi, T., Sekido, Y., Abiko, I., Nihei, K. (1988). Difference in electroluminescent ZnS:Tb,F thin films prepared by electron-beam evaporation and RF magnetron sputtering. *Jpn. J. Appl. Phys.* **26**, L1205–1207.

Miura, N., Sasaki, T., Matsumoto, H., and Nakano, R. (1991). Strong ultraviolet-emitting ZnF_2:Gd thin film electroluminescent device. *Jpn. J. Appl. Phys.* **30**, L1815–L1816.

Miura, N., Ishikawa, T., Sasaki, T., Oka, T., Ohata, H., Matsumoto, H., and Nakano, R. (1992). Several blue-emitting thin-film electroluminescent device. *Jpn. J. Appl. Phys.* **31**, L46–L48.

Miura, N., Ogawa, K., Kobayashi, S., Matsumoto, H., and Nakano, R. (1994). Electroluminecence spectra of rare-earth-doped $ZnS_{1-x}Se_x$ thin films. *J. Cryst. Growth* **138**, 1046–1050.

Miura, N., Namiki, T., Matsumoto, H., and Nakano, R. (1997). Electroluminescent spectra for Tb^{3+} doped $Zn_{1-x}Cd_xS$ thin-films. *J. Lumin.* **72–74**, 999–1001.

Miyata, T., Minami, T., Takata, S., and Fukuda, I. (1991). New high-luminance multicolor TFEL devices using an oxide phosphor emitting layer. *SID 1991 Digest* **22**, 286–289.

Moffatt, D. M. (1994). Flat panel display substrates. *Mater. Res. Soc. Symp. Proc.* **345**, 163–174.

Moffatt, D. M. (1996). Glass substrates for flat panel displays. *MRS Bull.* **21**, 31–34.

Morton, D. C., and Williams, F. E. (1979). A new thin-film electroluminescent material — ZnF_2:Mn. *Appl. Phys. Lett.* **35**, 671–672.

Mueller-Mach, R., Mueller, G. O., Alinsog, E., and Helbing, R. (1996a). High luminance from thin film electroluminescence devices. In *Inorganic and Organic Electroluminescence* (Mauch, R. H., and Gumlich, H.-E., eds.), Wissenschaftliche & Technik Verlag, Berlin, pp. 381–384.

Mueller-Mach, R., Mueller, G. O., and Nauka, C. (1996b). 4f–4f Emitters in ZnS thin film electroluminescence devices. *2nd Int. Conf. on the Science and Technology of Display Phosphors, Extended Abstracts*, San Diego, CA, pp. 191–194.

Mueller-Mach, R., Fouassier, C., Fan, X. W., Mueller, G. O., Zhong, G. Z., Garcia, A., Zhao, L. J., and Sun, J. M. (1997). Ho^{3+}, a multi-color emitter in sulfides. *3rd Int. Conf. on the Science and Technology of Display Phosphors, Extended Abstracts*, Huntington Beach, CA, pp. 45–48.

Müller, G. O. (1984). Basics of Electron-Impact-Excited Luminescence Devices. *Phys. Status Solidi A* **81**, 597–608.

Müller, G. O. (1995). Efficiency of color electroluminescence. *Proc. 7th Int. Workshop on Electroluminescence*, Beijing, pp. 7–21.

Müller, G. O., and Mach, R. (1988). Physics of electroluminescence devices. *J. Lumin.* **40–41**, 92–96.

Nakanishi, Y., Gurumurugan, K., Mitsui, T., Aoki, T., and Hatanaka, Y. (1997). Structural and luminescent properties of $Zn_{1-x}Mg_xS$:Mn thin films including oxygen. *Int. Display Research Conference, Conference Record*, Toronto, pp. 346–349.

Nakayama, T., Onisawa, K., Fuyama, M., and Hanazono, M. (1992). TiO_2/SiO_2 Multilayer insulating films for ELDs. *J. Electrochem. Soc.* **139**, 1204–1206.

Naman, A., Pathagney, B., Li, J. H., Jones, K. S., Holloway, P. H., Sun, S. S., and Dennis, W. M. (1996). The effect of gallium doping on the microstructure of SrS:Ce,F electroluminescent thin films. *2nd Int. Conf. on the Science and Technology of Display Phosphors*, San Diego, CA, pp. 37–38.

Neyts, K. A., Corlatan, D., De Visschere, P., and Van den Bossche, J. (1994). Observation and simulation of space-charge effects and hysteresis in ZnS:Mn AC thin-film electroluminescent devices. *J. Appl. Phys.* **75**, 5339–5346.

Nire, T., Matsuno, A., Wada, F., Fuchiwaki, K., and Miyakoshi, A. (1992). Multicolor TFEL display panel with a double-heterointerface-structured active layer *SID 1992 Digest* **23**, 352–355.

Nire, T., Matsuno, A., Miyakoshi, A., and Ohmi, K. (1994). ZnS:Mn electroluminescent thin films prepared by multisource deposition under controlled sulfur vapor pressure. *Jpn. J. Appl. Phys.* **33**, 2605–2612.

Nykänen, E., Lehto, S., Leskelä, M., Niinistö, L., and Soininen, P. (1992). Blue electroluminescence in Pb^{2+} doped CaS and SrS thin films. *Proc. 6th Int. Workshop on Electroluminescence*, El Paso, TX, pp. 199–204.

Nykänen, E., Soininen, P., Niinistö, L., Leskelä, M., and Rauhala, E. (1995). Electroluminescent SrS:Ce,F thin films deposited by the atomic layer epitaxy process. *Proc. 7th Int. Workshop on Electroluminescence*, Beijing, pp. 437–444.

O'Brien, T. A., Rack, P. D., Holloway, P. H., and Zerner, M. C. (1998). Crystal field and molecular orbital calculation of the optical transition in Ce doped alkaline earth sulfide (MgS, CaS, SrS, and BaS) phosphors. *J. Lumin.* **78**, 245–257.

Oberacker, T. A., and Schock, W. H. (1995a). The influence of halides on the luminescent properties of $SrS:CeX_3$ ($X = Cl$, F) TFEL devices. *Proc. 7th Int. Workshop on Electroluminescence*, Beijing, pp. 244–250.

Oberacker, T. A., Velthaus, K. O., Mauch, R. H., Schock, H. W., and Tuenge, R. T. (1995b). Growth mechanism of $SrGa_2S_4:CeCl_3$ films for thin film electroluminescent devices. *Proc. 7th Int. Workshop on Electroluminescence*, Beijing, pp. 160–161.

Oberacker, T. A., and Schock, W. H. (1996a). Investigations of the influence of halides on the properties of CeX_3-doped ($X = Cl$, F) strontium sulfide thin film electroluminescent devices. *J. Cryst. Growth* **159**, 935–938.

Oberacker, T. A., Schlotterbeck, G., Bilger, G., Braunger, D., and Schock, H. W. (1996b). Field quenching effects in polycrystalline SrS:Pb and SrS:Ce,Pb thin films for electroluminescent devices. *J. Appl. Phys.* **80**, 3526–3531.

Oberacker, T. A., Troppenz, U., Huettl, B., Gaerther, T., Herrmann, R., and Velthaus, K. O. (1997). High efficiency blue from SrS:Ce/ZnS:Mn "Color by White" thin film electroluminescence devices. *Conference Records of the 1997 International Display Research Conference*, Toronto, pp. 297–300.

Ogura, T., Mikami, A., Tanaka, K., Taniguchi, K., Yoshida, M., and Nakajima, S. (1986). High-brightness green-emitting electroluminescent devices with ZnS:Mn,F active layers. *Appl. Phys. Lett.* **28**, 1570–1571.

Ohmi, K., Tanaka, S., Kobayashi, H., and Nire, T. (1992a). Electroluminescent devices with $(Y_2O_3:Tb/ZnS)_n$ multilayered phosphor thin films prepared by multisource deposition. *Jpn. J. Appl. Phys.* **31**, L1366–L1369.

Ohmi, K., Tanaka, S., Yamano, Y., Fujimoto, K., Kobayashi, H., Mauch, R. H., Velthaus, K. O., and Schock, H. W. (1992b). White light emitting electroluminescent devices with $(SrS:Ce/ZnS:Mn)_n$ multilayered thin films. *Japan Displays '92*, pp. 725–728.

Ohmi, K., Hirose, T., Harada, M., Tanaka, S., and Kobayashi, H. (1997). Imrovements in aging characteristics by Zn doping of electron-beam evaporated SrS:Ce thin-film electroluminescent devices. *Jpn. J. Appl. Phys.* **36**, L33–L36.

Ohmi, K., Yamabe, K., Fukada, H., Fujiwara, T., Tanaka, S., and Kobayashi, H. (1998). Blue emitting SrS:Cu thin-film electroluminescent devices prepared by hot-wall deposition technique. *4th Int. Conf. on the Science and Technology of Display Phosphors, Extended Abstracts*, Bend, OR, pp. 219–222.

Ohnishi, H., and Okuda, T. (1989). Blue-green color TFEL device with sputtered SrS:Ce thin films. *SID 1989 Digest* **20**, 317–320.

Ohnishi, H., Iwase, R., and Yamasaki, Y. (1988), Red-color ACTFEL devices using sputtered CaS:Eu thin films. *Proc. SID* **29**, 311–315.

Ohwaki, J., Yamauchi, N., Kozawaguchi, H., and Tsujiyama, B. (1987). The role of stacked insulating layers on thin-film electroluminescent devices. *Jpn. J. Appl. Phys.* **26**, 1064–1068.

Ohwaki, J., Kozawaguchi, H., and Tsujiyama, B. (1990). Stacked insulator structure thin-film electroluminescent display devices. *J. Electrochem. Soc.* **137**, 340–342.

Okamoto, K., and Miura, S. (1986). Excitation mechanism in thin-film electroluminescent devices. *Appl. Phys. Lett.* **49**, 1596–1598.

Okamoto, K., Nasu, Y., and Hamakawa, Y. (1981). Low-threshold-voltage thin-film electroluminescent devices. *IEEE Trans. Electron Devices* **ED-28**, 698–702.

Okamoto, K., Yoshimi, T., and Miura, S. (1988). TbOF complex centers in ZnS thin-film electroluminescent devices. *Appl. Phys. Lett.* **53**, 678–680.

Okamoto, K., Yoshimi, T., Nakamura, K., Kobayashi, T., Sato, S., and Miura, S. (1989a). ZnS:TbOF thin-film green electroluminescent panel fabricated by two-target sputtering. *Jpn. J. Appl. Phys.* **28**, 1378–1384.

Okamoto, S., Nakazawa, E., and Tsuchiya Y. (1989b), Electroluminescence of rare-earth activated SrS thin-films. *Electroluminescence, Springer Proc. Phys.* Vol. 38 (Shionoya, S., and Kobayashi, H. eds.), Springer-Verlag, Berlin, pp. 195–198.

Okamoto, S., Nakazawa, E., and Tshuchiya, Y. (1990a). White-emitting thin-film electroluminecesce devices with SrS phosphor doubly activated with rare-earth ions. *Jpn. J. Appl. Phys.* **29**, 1987–1990.

Okamoto, S., Nakazawa, E., Kuki, T., and Tshuchiya, Y. (1990b). Effects of sulfur-cracking in the fabrication of SrS thin film electroluminescence devices. *Acta Polytechn. Scand., Ser. Appl. Phys.* **Ph 170**, 203–206.

Okamoto, S., Kuki, T., and Suzuki, T. (1993). SrS:Ce thin-film electroluminescent devices fabricated by post-annealing technique and their electrical properties. *Jpn. J. Appl. Phys.* **32**, 1672–1680.

Okamoto, S., and Nakazawa, E. (1995). Transient emission mechanisms in thin-film electroluminescent devices with rare-earth-ion-activated SrS phosphor layers. *Jpn. J. Appl. Phys.* **34**, 521–526.

Okibayashi, K., Ogura, T., Terada, K., Taniguchi, K., Yamashita, T., Yoshida, M., and Nakajima, S. (1991). High-brightness multi-color EL display panels with black-matrix filter. *SID 1991 Digest* **22**, 275–278.

Onisawa, K., Fuyama, M., Taguchi, K., Tamura, K., and Ono, Y. A. (1988). Luminance improvement of blue-green emitting SrS:Ce EL cell by controlling vacuum conditions with sulfur addition. *J. Electrochem. Soc.* **135**, 2631–2634.

Onisawa, K., Taguchi, K., Fuyama, M., Tamura, K., Abe, Y., and Ono, Y. A. (1989). Luminescence improvement of red-emitting CaS:Eu thin film electroluminescent cells prepared by electron beam evaporation. *J. Electrochem. Soc.* **136**, 2736–2740.

Onisawa, K., Abe, Y., Nakayama, T., and Hanazono, M. (1991a). Effects of substrate temperature during phosphor layer deposition on luminance of SrS:Ce blue-green-emitting thin-film electroluminescent devices. *Jpn. J. Appl. Phys.* **30**, 314–319.

Onisawa, K., Abe, Y., Tamura, K., Nakayama, T., Hanazono, M., and Ono, Y. A. (1991b). Oxygen in SrS phosphor powder and its effects on performance of thin film electroluminescent devices. *J. Electrochem. Soc.* **138**, 599–601.

Ono, A. (1990). Different techniques employed in deposition of TFEL devices. *Acta Polytechn. Scand., Ser. Appl. Phys.* **Ph 170**, 41–48.

Ono, A. (1992). In *Electroluminescence, Encyclopedia of Applied Physics*, Vol. 5 (Trigg, G. L. ed.), VCH Publishers, New York, pp. 1–31.

Ono, A. (1995). In *Electroluminescent Displays*, World Scientific Publishing Co., Singapore.

Ono, A. (1997). Materials for full-color electroluminescent displays. *Annu. Rev. Mater. Sci.* **27**, 283–303.

Ono, Y. A., Fuyama, M., Onisawa, K., Tamura, K., and Ando, M. (1989). White-light emitting thin film electroluminescent devices with stacked SrS:Ce/CaS:Eu active layers. *J. Appl. Phys.* **66**, 5564–5571.

Oseto, S., Kageyama, Y., Takahashi, M,. Deguchi, H., Kameyama, K., and Fujimura, I. (1989). SrSe:Ce thin film electroluminescent devices. *Electroluminescence, Springer Proc. Phys.* Vol. 38 (Shionoya, S., and Kobayashi, H. eds.), Springer-Verlag, Berlin, pp. 191–194.

Park, W., Jones, T. C., Mohammed, E., Summers, C. J., and Sun, S.-S. (1997). Luminescence properties of a new blue electroluminescent phosphors, SrS:Cu. *3rd Int. Conf. on the Science and Technology of Display Phosphors, Extended Abstracts*, Huntington Beach, CA, pp. 215–218.

Park, W., Jones, T. C., Mohammed, E., and Summers, C. J. (1998). Luminescence properties of SrS:Cu,Ag thin film electroluminescent phosphors. *4th Int. Conf. on the Science and Technology of Display Phosphors, Extended Abstracts*, Bend, OR, pp. 215–218.

Peters, T. E., and Baglio, J. A. (1972). Luminescence and structural properties of thiogallate phosphors Ce^{3+} and Eu^{2+}-activated phosphors. Part I. *J. Electrochem. Soc.* **119**, 230–236.

Petre, D., Pintilie, I., Ciurea, M. L., and Botila, T. (1995). Interface trapping states in MISIM structures with ZnS:Mn. *Thin Solid Films* **260**, 54–57.

Pham-Thi, M., Ruelle, N., Benalloul, P., Barthou, C., Xian, H., Benoit, J., Bourdon, A. (1992). Radiative properties of Eu^{2+} in CaS. *Proc. 6th Int. Workshop on Electroluminescence*, El Paso, TX, pp. 240–245.

Poelman, D., van Meirhaege, R. L., Laflere, W. H., and Cardon, F. (1992). The influence of Se co-evaporation on the electroluminescent properties of SrS:Ce thin films. *J. Lumin.* **52**, 259–264.

Poelman, D., Vercaemst, R., van Meirhaege, R. L., Laflere, W. H., and Cardon, F. (1993). Effect of moisture on performance of SrS:Ce thin film electroluminescent devices. *Jpn. J. Appl. Phys.* **32**, 3477–3480.

Poelman, D., Vercaemst, R., Van Meirhaeghe, R. L., Laflere, W. H., and Cardon, F. (1995). The influence of Se-coevaporation on the emission spectra of CaS:Eu and SrS:Ce thin film electroluminescent devices. *J. Lumin.* **65**, 7–10.

Rack, P. D., and Holloway, P. H. (1998). The structure, device physics and materials properties of thin film electroluminescent devices. *Mater. Sci. Eng. Repts.* **21**, 171–219.

Rack, P. D., Holloway, P. H., and Sun, S.-S. (1995). Effects of oxygen doping on the chemical state and radiative transition of Ce in $Ca_xSr_{1-x}Ga_2S_4$:Ce blue electroluminescent phosphor. *1st Int. Conf. on the Science and Technology of Display Phosphors, Extended Abstracts*, San Diego, CA, pp. 79–81.

Ranta-aho, T., and Yliammi, M. (1993). Blue emitting fluoride thin films. *13th Int. Display Research Conference, Eurodisplay '93, Conference Proceedings*, Strasbourg, pp. 507–510.

Ryu, J. H., Lim, S., and Wager, J. F. (1994). Alternating-current thin-film electroluminescent devices with multiple dielectric layers. *Thin Solid Films* **248**, 63–68.

Saanila, V., Ihanus, J., Ritala. M., and Leskelä, M. (1998). Atomic layer epitaxy growth of BaS and BaS:Ce thin films from in situ synthesized Ba(thd)$_2$. *Chem. Vap. Deposition* **4**, 227–233.

Sakama, H., Ohmura, M., Tonouchi, M., and Miyasato, T. (1993). *Jpn. J. Appl. Phys.* **32**, 1681–1690.

Samuels, J. A., Smith, D. C., Sieblein, K. N., Salazar, K., Tuenge, R. T., Schaus, C. F., Le, H., Hitt, J., Thuemler, R. L., and Wager, J. F. (1996). MOCVD of SrS and SrS:Ce thin films for electroluminescent flat panel displays. *Mater. Res. Soc. Symp. Proc.* **415**, 15–20.

Sasakura, H., Kobayashi, H., Tanaka, D., Mita, J., Tanaka, T., and Nakayama, H. (1981). The dependences of electroluminescent characteristics of ZnS:Mn thin films upon their device parameters. *J. Appl. Phys.* **52**, 6901–6906.

Satoh, T., Fujikawa, H., Ishii, M., Ohwaki, T., and Taga, Y. (1997). Interfacial stability between

Ta-Sn-O films and indium tin oxide electrodes. *Jpn. J. Appl. Phys.* **36**, L1699–L1701.

Saunders, A., and Vecht, A. (1989). The role of chemical vapour deposition in the fabrication of high field electroluminescent displays. *Electroluminescence, Springer Proc. Phys.* Vol. 38 (Shionoya, S., and Kobayashi, H., eds.), Springer-Verlag, Berlin, pp. 210–217.

Schön, S., Chaichimansour, M., Park, W., Yang, T., Wagner, B. K., Summers, C. J. (1997). Homogeneous and δ-doped ZnS:Mn grown by MBE. *J. Cryst. Grow.* **175–176**, 598–602.

Shanker, V., Chatterjee, S., and Ghosh, P. K. (1992). Electroluminescence in Tb-doped Ga_2O_2S phosphor. *J. Appl. Phys.* **72**, 5416–5419.

Shannon, R. D. (1976). Revised effective ionic radii and systematic studies of interatomic distances in halides and chalcogenides. *Acta Crystallog.* **A32**, 751–767.

Skarp, J. (1984). Combination film, in particular for thin film electroluminescent displays. U.S. Patent 4,486,487.

Soenen, B., De Visschere, P., Ihanus, J., Ritala, M., and Leskelä, M. (1997). Aging of EL Devices based on ZnS Deposited with ALE from Various Precursors. *3rd Int. Conf. on the Science and Technology of Display Phosphors, Extended Abstracts,* Huntington Beach, CA, pp. 49–52.

Sohn, S. H., and Hamakawa, Y. (1992). A model for emission from $ZnS:Ce^{3+}$ and $SrS:Ce^{3+}$ thin film electroluminescent devices. *Jpn. J. Appl. Phys.* **31**, 3901–3906.

Sohn, S. H., Hyun, D. G., Deguchi, K., and Hamakawa, Y. (1993). Electroluminesce in $ZnS_{1-x}Te_x$:Ce thin-film devices. *J. Appl. Phys.* **73**, 4092–4094.

Soininen, E. (1998). Color TFEL. *4th Int. Conf. on the Science and Technology of Display Phosphors, Extended Abstracts,* Bend, OR, pp. 165–166.

Soininen, E., Härkönen, G., and Vasama, K. (1998). A novel atomic layer epitaxy ZnS:Mn process for improved electroluminescent display performance. *J. SID.* In press.

Soininen, P., Leskelä, M., Niinistö, L., Nykänen, E., and Rauhala, E. (1992). SrS:Mn as green electroluminescent material. *Proc. 6th Int. Workshop on Electroluminescence,* El Paso, TX, pp. 217–221.

Soininen, P., Nykänen, E., Leskelä, M., and Niinistö, L. (1993). Blue electroluminescence in $SrS:Ce,SiCl_4$ thin film grown by atomic layer epitaxy. *13th International Display Research Conference, Eurodisplay'93, Conference Proceedings,* Strasbourg, pp. 511–514.

Soininen, P. J., Nykänen, E., Niinistö, L. and Leskelä, M. (1996). Improved luminance from electroluminescent SrS:Ce thin films deposited by the atomic layer epitaxy. In *Inorganic and Organic Electroluminescence* (Mauch, R. H., and Gumlich, H.-E., eds.), Wissenschaftliche & Technik Verlag, Berlin, pp. 149–152.

Sowa, K., Tanabe, M., Furukawa, S., Nakanishi, Y., and Hatanaka, Y. (1992). Characteristics of Y_2O_3:Eu/ ZnS/Y_2O_3:Eu red light emitting electroluminescent devices. *Jpn. J. Appl. Phys.* **31**, 3598–3602.

Sowa, K., Tanabe, M., Furukawa, S., Nakanishi, Y., and Hatanaka, Y. (1993). Growth of Y_2O_2S:Eu thin films by reactive magnetron sputtering and electroluminescent characteristics. *Jpn. J. Appl. Phys.* **32**, 5601–5602.

Sreeram, A. N., Riddle, G. H. N., Sun, S.-S., Morton, D. C., and Mahdi, K. (1997). EL displays on multilayer ceramic-on-metal circuit boards. *Conference Records of the 1997 Int. Display Research Conference,* Toronto, pp. 362–365.

Su, S. H., Yokoyama, M., and Su, Y. K. (1997). 9 inch diagonal ZnS and ZnS:Mn films fabricated by metalorganic chemical vapor deposition. *J. Electrochem. Soc.* **144**, 4310–4313.

Sun, S. C., and Chen, T. F. (1997). Effects of electrode materials and annealing ambients on the electrical properties of TiO_2 thin films by metalorganic chemical vapor deposition. *Jpn. J. Appl. Phys.* **36**, 1346–1350.

Sun, S.-S. (1990). TFEL device having multiple layer insulators. U.S. Patent 4,897,319.

Sun, S.-S. (1998a). Blue emitting SrS:Ag,Cu TFEL development. *Conference Records of the 1998 International Display Research Conference, Asia Display '98*, Seoul, Korea, CD-ROM.

Sun, S.-S. (1998b). Personal communication.

Sun, S.-S., and Khormaei, R. (1992). Variation of the characteristics of ZnS:Mn AC thin film electroluminescent devices with insulators. *High-Resolution Displays and Projection Systems, Proc. SPIE* **1664**, 48–56.

Sun, S.-S., Tuenge, R. T., Kane, J., and Liang, M. (1994). Electroluminescence and photoluminescence of cerium-activated alkaline earth thiogallate thin films and devices. *J. Electrochem. Soc.* **141**, 2877–2883.

Sun, S.-S., Dickey, E., Kane, J. and Yocom, P. N. (1997a). A bright and efficient new blue TFEL phosphor. *Conference Records of the 1997 International Display Research Conference*, Toronto, pp. 301–304.

Sun, S.-S., Dickey, E., Tuenge, R. T., Wentross, R., and Kane, J. (1997b). High performance alkaline earth thiogallate blue emitting ACTFEL devices. *1st International Conference on the Science and Technology of Display Phosphors, Extended Abstracts*, San Diego, CA, pp. 119–121.

Suyama, T., Okamoto, K., and Hamakawa, Y. (1982). New Type of Thin-Film Electroluminescent Device Having a Multilayer structure. *Appl. Phys. Lett.* **41**, 462–464.

Swiatek, K., Suchocki, A., Stapor, A., Niinistö, L., Leskelä, M. (1989). Optical excitation and recombination mechanisms of Tb^{3+}-doped zinc sulphide thin films. *J. Appl. Phys.* **66**, 6048–6051.

Takahashi, K., Utsumi, K., Ohnuki, Y., and Kondo, A. (1990). $Sr_{1-x}Zn_xS$:Ce,F phosphor for thin film electroluminescent devices. *Electroluminescence, Springer Proc. Phys.*, Vol. 38 (Shionoya, S., and Kobayashi, H., eds), Sprigner-Verlag, Berlin, pp. 187–190.

Takahashi, K., Shibuya, K., and Kondo, A. (1992). Degradation induced by the Sr^{2+} vacancies in SrS:Ce electroluminescent devices. *J. Cryst. Growth* **117**, 979–982.

Takeda, M., Kakihara, Y., Yoshida, M., Nakata, Y., Kawaguchi, M., Kishishita, H., Yamauchi, Y., Inoguchi, T., and Mito, S. (1975). Inherent memory effects in ZnS:Mn thin-film EL devices. *Proc. 6th Conf. on Solid State Devices, Tokyo, 1974; Supplement to the J. Japan Soc. App. Phy.* **44**, 103–108.

Takeda, M., Kanatani, Y., Kishishita, H., and Uede, H. (1983). *Advances in Display Technology III, Proc. SPIE* **386**, p. 34.

Tammenmaa, M., Leskelä, M., Koskinen, T., and Niinistö, L. (1986). Zinc sulphide thin films doped with rare earth ions. *J. Less-Commom Metals* **126** 209–214.

Tanaka, S., Kobayashi, H., Sasakura, H., and Hamakawa, Y. (1976a). Evidence for the direct impact excitation of Mn centers in electroluminescent ZnS:Mn films. *J. Appl. Phys.* **47**, 5391–5393.

Tanaka, S., Maruyama, Y., Kobayashi, H., and Sasakura, H. (1976b). Electroluminescence in rare earth doped Y_2O_3, La_2O_3, and Y_2O_2S powder layers. *J. Electrochem. Soc.* **123**, 1917–1918.

Tanaka, S., Kobayashi, H., Shiiki, M., Kunou, T., Shanker, V., and Sasakura, H. (1984). Electroluminescence due to allowed 5d–4f transition of Ce^{3+} in a ZnS Thin Film. *J. Lumin.* **31& 32**, 945–947.

Tanaka, S., Shanker, V., Siiki, M., Deguchi, H., and Kobayashi, H. (1985). Multicolor electroluminescence in alkaline-earth sulfide thin-film devices. *Proc. SID* **26**, 255–258.

Tanaka, S., Yoshiama, H., Mikami, Y., Nishiura, J., Ohshio, S., and Kobayashi, H. (1988a). Excitation mechanism in red CaS:Eu and blue SrS:Ce,K TFEL devices. *Proc. SID* **29**, 77–81.

Tanaka, S., Yoshiyama, H., Nishiura, J., Ohshio, S., Kawakami, H., and Kobayashi, H. (1988b). Bright white-light electroluminescent devices with new phosphor thin film based on SrS. *Proc. SID* **29**, 305–310.

Tanaka, S., Morita, H., Yamada, K., and Kobayashi, H. (1991). Luminance improvement of blue- and white-emitting SrS TFEL devices by annealing in Ar-S atmosphere. *Conference Records of the 1991 Int. Display Research Conference*, San Diego, CA, pp. 137–140.

Tanaka, S., Ohmi, K., Yamada, K., Yamano, Y., and Kobayashi, H. (1992). Blue emitting electroluminecet device with $(ZnS:Tm^{3+}/SrS)_n$ multilayered phosphor thin-films. *Proc. 6th International EL Workshop*, El Paso, TX, pp. 193–198.

Tanaka, S., Ohmi, K., Fujimoto, K., Kobayashi, H., Nire, T., Matsuno, A., and Miyakoshi, A. (1993). Electrical and luminescent characterization of white light emitting (ZnS:Mn/SrS:Ce/ZnS:Mn) multilayered thin film electroluminescent devices. *13th International Display Research Conference, Eurodisplay '93, Conference Proceedings*, Strasbourg, pp. 237–240.

Tanda, S., Miyakoshi, A., and Nire, T. (1989). Bright SrS TFEL device prepared by multisource deposition. *Electroluminescence, Springer Proc. Phys.*, Vol. 38 (Shionoya, S., and Kobayashi, H., eds.), Springer-Verlag, Berlin, pp. 180–182.

Taniguchi, K., Tanaka, K., Ogura, T., Kakihara, Y., Nakajima, S., and Inoguchi, T. (1984). EL device with stable memory effect. *Proc. 4th Display Research Conference*, Paris, pp. 89–92.

Theis, D., Oppolzer, H., Ebbinghaus, G., and Schild, S. (1983). Cross-sectional transmission electron microscopy of electroluminescent thin films fabricated by various deposition methods. *J. Cryst. Growth* **63**, 47–57.

Thomas C. B., and Cranton, W. M. (1993). High efficiency ZnS:Mn ac thin film electroluminescent device structure. *Appl. Phys. Lett.* **63**, 3119–3120.

Tiku, S. K., and Rustomji, S. H. (1989). Dielectrics for bright EL displays. *IEEE Trans. Electron Devices* **36**, 1947–1952.

Tiku, S. K., and Smith, G. C. (1984). Choice of dielectrics for TFEL displays. *IEEE Trans. Electron Devices* **ED-31**, 105–108.

Tohda, T., Fujita, Y., Matsuoka, T., and Abe, A. (1986). New efficient phosphor material ZnS:Sm,P for red electroluminescent devices. *Appl. Phys. Lett.* **48**, 95–96.

Tohda, T., Okajima, M., Yamamoto, M., and Matsuoka, T. (1991). Blue-green emitting SrS:Ce,Cl thin-film electroluminescent devices prepared by reactive evaporation using H_2S gas. *Jpn. J. Appl. Phys.* **30**, 2786–2790.

Tong, W., Yang, T., Park, W., Chaiichimansour, M., Schön, S., Wagner, B. K., and Summers, C. J. (1997a). Gas source molecular beam epitaxy of SrS:Ce for flat panel displays. *J. Electron. Mater.* **26**, 728–731.

Tong, W., Zhang, L., Park, W., Chaichimansour, M., Wagner, B. K., and Summers, C. J. (1997b). Charge compensation study of molecular beam epitaxy grown SrS:Ce. *Appl. Phys. Lett.* **71**, 2268–2270.

Törnqvist, R. (1983). Manganese concentration dependent saturation in ZnS:Mn thin film electroluminescent devices. *J. Appl. Phys.* **54**, 4110–4117.

Törnqvist, R. (1985). Properties and performance of TFEL structures. *J. Cryst. Growth* **72**, 538–544.

Törnqvist, R. (1992a). Multicolor thin film electroluminescent displays. *Proc. 6th Int. Workshop on Electroluminescence*, El Paso, TX, pp. 329–336.

Törnqvist, R. (1992b). Thin-film electroluminescence: high performance without complexity. *Displays* **13**, 81–88.

Törnqvist, R. (1997). TFEL "Colour by white." *SID Digest* **28**, 855–858.

Törnqvist, R., and Korpela, S. (1982). On the aging of ZnS:Mn electroluminescent thin films grown by the atomic layer epitaxy technique. *J. Cryst. Growth* **59**, 395–398.

Törnqvist, R., Antson, J., Skarp, J., and Tanninen, V. (1983). How the ZnS:Mn layer thickness contributes to the performance of AC thin-film EL devices grown by atomic layer epitaxy (ALE). *IEEE Trans. Electron Devices* **ED-30**, 468–471.

Törnqvist, R. O., Harju, T. T., Honkala, J. T., Viljanen, J. H., Åberg, M. H., and Kattelus, H. P. (1991). Low-power thin-film electroluminescent display. *SID 1991 Digest* **22**, 63–65.

Troppenz, U., Huettl, B., Pohl, V., Velthaus, K.-O., Sun, S.-S., and Tuenge, D., (1998). Private communication.

Tuenge, R. T. (1989). Thin film electroluminescent phosphors for patterned full-color displays. *Electroluminescence, Springer Proc. Phys.* Vol. 38 (Shionoya, S., and Kobayashi, H., eds.), Springer-Verlag, Berlin, pp. 132–138.

Tuenge, R. T. (1992). Current status of color TFEL phosphors. *Proc. 6th Int. Workshop on Electroluminescence,* El Paso, TX, pp. 173–178.

Tuenge, R. T., Coovert, R. E., and Barrow, W. A. (1981). *159th Electrochem. Soc. Meeting, Extended Abstracts,* p. 422.

Vecht, A., Waite, M., Higton, M., and Ellis, R. (1981). DC electroluminescence in alkaline earth sulphides. *J. Lumin.* **24–25**, 917–920.

Velthaus, K. O., Mauch, R. H., Schock, H. W., Tanaka, S., Yamada, K., Ohmi, K., and Kobayashi, H. (1992). Efficient electroluminescent devices based on ZnS/SrS:Ce multilayered phosphors. *Proc. 6th Int. Workshop on Electroluminescence,* El Paso, TX, pp. 187–192.

Velthaus, K. O., Mauch, R. H., and Schock, H. W. (1994). The role of ZnS in ZnS/SrS:Ce,Cl multilayered TFEL devices. *Adv. Mater. Opt. Electron.* **3**, 111–116.

Velthaus, K. O., Troppenz, U., Hüttl, B., Herrmann, R., and Mauch, R. H. (1995). Novel anion precursors for alkaline-earth based thin film electroluminecence. *Proc. 7th Int. Workshop on Electroluminescence,* Beijing, pp. 451–457.

Velthaus, K. O., Hüttl, B., Troppenz, U., Gaertner, T., and Bilger, G. (1998). Low temperature deposition for SrS:Cu,Ag EL-phosphor. *4th Int. Conf. on the Science and Technology of Display Phosphors, Extended Abstracts,* Bend, OR, pp. 231–234.

Velthaus, K. O., Hüttl, B., Troppenz, U., Herrmann, R., and Mauch, R. H. (1997). New deposition process for very blue and bright SrS:Ce,Cl TFEL devices. *SID 97 Digest* **28**, 411–414.

Venghaus, H, Theis, D., Oppolzer, H., and Schild, S. (1982). Microstructure and light emission of ac-thin film electroluminescent devices. *J. Appl. Phys.* **53**, 4146–4151.

Vercaemst, R., Poelman, D., van Meirhaege, R. L., Fiermans, L., Laflere, W. H., and Cardon, F. (1995). An XPS study of the dopants' valence state and the composition of $CaS_{1-x}Se_x$:Eu and $CaS_{1-x}Se_x$:Ce thin film electroluminescent devices. *J. Lumin.* **63**, 19–30.

Vlasenko, N. A., Zynio, S. A., and Kopytko, Yu. V. (1975). The effect of Mn concentration on ZnS:Mn electroluminescence decay. *Phys. Status Solidi A* **29**, 671–676.

Vlasenko, N. A., Kononets, Ya. F., Beletskii, A. I., Denisova, Z. I., Kopytko, Yu. F., Soininen, E. L., Törqvist, R. O., and Vasama, K. M. (1997). Characterization and aging behaviour of ZnS:Mn TFEL devices grown by two different ALE processes. *3rd Int. Conf. on the Science and Technology of Display Phosphors, Extended Abstracts,* Huntington Beach, CA, pp. 53–56.

Wager, J. F., and Keir, P. D. (1997). Electrical characterization of thin-film electroluminescent devices. *Annu. Rev. Mater. Sci.* **27**, 223–248.

Wang, C. W., Sheu, T. J., Su, Y. K., and Yokoyama, M. (1997). The study of aging mechanism in ZnS:Mn thin-film electroluminescent devices grown by MOCVD. *Appl. Surf. Sci.* **113–114**, 709–713.

Wright, P. J., Cockayne, B., Cattell, A. F., Dean, P. J., and Pitt, A. D. (1982). Manganese doping of ZnS and ZnSe epitaxial layer grown by organometallic chemical vapour deposition. *J. Cryst. Growth* **59**, 155–160.

Wu, X. (1996). Hybrid EL displays. In *Inorganic and Organic Electroluminescence* (Mauch, R. H., and Gumlich, H.-E., eds). Wissenschaftliche & Technik Verlag, Berlin, pp. 285–289.

Xian, H., Zhong, G., Tanaka, S., and Kobayashi, H. (1989). Electroluminescence and photoluminescence in Eu-doped SrS thin-films. *Jpn. J. Appl. Phys.* **28**, L1019–L1021.

Xiao, T., and Kitai, A. H. (1997). New electroluminescent materials based on alkaline earth gallates. *Proc. SID*, Toronto, pp. 310–313.

Xiao, T., Kitai, A. H., Liu, G., and Nakua, A. (1997). Bright green oxide phosphors for EL displays. *SID 1997 Digest* **28**, 415–418.

Yamashita, N. (1991). Photoluminescence properties of Cu^+ centers in MgS, CaS, SrS. and BaS. *Jpn. J. Appl. Phys.* **30**, 3335–3340.

Yamashita, N., Ohira, T., Mizuochi, H., and Asano, S. (1984). Luminescence of Pb^{2+} centers in SrS and SrSe phosphors. *J. Phys. Soc. Jpn.* **53**, 419–426.

Yamashita, N., Michitsuji, Y., and Asano, S. (1987). Photoluminescence spectra and vibrational characteristics of $SrS:Ce^{3+}$ and $SrSe:Ce^{3+}$ phosphors. *J. Electrochem. Soc.* **134**, 2932–2934.

Yamashita, N., Ebisumori, K., and Nakamura, K. (1993a). Luminescence from the aggregated Cu^+ centers in $SrS:Cu^+$. *Jpn. J. Appl. Phys.* **32**, 3846–3850.

Yamashita, N., Fukumoto, S., Ibuki, S., and Ohnishi, H. (1993b). Photoluminescence of Eu^{2+} and Eu^{3+} centers in CaS:Eu,Na phosphors. *Jpn. J. Appl. Phys.* **32**, 3135–3139.

Yamashita, N., Harada, O., and Nakamura, K. (1995). Photoluminescence spectra of Eu^{2+} centers in Ca(S,Se):Eu and Sr(S,Se):Eu. *Jpn. J. Appl. Phys.* **34**, 5539–5545.

Yang, H., Park, Y. K., Ju, S. H., and Chang, B. H. (1995). Effect on anion vacancy on the chromaticity of luminescence for Sr(S,Se):Ce,Cl thin film electroluminescent devices. *Appl. Phys. Lett.* **67**, 70–72.

Yang, T., Chaichimansour, M., Park, W., Wagner, B. K., Schaus, C., Sun, S.-S., Thomas, E., Wang, Z. L., and Summers, C. J. (1996). MBE growth and characterization of SrGa2S4:Ce blue phosphor for thin-film electroluminescence. *J. SID* **4**, 311–313.

Yeom, T. H., Lee, Y. H., Hahn, T. S., Oh, M. H., and Choh, S. H. (1996). Electron-paramagnetic-resonance study of the Mn^{2+} luminescence center in ZnS:Mn powder and thin films. *J. Appl. Phys.* **79**, 1004–1007.

Ylilammi, M., and Ranta-aho, T. (1992). Growth of ZnS by ALE without chloride compound. *Nordic Semiconductor Meeting*, Hämeenlinna, pp. 43–46.

Yoshida, M., Mikami, A., Ogura, T., Tanaka, K., Taniguchi, K., and Nakajima, S. (1986). Stability improvement of red-emitting CaS:EuS thin-film electroluminescence. *SID 1986 Digest* **17**, 41–43.

Yoshida, M., Yamashita, T., Taniguchi, K., Tanaka, K., Ogura, T., Mikami, A., Nakaya, H., Yamaue, S., and Nakajima, S. (1989). ZnS:Mn electroluminescent devices with high performance using $SiO_2/Ta_2O_5/SiO_2$ insulating layer. *Electroluminescence, Springer Proc. Phys.*, Vol. 38 (Shionoya, S., and Kobayashi, H., eds.), Springer-Verlag, Berlin, pp. 273–276.

Yoshino, H., Ohura, M., Kurokawa, S., and Ohnishi, H. (1992). Improved luminance in green-color TFEL device with sputtered ZnS:TbOF. *Japan Display 1992*, pp. 737–740.

Yoshioka, T., Sano, Y., Numomura, K., and Tani, C. (1989). Characteristics of red electroluminescent devices using $CaS_{1-x}Se_x$:Eu phosphor layers, *SID 1989 Digest* **20**, 313–316.

Yun, S. J., Kim, Y. S., Kang, J. S., Park S.-H. K., Cho, K. I., and Ma, D. S. (1999). High-luminance blue-emitting CaS:Pb electroluminescent devices fabricated using atomic layer deposition. *SID 1999 Digest* **30**, 1142–1145.

Zhong, G.-Z., Li, C.-H., Tao, S.-W., Wang, S.-Q., and Zhang, C.-W. (1988). A full-color ACEL device using ZnS:Ho thin film and color filters. *SID 1988 Digest* **19**, 287–288.

CHAPTER 4

Microcavities for Electroluminescent Devices

Kristiaan Neyts

ELECTRONICS AND INFORMATION SYSTEMS DEPARTMENT
UNIVERSITY OF GHENT
GHENT, BELGIUM

I.	INTRODUCTION	183
II.	THEORETICAL DESCRIPTION	185
	1. *Light Waves in a Microcavity*	186
	2. *Dipole Antenna in a Microcavity*	188
	3. *Dipole Transitions in a Microcavity*	198
III.	BASIC EXAMPLES	200
	1. *Mirror Characteristics*	200
	2. *Radiation from Dipole Antennas*	202
	3. *Dipole Antennas Near a Mirror*	204
	4. *Semitransparent Mirror Reflectivity*	207
	5. *Double Mirror Microcavities*	210
IV.	APPLICATIONS	214
	1. *Inorganic EL Devices*	215
	2. *Light-Emitting Diodes*	220
	3. *Organic Electroluminescent Devices*	226
V.	CONCLUSION	230
	REFERENCES	232

I. Introduction

In the electroluminescent (EL) devices discussed in this book, the light is generated in a layer, which is embedded in a stack of thin films. Typically the thickness of the region where the light is generated and the thickness of the other layers in the stack are of the order of 1 μm, which is in the same range as the wavelength of the emitted light. The fact that the dimension of the thin film structures and the wavelength of visible light are in the same range is a coincidence that has enabled extensive use of optical characterization methods such as interference, photolithography, and ellipsometry. As the light is generated inside the thin film structure and there are total or partial reflections on the thin film interfaces, some interference effects can usually be observed in the emitted light. If the optical structure of the EL device is designed in such a way that interference effects strongly influence the emission characteristics, then the structure is called a microcavity. The reflection from a rough surface, such as a white sheet of paper, or the emission from a rough powder phosphor as used in a cathode ray tube, is normally Lambertian, which means that the observed brightness is independent of the angle of observation. For a microcavity, the light intensity will, in general, depend on the direction of the emission, and in addition, the angular distribution of the emission is different for every wavelength.

It has been known for quite some time that the probability for an excited state to emit a photon in a certain direction can be influenced by the optical environment. Drexhage et al. (1968) observed that the decay time of excited states is slightly modified when the decay takes place in the neighborhood of a metallic mirror. The modification in the decay time was much larger in the experiments where excited Rydbergh atoms with emission at radio frequencies were placed inside a microcavity, consisting of a vacuum gap between two metallic mirrors (Haroche and Kleppner, 1989). More recently, experiments based on photoluminescence, cathodoluminescence, or electroluminescence have revealed many aspects of microcavity effects in thin films.

The origin of the microcavity effect is related to the quantum-mechanical explanation of spontaneous radiative decay. In radiative decay, there is a quantum-mechanical transition of an excited state to a ground state, with simultaneous creation of a photon. Different ground states may be available and similarly, different photon states can be excited. In free space, the probability of emitting a photon in a certain direction is isotropic. But when the optical environment is modified, the density of photon states changes, and the probability that a photon will be emitted into a certain direction is modified accordingly. When a microcavity is designed in such a way that the total density of photon states is reduced, then the total probability that a photon is emitted will reduce by the same factor. If radiative decay is the only process, we can expect the decay time to increase.

The practical use of microcavities in electroluminescent devices has been investigated for the three kinds of EL devices: organic electroluminescent devices (Grüner *et al.*, 1996; Nakayama *et al.*, 1993; Saito *et al.*, 1993), light-emitting diodes (Bjork *et al.*, 1991; Hunt *et al.*, 1992; Rogers *et al.*, 1990; Yokoyama, 1992), and inorganic electroluminescent devices (Mauch, 1989; Mueller *et al.*, 1996; Neyts *et al.*, 1997; Vlasenko *et al.*, 1970). In general, the aim of these investigations is to increase the light output of the devices or to enhance the emission characteristics for specific applications. In most cases, the lateral dimensions of the EL devices are much larger than a few micrometers, and the structure is thus practically one-dimensional.

The most important parameter is then the angle with the normal on the thin film structure. Each microcavity structure leads to a certain wavelength and angular dependence, and a change in the radiative decay time.

In this chapter, we introduce a theoretical description for the emission from a one-dimensional planar microcavity, where the light is generated in a thin layer. The next section discusses the characteristics of metallic and dielectric mirrors, interference effects near a single mirror, and interference related to Fabry–Pérot modes. Section IV is devoted to the use of microcavity effects in practical electroluminescent devices: inorganic EL, organic EL devices, and light-emitting diodes.

II. Theoretical Description

The aim of this section is the mathematical description and the numerical calculation of the interference and cavity effects in thin film structures. We use the equivalence between light emission from a dipole transition and the power radiated by a classical elementary dipole antenna, as described in the papers of Lukosz (1980, 1981) and Lukosz and Kunz (1977, 1979) and more recently in Neyts (1998). With this theory, the radiated power of a dipole antenna can be calculated as a function of the location and orientation. The power can then be related to the probability for the emission of a photon by a dipole transition. With this method, the emission and change in decay time has been simulated successfully for many microcavity systems (Chance *et al.*, 1974; Neyts *et al.*, 1995, 1997; Neyts, 1996; So *et al.*, 1999). The basic formulas for the calculation of the radiated power of an electrical dipole antenna are given in Subsection 2 of Section II.

Other approaches for the simulation of microcavity effects — for example, the expansion in vacuum field modes — have been described in the literature. In the first part of such a calculation, orthonormal modal fields must be calculated and normalized for the given structure. In the second part, the coupling between the dipole radiator and the field modes is calculated. For the modal fields that correspond to plane waves in the half infinite medium outside of the cavity, the calculation is relatively straightforward (Bjork *et al.*, 1991; De Martini *et al.*, 1991; Ho *et al.*, 1996). It can also be performed when guided modes are taken into account (Brorsen and Skovgaard, 1996; Marcuse, 1972; Rigneault *et al.*, 1996), but, up to now, not when absorbing media are present in the stack. This method is conceptually close to the quantum-mechanical description of an excited state, which can emit a photon into the modes to which it is coupled, but has the disadvantage that it is difficult to use for absorbing media.

1. Light Waves in a Microcavity

In this section, we introduce the thin film structure shown in Fig. 1, where the emitting medium with index of refraction n_e and thickness d_e is located between two stacks of layers. The surrounding layers have thicknesses d_i and indices of refraction n_i. The z axis is chosen to be perpendicular to the plane of the thin films, which is parallel to the xy plane. For large and small values of z, the stack ends in half-infinite media with indices of refraction n_+ and n_- in the z and $-z$ directions, respectively. Throughout this chapter, we assume that the material of the emitting layer is nonabsorbing and thus that n_e is real. The other layers can be either transparent, with a real index of refraction, or absorbing, in which case the index has a positive imaginary part.

We consider light with frequency v and wavelength $\lambda = c/v$ in vacuum, where c is the velocity of light. In a material with index of refraction n_i, the amplitude of the wave vector is given by

$$k_i = \frac{2\pi n_i}{\lambda} \tag{1}$$

In the emitting medium and other loss-free media, this amplitude is real; in absorbing media, it is complex. In a source-free region, any oscillating field with time dependence $\exp(-j2\pi vt)$ can be written as a Fourier integral over the real wave vector components $k_{x,y}$ and $k_{y,i}$, which are parallel with the xy plane. In this integration, TE (or s) and TM (or p) polarization (see Fig. 2) and waves moving in the z and in the $-z$ directions are taken into account. Because practically all thin film deposition methods yield layers that do not have a preferential orientation in the plane of the layer, we

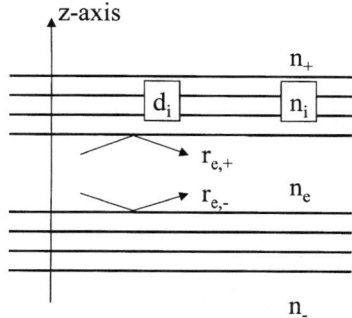

FIG. 1. Structure of a light-emitting thin film device: emitting loss-free medium with index of refraction n_e and thickness d_e, intermediate layers are labeled with i, and half infinite media are labeled $+$ and $-$. Amplitude reflection coefficients are $r_{e,+}$ and $r_{e,-}$.

4 MICROCAVITIES FOR ELECTROLUMINESCENT DEVICES

further assume that the emission from the microcavity has rotational symmetry around the z axis. Instead of the components k_x and k_y in the xy plane, we use the variable κ, which runs from zero to infinity:

$$\kappa = (k_{x,i}^2 + k_{y,i}^2)^{1/2} \tag{2}$$

Then the z component of the wave vector (in the first quadrant) is

$$k_{z,i} = (k_i^2 - \kappa^2)^{1/2} \tag{3}$$

In this expression, $k_{z,i}$ is the complex square root in the first quadrant of the complex plane. In the case where $k_{z,i}$ is real, the contributions are plane waves with a real wave vector. If $k_{z,i}$ is complex, the waves are evanescent and decrease in amplitude along the z direction.

When k_i is real and larger than κ, we have a plane wave propagating at an angle α_i with the z axis, with

$$\kappa = k_i \sin \alpha_i \tag{4}$$

where $\alpha_i < \pi/2$ for waves traveling in the z direction and $\alpha_i > \pi/2$ for waves in the $-z$ direction, as shown in Fig. 3.

In the microcavity structure, plane and evanescent waves with TM or TE polarization can be reflected or transmitted. The sign conventions for TE and TM are illustrated in Fig. 2, the complex Fresnel coefficients for reflection of a wave in medium n_i on an adjacent medium with index n_{i+1} are

$$r_{i,i\pm1}^{TM} = \frac{k_{z,i}/n_i^2 - k_{z,i\pm1}/n_{i\pm1}^2}{k_{z,i}/n_i^2 + k_{z,i\pm1}/n_{i\pm1}^2} \qquad r_{i,i\pm1}^{TE} = \frac{k_{z,i} - k_{z,i\pm1}}{k_{z,i} + k_{z,i\pm1}} \tag{5}$$

The amplitude transmission coefficients for TE or TM polarization are given by $t_{i,i\pm1} = 1 \pm r_{i,i\pm1}$.

If there are many layers between the emitting medium and the half-infinite medium in the z or $-z$ directions, then all interfaces have to be taken into account for the calculation of the total reflection and transmission coefficients $r_{e,\pm}^{TM,TE}$, $t_{e,\pm}^{TM,TE}$. Starting from the outermost layer, the total reflection

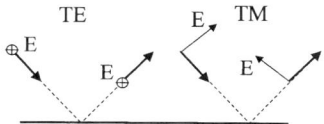

FIG. 2. Sign convention of the reflection coefficients for TE and TM polarization.

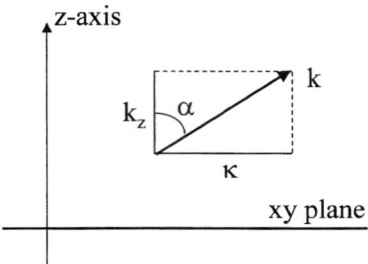

FIG. 3. The k-vector of a plane wave has a perpendicular contribution k_z and one parallel to the layers κ. The definition of the angle α is also given.

and transmission coefficients can be found by iterative calculation:

$$r_{i,j} = \frac{r_{i,i\pm 1} + r_{i\pm 1,j}\exp(2jk_{z,i+1}d_{i+1})}{1 + r_{i,i\pm 1} \cdot r_{i\pm 1,j}\exp(2jk_{z,i+1}d_{i+1})} \quad (6)$$

$$t_{i,j} = \frac{t_{i,i\pm 1}t_{i\pm 1,j}\exp(jk_{z,i+1}d_{i+1})}{1 + r_{i,i\pm 1}r_{i\pm 1,j}\exp(2jk_{z,i+1}d_{i+1})} \quad (7)$$

The energy reflection and transmission coefficients are then given by

$$R_{e,\pm}^{\text{TM,TE}} = |r_{e,\pm}^{\text{TM,TE}}|^2 \quad (8)$$

and

$$\begin{cases} T_{e,\pm}^{\text{TM}} = |t_{e,\pm}^{\text{TM}}|^2 \dfrac{n_e^2}{n_\pm^2} \dfrac{k_{z,\pm}}{|k_{z,e}|} \\ T_{e,\pm}^{\text{TE}} = |t_{e,\pm}^{\text{TE}}|^2 \dfrac{k_{z,\pm}}{|k_{z,e}|} \end{cases} \quad (9)$$

for $\mathscr{I}m(k_{z,\pm}) = 0$, so that for a loss-free mirror, the sum of reflection and transmission coefficients is equal to 1.

2. Dipole Antenna in a Microcavity

In this section, we evaluate the radiation of an elementary electrical dipole antenna in a microcavity with oscillating dipole moment $\bar{p}_0$ [leaving out the time behavior $\exp(-j2\pi vt)$], which is located in a loss-free emitting medium with index of refraction n_e. In realistic electroluminescent devices, dipole antennas with different locations and frequencies must be considered. In an infinite medium n_e, the total power emitted by the monochromatic dipole antenna is given by

$$L_e = \frac{v \cdot k_e^3}{6n_e^2 \varepsilon_0} p_0^2 \qquad (10)$$

In the following, we scale all powers with this factor and use the dimensionless power F, which is the actual generated power, divided by the power that would be emitted in an infinite medium n_e.

The calculation of the electromagnetic field caused by an elementary electric dipole antenna in a thin film structure is rather complex. By using the superposition of elementary plane and evanescent waves the total power F supplied by the dipole antenna (including dissipation in absorbing media and radiation into the far field) can be written as an integral:

$$F = \int_0^\infty K(\kappa) \, d\kappa^2 \qquad (11)$$

with K the power density, per unit $d\kappa^2$ (Lukoy, 1980, 1981; Lukosz and Kunz, 1977, 1979; Neyts, 1998).

The power density can be separated into different components: according to the polarity we distinguish K^{TE} and K^{TM}; according to the direction of the emission along the z axis we have K_+ and K_-; according to whether the light is absorbed in the structure or emitted into the far field we have contributions K_A and K_T; and also combinations with multiple indices are considered, as shown in Fig. 4.

In the case of plane waves, an emission angle α_i can be associated with the radiation according to Eq. (4). The power density K per interval $d\kappa^2$ can

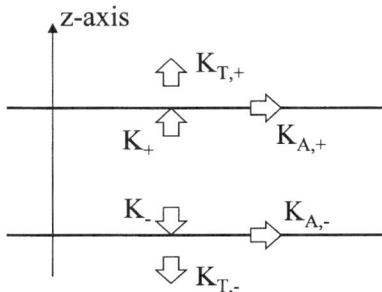

FIG. 4. The radiation from the dipole antenna has contributions in the $z(K_+)$ and $-z$ directions (K_-). Each contribution can be separated in a part that is absorbed (e.g., $K_{A,+}$) by the intermediate layers and a part that is transmitted $(K_{T,+})$.

then also be transformed to a power density P per solid angle $d\Omega$. For example, for the radiated power in medium n_+, assumed to be loss-free, the angular distribution is related to the transmitted power $K_{+,T}$ by

$$P_+(\alpha_+) = \frac{k_+^2 \cos \alpha_+}{\pi} K_{+,T}(k_+ \sin \alpha_+) \tag{12}$$

a. *Total Power*

In the case where the elementary dipole is located at a distances z_+ and z_- from the interfaces of the emitting medium, as shown in Fig. 5, the power densities for a dipole that is oriented parallel to the z axis are given by (Lukosz, 1980; Neyts, 1998):

$$K_\perp^{TM} = \frac{3}{4} \mathcal{R}e \left\{ \frac{\kappa^2}{k_e^3 \cdot k_{z,e}} \frac{(1 + a_+^{TM})(1 + a_-^{TM})}{1 - a^{TM}} \right\} \qquad K_\perp^{TE} = 0 \tag{13}$$

and for a dipole parallel to the xy plane, the power density is

$$\begin{aligned} K_\parallel^{TM} &= \frac{3}{8} \mathcal{R}e \left\{ \frac{k_{z,e}}{k_e^3} \frac{(1 - a_+^{TM})(1 - a_-^{TM})}{1 - a^{TM}} \right\} \\ K_\parallel^{TE} &= \frac{3}{8} \mathcal{R}e \left\{ \frac{1}{k_e \cdot k_{z,e}} \frac{(1 + a_+^{TE})(1 + a_-^{TE})}{1 - a^{TE}} \right\} \end{aligned} \tag{14}$$

where $\mathcal{R}e$ stands for the real part of the complex number and $a_\pm^{TM,TE}$ is the reflection coefficient with respect to the location of the dipole antenna:

$$a_+^{TM,TE} = r_{e,+}^{TM,TE} \exp(2jk_{z,e}z_+)$$
$$a_-^{TM,TE} = r_{e,-}^{TM,TE} \exp(2jk_{z,e}z_-)$$

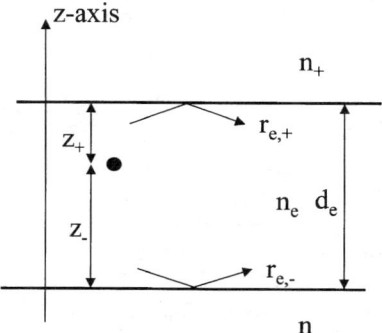

FIG. 5. The emitting layer has a thickness d_e and the radiating dipole antenna (or excited state with dipole transition) is located at distances z_- and z_+ from the interfaces.

4 MICROCAVITIES FOR ELECTROLUMINESCENT DEVICES

and $a^{\mathrm{TM,TE}}$ is the product of the two:

$$a^{\mathrm{TM,TE}} = a_+^{\mathrm{TM,TE}} a_-^{\mathrm{TM,TE}} = r_{e,+}^{\mathrm{TM,TE}} r_{e,+}^{\mathrm{TM,TE}} \exp(2jk_{z,e}d_e) \qquad (15)$$

The four expressions for K are of the same shape, namely:

$$K^\beta = \mathrm{cst} \cdot \mathcal{R}e \left\{ \frac{\beta^2}{k_e^3 k_{z,e}} \frac{(1 \pm a_+)(1 \pm a_-)}{1-a} \right\} \qquad (16)$$

with $\mathrm{cst}^e = 3/4, 0$ or $3/8$, β equal to κ, $k_{z,e}$, or k_e and using the $+$ or the $-$ sign, depending on the polarization and orientation of the dipole radiator.

In Eqs. (13) and (14), the effect of the microcavity structure is concentrated in the factor $(1 \pm a_+)(1 \pm a_-)/(1-a)$, which can be interpreted as an interference factor. This factor is unity in the case of the infinite medium n_e, because there are no reflections. The factor $1 \pm a_\pm$ in the numerator describes the wide-angle interference between directly emitted and reflected radiation, as represented in Fig. 6. The factor $1-a$ in the denominator describes multiple beam interference, which occurs when radiation is repeatedly reflected between two interfaces, yielding an infinite sum: $1 + a + a^2 + a^3 + \cdots$.

For an electrical dipole antenna making an angle θ with the z axis, the radiated power is simply the sum of two contributions:

$$K_\theta^{\mathrm{TM}} = K_\perp^{\mathrm{TM}} \cos^2\theta + K_\parallel^{\mathrm{TM}} \sin^2\theta \qquad K_\theta^{\mathrm{TE}} = K_\perp^{\mathrm{TE}} \cos^2\theta + K_\parallel^{\mathrm{TE}} \sin^2\theta \qquad (17)$$

For a randomly oriented dipole antenna, with equal probability for all

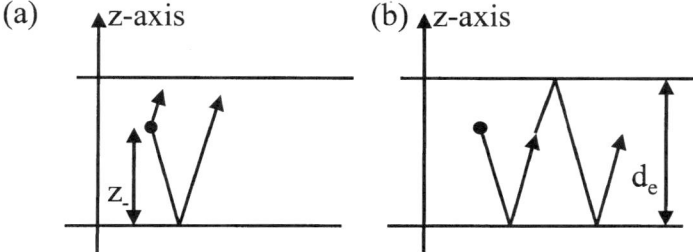

FIG. 6. (a) For wide-angle interference, directly emitted and reflected are added and the distance z_- is essential. (b) Multiple beam interference occurs when the radiation is reflected back and forth between the two parallel reflectors. The thickness of the layer d_e is the essential parameter.

directions in space, we have

$$K_{RND}^{TM} = \frac{1}{3} K_{\perp}^{TM} + \frac{2}{3} K_{\parallel}^{TM} \qquad K_{RND}^{TE} = \frac{1}{3} K_{\perp}^{TE} + \frac{2}{3} K_{\parallel}^{TE} \qquad (18)$$

b. Power in Plane Waves

In the case of plane waves in the emitting medium, thus when $k_{z,e}$ is real, the total power supplied by the dipole antenna can be separated into different contributions. In Eq. (22), the factor $\beta^2/(k_e^3 \cdot k_z)$ is then real and can be placed in front. The factor under the $\mathcal{R}e$ sign can be decomposed into contributions to the z and $-z$ directions, because

$$\mathcal{R}e \left\{ \frac{(1 \pm a_+)(1 \pm a_-)}{1 - a} \right\} = \frac{1}{2} \frac{|1 + a_-|^2}{|1 - a|^2} (1 - |a_+|^2) + \frac{1}{2} \frac{|1 \pm a_+|^2}{|1 - a|^2} (1 - |a_-|^2) \qquad (19)$$

Here the first term describes the net power propagating from the antenna into the z direction: it is the difference between the total power in the z direction and the power reflected on the interface at the positive side.

In the case of electroluminescent devices, the refractive index of the medium where the light is generated is larger than the index of glass or air. Because of this, we use the angle in the emitting medium α_e to characterize the light generation and emission (for the light that is trapped in the structure, the angle α_+ would be imaginary). For the power density P transmitted in the positive direction we find the following contributions for a perpendicular dipole antenna (Neyts, 1998):

$$P_{\perp}^{TM}(\alpha_e) = \frac{3}{8\pi} \sin^2 \alpha_e \frac{|1 + a_-^{TM}|^2}{|1 - a^{TM}|^2} T_{e,+}^{TM} \qquad P_{\perp}^{TE}(\alpha_e) = 0; \qquad (20)$$

and for a dipole parallel to the xy plane:

$$P_{\parallel}^{TM}(\alpha_e) = \frac{3}{16\pi} \cos^2 \alpha_e \frac{|1 - a_-^{TM}|^2}{|1 - a^{TM}|^2} T_{e,+}^{TM} \qquad P_{\parallel}^{TE}(\alpha_e) = \frac{3}{16\pi} \frac{|1 + a_-^{TE}|^2}{|1 - a^{TE}|^2} T_{e,+}^{TE}$$

c. Power in Discrete Modes

Discrete modes can appear when planes waves in medium e encounter total internal reflection on both interfaces: $|r_{e,+}^{TE,TM}| = |r_{e,-}^{TE,TM}| = 1$. This happens when there are no absorbing materials in the considered structure (all n_i real) and $k_-, k_+ < \kappa < k_e$ (thus $k_{z,e}$ is real). From Eq. (19) it seems that the radiated power becomes zero in this case, because $|a_+^{TE,TM}| = |a_-^{TE,TM}| = 1$; but the denominator can also vanish. Equation (19) becomes singular at certain values κ_m when $a^{TE,TM} = 1$, which corresponds to the discrete TE and TM guided modes of the optical system. The singularity in the power can be integrated when the expression $a = |a|e^{j\phi}$ is used for a in the neighborhood of the singularity (with $|a| \to 1$):

$$\int_{-\varepsilon}^{\varepsilon} \mathscr{R}e \left\{ \frac{(1 \pm a_+)(1 \pm a_-)}{1-a} \right\} d\phi = 2\pi [1 \pm \mathscr{R}e(a_+)] \tag{21}$$

The total power dissipated into a certain mode with parameter κ^m is given by (Neyts, 1998)

$$\int_{\kappa_m^2 - \varepsilon}^{\kappa_m^2 + \varepsilon} K^\beta(\kappa) \, d\kappa^2 = 2\pi \, \text{cst}^e \frac{\beta_m^2}{k_e^3 \cdot (k_{z,e})_m} \frac{1 \pm \mathscr{R}e(a_{+,m})}{\left|\dfrac{d\phi}{d\kappa^2}\right|} \tag{22}$$

The power radiated into a certain mode varies as a sine function of the location (z_+ and z_-) of the antenna in the layer. The positions of the minima and maxima correspond with the positions where the modes have their respective nodes and antinodes.

Ideally, discrete guided modes are not absorbed or emitted: the light remains trapped in the structure until it eventually reaches the lateral edge of the device. Often there is some small absorption in the structure, enough to dissipate all the radiation before the edge of the device is reached.

d. Power in Evanescent Waves

In the case of evanescent waves in the emitting layer, a similar separation of contributions can be carried out. In the case $k_{z,e}$ is imaginary, the relevant quantity can be separated as follows:

$$\pm \mathscr{I}m \left\{ \frac{(1 \pm a_+)(1 \pm a_-)}{1-a} \right\} = \frac{|1 \pm a_-|^2}{|1-a|^2} \mathscr{I}m(a_+) + \frac{|1 \pm a_+|^2}{|1-a|^2} \mathscr{I}m(a_-) \tag{23}$$

where the two terms describe the net power propagating in the z and $-z$ directions. In this case, $\mathscr{I}m(a_+)$ is different from zero only when there is an absorbing medium on the positive side of the emitting medium, or when $k_{z,+}$ is real and the evanescent waves in medium n_e can be transformed into plane waves in medium n_+. For the power supplied by the antenna in the positive direction, we find the following positive contribution (Neyts, 1998):

$$K_+^\beta = \text{cst} \frac{|\beta|^2}{k_e^3 \cdot |k_{z,e}|} \frac{|1 \pm a_-|^2}{|1 - a|^2} \mathscr{I}m(a_+) \tag{24}$$

which can also be separated into a transmission part and an absorption part.

e. *Optimizing the Perpendicular Emission*

For the emission in the perpendicular direction (z_+), the enhancement factor due to the microcavity is independent of the kind of dipole or the type of polarization that is considered:

$$P(\alpha_e = 0) = P_{\inf}(\alpha_e = 0) \frac{|1 + a_-^{TE}|^2}{|1 - a|^2} T_{e,+} \tag{25}$$

As the sign convention is different for TM and TE, we have specified the factor a for the most commonly used TE polarization convention. The distances z_+ and z_- between the dipole antenna and the mirrors can be optimized to reach the highest possible value for the power P. This is obtained when the phase of the factors a_-^{TE} and a_+^{TE} (and the phase of a) are equal to zero. For mirrors with given complex reflection coefficients, the optimum distances from the interfaces are, with m_+ and m_- integers,

$$z_- = \frac{\lambda}{2n_e}\left(m_- - \frac{\varphi(r_{e,-}^{TE})}{2\pi}\right) \tag{26}$$

$$z_+ = \frac{\lambda}{2n_e}\left(m_+ - \frac{\varphi(r_{e,+}^{TE})}{2\pi}\right) \tag{27}$$

$$d_e = z_+ + z_- = \frac{\lambda}{2n_e}\left(m_+ + m_- - \frac{\varphi(r_{e,+}^{TE})}{2\pi} - \frac{\varphi(r_{e,-}^{TE})}{2\pi}\right) \tag{28}$$

The power in the perpendicular direction is then (Hunt et al., 1993)

$$P(\alpha_e = 0) = P_{\inf}(\alpha_e = 0) \frac{(1 + \sqrt{R_-})^2}{(1 - \sqrt{R_- R_+})^2} T_{e,+} \tag{29}$$

This formula can be used to optimize the mirror reflectivities. The value for R_- should be as high possible to enhance the intensity in the z direction. The mirror in the z direction should be loss free: $T_+ = 1 - R_+$. The optimum reflectivity R_+ is then obtained when it is equal to R_-. When the mirror R_+ is not loss-free, the optimum transmission and reflection of the front mirror can be determined numerically.

f. Microcavity Modes and Effective Thickness

When the factor a in Eq. (19) is close to unity, multiple beam interference leads to a large enhancement of the intensity. The range of wavelengths and angles for which the enhancement is considerable is called a mode of the microcavity. By separating the amplitude and the phase of the factor $a = \sqrt{R_{e,+} R_{e,-}}\, e^{j\phi}$, the multiple beam interference factor in Eq. (19) becomes (Dodabalapur et al., 1994, 1996; Grüner et al., 1996; Neyts, 1998)

$$\frac{1}{|1-a|^2} = \frac{1}{1 + R_{e,+} R_{e,-} - 2\sqrt{R_{e,+} R_{e,-}} \cos\phi} \tag{30}$$

According to Eq. (15), the phase factor contains contributions due to the thickness of the emitting medium and to the mirrors:

$$\phi = \frac{4\pi n_e d_e \cos\alpha_e}{\lambda} + \varphi(r_{e,+}) + \varphi(r_{e,-}) \tag{31}$$

This equation shows that for a thick cavity the phase ϕ is very sensitive to the emission angle α.

In a good approximation, the multiple beam interference factor will be largest when the phase ϕ is a multiple of 2π. Let us assume that this happens at a given α_0 and λ_0 and that the corresponding value for a is a_0. In the neighborhood of this maximum the phase can be linearized as a function of α and λ:

$$\phi = m2\pi + \frac{4\pi n_e}{\lambda_0}\left((\cos\alpha_e - \cos\alpha_0) d_{\text{eff},\alpha} - \frac{\lambda - \lambda_0}{\lambda_0} \cos\alpha_0 d_{\text{eff},\lambda}\right) \tag{32}$$

with $d_{\text{eff},\alpha}$ and $d_{\text{eff},\lambda}$ the effectiveness thickness of the microcavity with respect to variations in, respectively, the angle and the wavelength:

$$d_{\text{eff},\alpha} = d_e + \frac{\lambda_0}{4\pi n_e}\left(\frac{d[\varphi(r_{e,+}) + \varphi(r_{e,-})]}{d\cos\alpha_e}\right)_{\alpha_0,\lambda_0} \tag{33}$$

$$d_{\text{eff},\lambda} = d_e + \frac{\lambda_0^2}{4\pi n_e \cos\alpha_0}\left(-\frac{d[\varphi(r_{e,+}) + \varphi(r_{e,-})]}{d\lambda}\right)_{\alpha_0,\lambda_0} \tag{34}$$

In general these two effective thicknesses are different and their values are related to the complex mirror reflection coefficients. For TM and TE polarization, different effective thicknesses are obtained, except in the perpendicular direction. For metallic mirrors, the contribution of the phase factors is relatively small and the effective thickness is mainly determined by the thickness of layer d_e. In the case of distributed dielectric mirrors, the contribution of the phase factors is usually very important.

When the angle α is increased, the same value of ϕ is obtained for a smaller value of the wavelength. This means that the microcavity mode shifts toward the blue side of the spectrum at larger angles. The shift follows from Eq. (32) and is proportional with the ratio of the effective thicknesses:

$$\frac{d \cos \alpha_e}{d\lambda} = \frac{\cos \alpha_0}{\lambda_0} \frac{d_{\text{eff},\lambda}}{d_{\text{eff},\alpha}} \qquad (35)$$

In the vicinity of the maximum $\phi = 0$, the factor $\cos \phi$ in Eq. (30) can be substituted for by its small-angle series expansion and $\sqrt{R_{e,+}R_{e,-}}$ is approximately constant. The full width at half maximum of the interference peak is then $\Delta\phi_{\text{FWHM}} = 2(1 - \sqrt{R_{e,+}R_{e,-}})/\sqrt[4]{R_{e,+}R_{e,-}}$, which yields for the angular and spectral width of the microcavity modes (Grüner et al., 1996; Hunt et al., 1992)

$$\Delta \cos \alpha_{\text{FWHM}} = \frac{1 - \sqrt{R_{e,+}R_{e,-}}}{\sqrt[4]{R_{e,+}R_{e,-}}} \frac{\lambda_0}{2\pi n_e d_{\text{eff},\alpha}} \qquad (36)$$

$$\Delta\lambda_{\text{FWHM}} = \frac{1 - \sqrt{R_{e,+}R_{e,-}}}{\sqrt[4]{R_{e,+}R_{e,-}}} \frac{\lambda_0^2}{2\pi n_e d_{\text{eff},\lambda} \cos \alpha_0} \qquad (37)$$

This illustrates the importance of the effective thickness of the microcavity: a larger effective thickness will reduce the width of the angular and spectral region where the emission is enhanced. The multiple beam interference factor in Eq. (30) is then

$$\frac{1}{|1-a|^2} = \frac{1}{(1 - \sqrt{R_{e,+}R_{e,-}})^2}$$
$$\times \frac{1}{1 + \{[2(\cos \alpha_e - \cos \alpha_0)/\Delta \cos \alpha_{\text{FWHM}}] - [2(\lambda - \lambda_0)/\Delta\lambda_{\text{FWHM}}]\}^2} \qquad (38)$$

For monochromatic light, more than half of the intensity in the mode around $\phi = 0$ is emitted in the interval $(\cos\alpha_0 \pm 0.5\Delta\cos\alpha_{\text{FWHM}})$, which corresponds to a solid angle (Neyts, 1997):

$$\Delta\Omega_e = \frac{1 - \sqrt{R_{e,+}R_{e,-}}}{\sqrt[4]{R_{e,+}R_{e,-}}} \frac{\lambda_0}{n_e d_{\text{eff},\alpha}} \tag{39}$$

g. *Averaging over Thick Layers*

There are many important applications where the dipole antennas are not located in a plane, but rather distributed over a region that is thicker than a few wavelengths. In this case, the formulas already given must be averaged out over the parameters z_+ and z_-. When the thickness of the region is sufficiently large, then the wide-angle interference effects will disappear, because in the formula for the radiated power density P into plane waves in the z direction Eq. (20), the averaging procedure yields:

$$|1 \pm a_-|^2 = 1 \pm 2\mathcal{R}e(a_-) + a_-|^2 \approx 1 + R_- \tag{40}$$

This averaging procedure can be carried out for most expressions where plane waves are involved. For waves that are evanescent in the medium n_e, the averaging procedure is not valid because there is an exponential relation with the coordinate z instead of a sinusoidal relation.

For the multiple beam interference factor the expression in Eq. (30) can be used. For randomly oriented dipoles, including contributions from perpendicular and parallel dipoles, the resulting intensity (TM or TE) becomes (Neyts, 1997):

$$P_+(\alpha_e) = \frac{1}{8\pi} \frac{(1 + R_{e,-})T_{e,+}}{1 + R_{e,+}R_{e,-} - 2\sqrt{R_{e,+}R_{e,-}}\cos\phi} \tag{41}$$

where R, T, and ϕ depend on the polarization, the angle, and the wavelength of the emission. For thick microcavities this expression can be integrated over the solid angle corresponding to the mode, for ϕ going from $-\pi$ to π, because $R_{e,+}R_{e,-}$ is approximately constant. The result is the total emission in the mode (TM or TE) (Brorsen and Skovgaard, 1996; Neyts, 1997)

$$\int P_+(\alpha_e) d\Omega_e = \frac{\lambda}{4n_e d_{\text{eff},\alpha}} \frac{(1 + R_{e,-})T_{e,+}}{2(1 - R_{e,+}R_{e,-})} \tag{42}$$

Roughly each mode contains the same fraction $\lambda/(4n_e d_{\text{eff},\alpha})$ of the total emission in medium n_e. The second factor indicates which fraction of the mode is emitted in the forward direction and is limited to unity. From this we conclude that large enhancements of the integrated intensity can be achieved only when the maximum of one single mode fits in the desired solid angle. The emission in a single mode in medium n_+ can be enhanced by reducing the effective thickness of the cavity (Neyts, 1997).

3. Dipole Transitions in a Microcavity

The radiation from the dipole antenna we have discussed is continuous, but in electroluminescent devices the excited states can emit only one photon each. Nevertheless, the results of the dipole antenna theory can be applied to excited states by replacing intensity of the radiation by probability of photon emission. The probability for the excited state to emit a photon with κ^2 in the interval $[\kappa_1^2, \kappa_2^2]$ is thus proportional to the power radiated by a dipole antenna with the same location and orientation, in the same κ^2 interval.

When a thin film cavity changes the total power radiated by a dipole antenna by a factor F, which can be smaller or larger then 1, the corresponding integrated probability to emit a photon is changed by the same factor. For an excited state that has both radiative and nonradiative decay channels, the decay rate of the excited state in the infinite medium $\Gamma_{\text{inf}}(s^{-1})$ will be modified by the cavity into Γ_{cav}, with

$$\Gamma_{\text{inf}} = \Gamma_{nr} + \Gamma_r \quad \text{and} \quad \Gamma_{\text{cav}} = \Gamma_{nr} + F\Gamma_r \qquad (43)$$

with F the modification factor due to the presence of the cavity, given in Eq. (11). For an excited state at time $t = 0$, the probability to emit a photon in the interval $t, t + dt]$ is given by $\Gamma_r \exp(-\Gamma_{\text{inf}} t) dt$ in the infinite medium and by $F\Gamma_r \exp(-\Gamma_{\text{cav}} t) dt$ for an atom inside the cavity (in this case the radiation dissipated in absorbing materials is included). The change in the decay time gives an additional factor in the emission intensity per solid angle for a dipole transition (Grüner et al., 1996; Neyts, 1998):

$$P_{\text{dipole}}(\alpha) = \frac{\Gamma_{nr} + \Gamma_r}{\Gamma_{nr} + F\Gamma_r} P(\alpha) \qquad (44)$$

In the context of this chapter, we are mainly interested in the effect of the microcavity on the emission. As a reference we set the total radiative emission $E_{\text{inf}} = 1$ for the emitting layer embedded in an infinite medium n_e. Using Eq. (44), the total radiative emission from a dipole transition in a microcavity (including transmission and dissipation in absorbing materials) is

$$E_{\text{tot}} = \frac{\Gamma_{nr} + \Gamma_r}{\Gamma_{nr} + F\Gamma_r} F \qquad (45)$$

In a specific application, only the light that is emitted into a certain solid angle or interval $[\kappa_1^2, \kappa_2^2]$ is considered as useful emission. This quantity is defined as E_{appl} and given by

$$E_{\text{appl}} = \frac{\Gamma_{nr} + \Gamma_r}{\Gamma_{nr} + F\Gamma_r} \int_{\kappa_1^2}^{\kappa_2^2} K(\kappa) \, d\kappa^2 \qquad (46)$$

This value can be larger than unity, which means that it is in principle possible to collect more light from the microcavity than is emitted in an infinite medium.

When the nonradiative transition is dominant ($\Gamma_{nr} \gg \Gamma_r$), the total decay time does not change and the factor appearing in the three above equations is equal to one. In this case, $P_{\text{dipole}} = P$ and $E_{\text{tot}} = F$, thus is equal to the values given for the dipole antenna. In the case where the radiative transition is dominant ($\Gamma_r \gg \Gamma_{nr}$), the total emission E_{tot} is equal to unity, because for each excited state there is always one photon emitted. In Eqs. (44), (45), and (46), the common factor is then equal to $1/F$.

Up to now we have assumed that the radiation of the dipole transition is monochromatic, and calculated the angular dependence and the decay rate at a fixed wavelength. The relaxation of the excited state will not always yield the same ground state, because there is a distribution of final states, with different energies. In addition, the transition to the final state may be connected with the creation of phonons, which reduce the energy and increase the wavelength of the emitted photon. If a given excited state has a spectral distribution $S_\lambda(\lambda)$ in the infinite medium with index of refraction n_e, it must be modeled by a corresponding ensemble of dipole antennas. The radiation of this ensemble of dipoles is then a function of the wavelength and $K(\kappa)$ and $P(a)$ must be replaced by $K(\kappa, \lambda) \cdot S_\lambda(\lambda)$ and $P(\alpha, \lambda) \cdot S_\lambda(\lambda)$ respectively. The factor F, which is important to estimate the change in the decay time according to Eq. (43), is then given by

$$F = \int_0^\infty \int_0^\infty K(\kappa, \lambda) S_\lambda(\lambda) \, d\kappa^2 \, d\lambda \qquad (47)$$

which replaces Eq. (11). The spectrum is normalized as follows:

$$\int_0^\infty S_\lambda(\lambda) \, d\lambda = 1 \qquad (48)$$

III. Basic Examples

Before discussing the applications of thin film microcavities in Section IV, we consider a number of basic microcavity structures to illustrate the effect of different parameters. The behavior of metallic mirrors and distributed Bragg reflectors, the difference between perpendicular and parallel dipole antennas, and the effects of wide-angle and multiple-beam interference are discussed in this section. All the examples are given for an emitting medium with index of refraction $n_e = 3$ and for a wavelength of $\lambda = 600$ nm in air, or 200 nm in the emitting medium.

1. Mirror Characteristics

In the electroluminescent devices, which are considered in the next section, two types of mirrors are used. The first type of mirror is a simple metallic layer, which can be semitransparent or completely opaque. Metals not only reflect, but also absorb part of the incident light, and are characterized by an index of refraction with real and imaginary components. The reflectivity of metallic mirrors is limited by the absorption, in the visible spectrum the best metallic mirror is silver, the commonly used aluminum is not as good (the reflectivities for 600 nm light incident from air are 98.4 and 90%, respectively).

A second type of mirrors are the distributed Bragg reflectors or DBRs, which consist of a stack of dielectric layers with alternatingly high and low indices of refraction. At each interface, a fraction of the light is reflected and the structure is designed in such a way that all the reflected contributions add up constructively, leading to a high reflectivity. When each layer has a thickness of a quarter wavelength, the total reflectivity in the perpendicular direction is optimized for the given wavelength. For waves incident at an angle or for other wavelengths, the reflectivity can be considerably smaller. An important advantage of DBRs is the fact that they do not absorb any light, so when the reflection coefficient of a DBR is the same as for a metallic mirror, the transmission of the DBR will always be larger.

As a reference for realistic metals, it is useful to consider an idealized metallic mirror with a very large index of refraction. The reflectivity of a perfect mirror does not depend on the adjacent medium: it is always equal to unity, and the phase of the reflection coefficient is $0°$ for TM and $-180°$ for TE polarization. Note that there is a $180°$ phase difference between reflection coefficients for TE and TM polarization, due to the sign convention given in Fig. 2. Ideal mirrors have no absorption, and their effective thickness [see Eqs. (33) and (34)] is zero.

Figure 7 shows the reflection coefficient R for an aluminum mirror adjacent to a material with index of refraction $n = 3$ as a function of the angle of the incident light. The values are given for a wavelength of 600 nm and for the complex index of refraction of the aluminum set to $n_{Al} = 1 + 6j$. In the perpendicular direction, the reflectivity R is 77%. The effective thickness of a metal mirror is typically only a fraction of the wavelength; for the given example, of an Al mirror, the values are

$$d_{\text{eff},\alpha}^{\text{TE}} = \left(\frac{d\varphi(r_{1,2}^{\text{TE}})}{d\cos\alpha_1}\right)_{\alpha = 0°, \lambda = 600\,\text{nm}} = 15.5\,\text{nm}$$

$$d_{\text{eff},\alpha}^{\text{TM}} = -15.5\,\text{nm} \tag{49}$$

$$d_{\text{eff},\lambda} = 0\,\text{nm}$$

The DBR in Fig. 8 consists of five layers with quarter wavelength thickness and alternating indices of refraction $n = 2$ and $n = 3$. With this sequence, the optimum reflectivity of $R = 70\%$ is obtained in the perpendicular direction. Typical for DBRs is the reduction of the reflectivity at larger angles, for Fig. 8 this is at angles of the order of 30°. The phase of the reflection coefficient is more sensitive to variations in angle, which leads to a larger effective thickness for the DBR:

$$d_{\text{eff},\alpha}^{\text{TE,TM}} = 143\,\text{nm}$$
$$d_{\text{eff},\lambda} = 80\,\text{nm} \tag{50}$$

As the effective thickness of a DBR is an important fraction of the

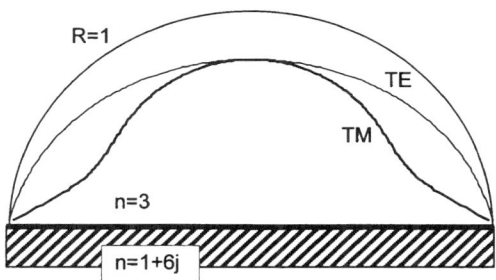

FIG. 7. Reflectivity of an aluminum mirror (with index of refraction $1 + 6j$) as a function of angle and polarization. The outer circle represents $R = 1$.

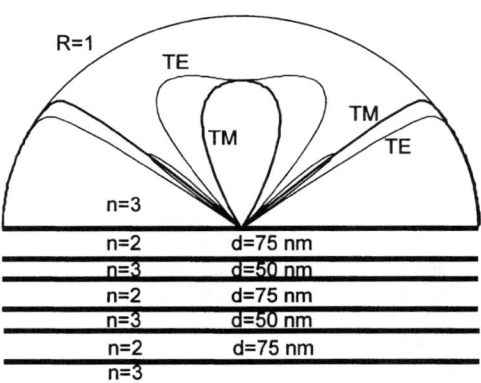

FIG. 8. Reflectivity of a DBR mirror as a function of angle and polarization. The outer circle represents $R = 1$.

wavelength, microcavities using DBRs will always have a relatively large effective thickness than metal mirrors, which has implications on the angle and wavelength dependence of the emission. This effect is even more important when the contrast between the different layers of the DBR is smaller than in the example given here.

For a DBR with 21 quarter-wavelength layers and the same contrast of 1.5, a reflectivity of 0.9995 is obtained. In this case, the effective thicknesses are only slightly larger then for the previous structure, because the first five layers contribute a lot to the reflectivity:

$$d_{\text{eff},\alpha}^{\text{TE,TM}} = 175\,\text{nm}$$

$$d_{\text{eff},\lambda} = 100\,\text{nm}$$

Sometimes useful mirror characteristics can be obtained by combining the reflectivities of metallic and DBRs, as shown in Fig. 9.

2. Radiation from Dipole Antennas

The intensity emitted from a single dipole antenna in an infinite medium depends on the angle θ with the dipole axis as $\sin^2\theta$, which is shown in Fig. 10. When the axis of the dipole antenna is perpendicular to the substrate, there is only emission with TM polarization [Eq. (13)]. For dipole antennas with an axis parallel to the substrate, there is a degree of freedom, namely

4 MICROCAVITIES FOR ELECTROLUMINESCENT DEVICES 203

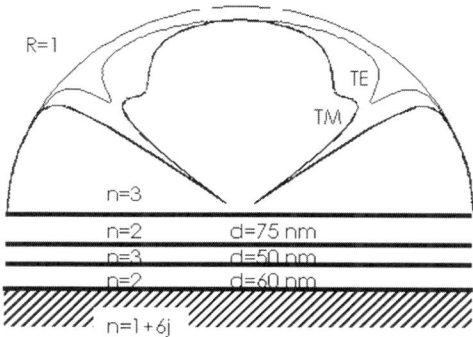

FIG. 9. Reflectivity of a combined aluminum–DBR mirror as a function of angle and polarization. The outer circle represents $R = 1$.

the azimuth angle. Because thin film deposition technologies usually do not yield anisotropy in the plane of the film, it is always assumed here that there is a random azimuth orientation. The resulting radiation in Fig. 10 is a combination of isotropic TE radiation and TM radiation, which is mainly perpendicular to the plane.

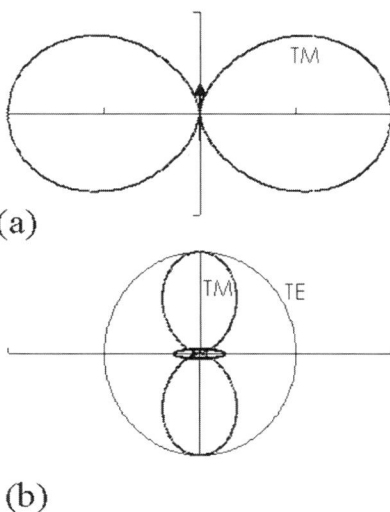

FIG. 10. (a) Distribution of radiation from an orthogonal dipole antenna. (b) Distribution of radiation from dipole antennas with random orientation in the xy plane.

For an ensemble of dipole antennas with random orientation in three directions, the emission is isotropic with equal contributions of TE and TM polarization. With the total intensity normalized to unity, the emission per unit of solid angle is $1/4\pi$, or $0.0796\,\mathrm{sr}^{-1}$.

3. Dipole Antennas Near a Mirror

When an electrical dipole antenna is located near a single perfect mirror, there will be wide-angle interference effects between the directly emitted and the reflected radiation. Instead of considering the reflections of the dipole antenna on the mirror, one can also consider the radiation emitted from the mirror image of the dipole antenna, which is located behind the mirror, with opposite polarity, as shown in Fig. 11. The emission from randomly oriented dipole antennas is enhanced by a factor of 4 in the perpendicular direction, when the distance to the mirror is a quarter wavelength. This is because the distance to the mirror image of half a wavelength compensates for the opposite polarity of the dipole. Figure 12 shows the total emission (TE + TM) as a function of the emission angle for a set of randomly oriented dipoles located at a distance of 50, 100, or 150 nm from the mirror, corresponding to $\lambda/4n_e$, $\lambda/2n_e$, and $3\lambda/4n_e$. For the distance of half a wavelength (100 nm) the emission in the perpendicular direction vanishes; in the other two cases, the emission in the perpendicular direction is enhanced by a factor of 4 and becomes $P_e = 1/\pi$ instead of $\frac{1}{4}\pi$. For the distance of 50 nm, the emission is enhanced by a factor of 2 or larger in almost all directions. The enhancement of a factor of 2 can be expected, because the light is emitted only in the upward direction, over a total solid angle of 2π. For larger distances from the mirror, there is an enhancement in some directions and an inhibition in others. When the emission of the dipole antenna is integrated over all angles, the total emission F in Fig. 12 is obtained. This value also represents the increase in the radiative decay rate for an electrical dipole transition (see Subsection 3 of Section II). Note

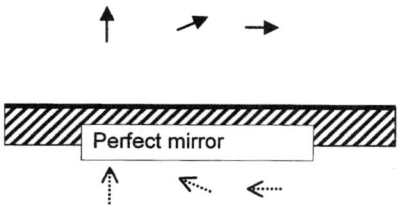

Fig. 11. Electrical dipoles (above) and their mirror images behind a perfect mirror (below).

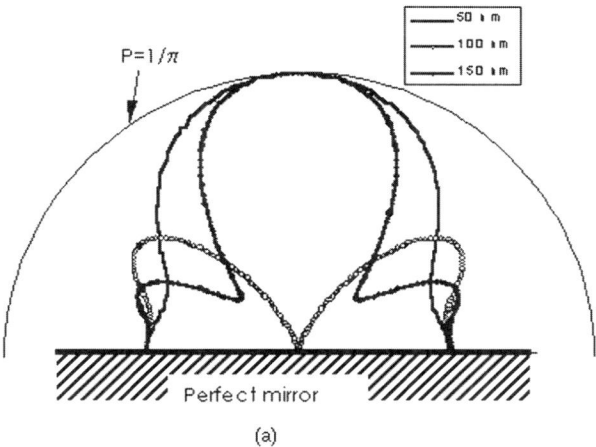

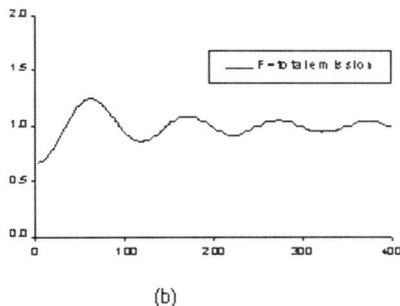

FIG. 12. (a) Radiation pattern (TM+TE) from a randomly oriented dipole antenna ($\lambda = 600$ nm) at distances of 50, 100, and 150 nm from a perfect mirror in a medium with $n_e = 3$. The outer circle corresponds with an intensity $P_e = 1/\pi$. (b) Integrated total emission F of the dipole antenna as a function of the distance from the perfect mirror.

that F oscillates around unity and at larger distances the effect of the mirror on F becomes negligible.

The concept of a mirror image of a dipole antenna is valid only for perfect mirrors, but can be a helpful tool for making estimations in the case of real mirrors. The calculated emission for randomly oriented dipole antennas near an aluminum mirror is given in Fig. 13. Because there is an additional

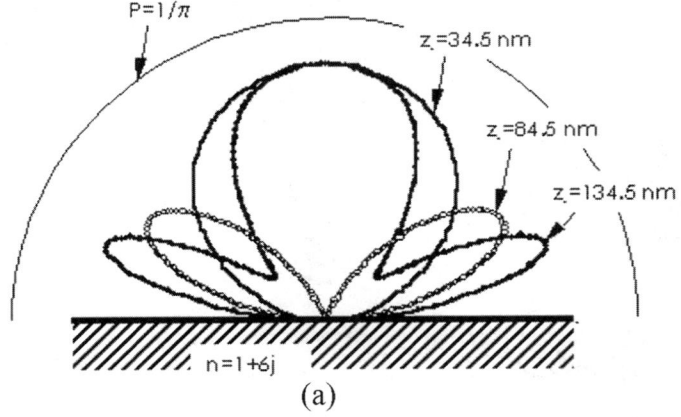

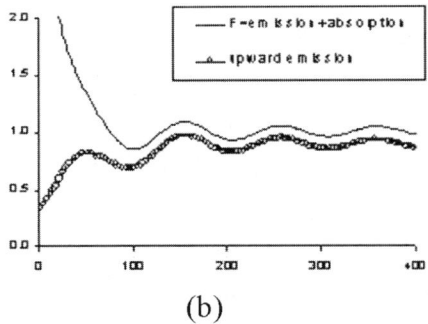

FIG. 13. (a) Radiation pattern (TM + TE) from a randomly oriented dipole antenna ($\lambda = 600$ nm) at distances 34.5, 84.5, and 134.5 nm from an aluminum mirror in a medium with $n_e = 3$. The outer circle corresponds with an intensity $P_e = 1/\pi$. (b) Integrated total emission F and net emission (not absorbed) of the dipole antenna as a function of the distance from the mirror.

phase shift of 52° at the aluminum mirror, the dipole antennas are placed 15.5 nm closer than for the perfect mirror. Because of absorption, the enhancement for the quarter-wavelength case is somewhat smaller than a factor 4 and the extinction is not complete for the half-wavelength case. The emission at angles close to 90°, or nearly parallel to the mirror, is strongly reduced compared to the case of perfect mirrors. This is due to the different phase of the reflection coefficient.

Figure 13b shows the integrated emission in the upward direction and the total emission F (including absorption in the Al mirror) as a function of the distance between the dipole antenna and the aluminum mirror. For very small distances, the total emission F becomes very large, because of the contribution of evanescent waves, which are very effectively absorbed in the aluminum mirror at close range. At larger distances, F becomes unity, as expected, and the integrated upward emission tends to the value corresponding with geometrical optics.

For the DBR, the enhancement in the perpendicular direction is obtained for a distance of half a wavelength, because the phase shift at the DBR differs from the phase shift at the ideal mirror by 180°. Due to the angle dependence of the reflectivity of the DBR, the angle dependence of the combined TM and TE emission is rather complex.

Because the DBR is semitransparent, a part of the emission is transmitted downward, through the DBR. The integrated upward emission, and the total emission F are shown in Fig. 14. There is again some oscillation, with the factor F approaching unity for sufficiently large distances.

Figure 15 shows the variation of the TM + TE emission for randomly oriented dipole antennas as a function of the distance from the mirror and the angle of the emission, for the ideal mirror, the aluminum mirror and the DBR already discussed.

4. SEMITRANSPARENT MIRROR REFLECTIVITY

When a second mirror is added to the microcavity, the modulation by wide-angle interference effects observed for the one-mirror cavity must be multiplied with an additional factor, which accounts for multiple-beam interference [see Eq. (19) or (20)]. In a microcavity with highly reflective mirrors, this modulation is much larger than for wide-angle interference. The resulting maxima due to multiple beam interference are often called Fabry–Pérot modes. The semitransparent mirror reflects a lot of radiation and inside the cavity the intensities are very high.

We now investigate the dependence on location and the reflectivity R_+ of the semitransparent mirror. For simplicity it is assumed that the phase changes at the front mirror and the back mirror are the same and equal to either π or $-\pi$. For a nonabsorbing front mirror, the factor in Eq. (20) depending on the front mirror, further called the multiple-beam interference

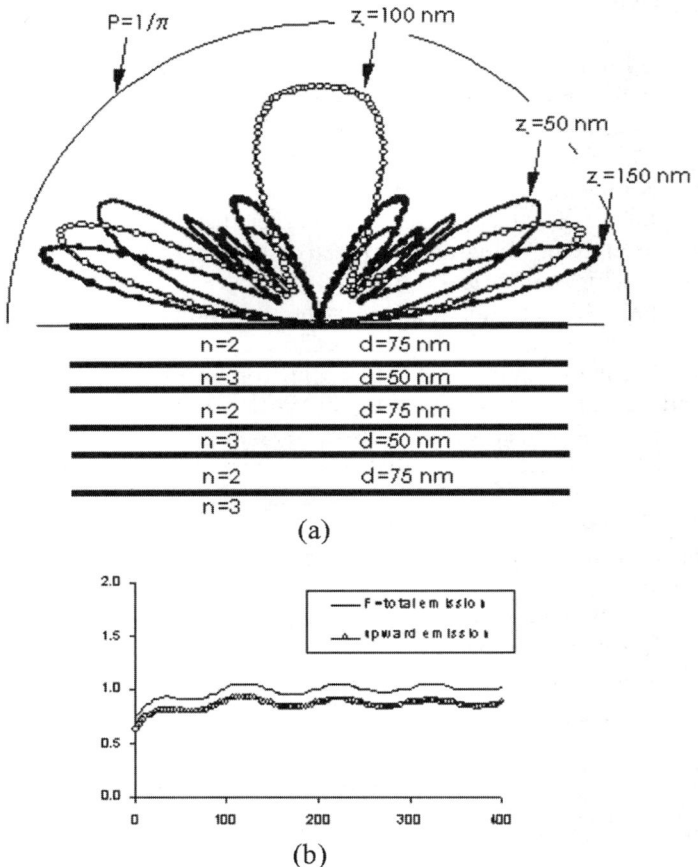

FIG. 14. (a) Upward radiation pattern (TM + TE) from a randomly oriented dipole antenna ($\lambda = 600$ nm) at distances of 50, 100, and 150 nm from a DBR mirror in a medium with $n_e = 3$. The outer circle corresponds with an intensity $P_e = 1/\pi$. (b) Integrated total emission F and reflected emission (not transmitted) of the dipole antenna as a function of the distance from the mirror.

factor, is given by

$$\frac{1 - |a_+|^2}{|1 - a|^2} = \frac{1 - R_+}{|1 - \sqrt{R_+ R_-} \exp(j 4\pi n_e d_e \cos \alpha / \lambda)|^2} \quad (51)$$

Figure 16 shows the multiple-beam interference factor in the perpendicular

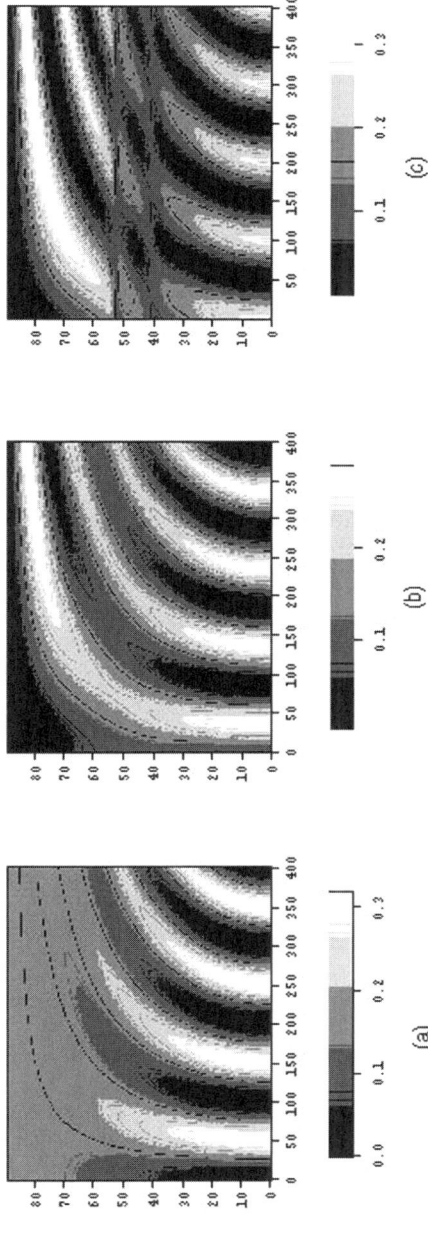

FIG. 15. Radiation P_e (TM + TE) in the upward direction for a randomly oriented dipole antenna ($\lambda = 600$ nm), as a function of the distance from the mirror and the angle of the emission for (a) the perfect mirror in Fig. 12; (b) the aluminum mirror in Fig. 13; and (c) the DBR in Fig. 14.

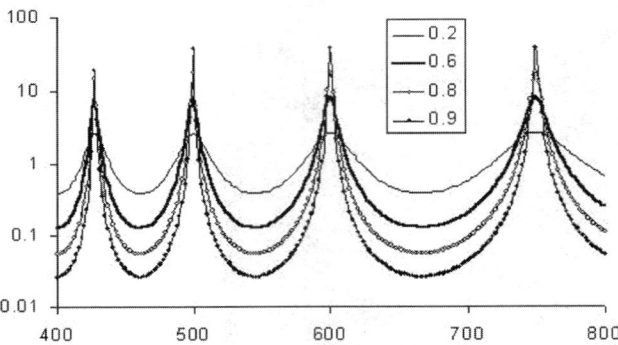

FIG. 16. Multiple-beam interference factor of Eq. (51) as a function of the wavelength λ for different reflectivities of the front mirror R_+. There is a perfect mirror on the other side: $R_- = 1$.

direction as a function of the wavelength for four different values of R_+. Other parameters of the structure are $R_- = 1$, $n_e = 3$, and $d_e = 500$ nm. Maxima occur when the optical thickness of the cavity ($n_e d_e = 1500$ nm) is a multiple of half of the wavelength. The maxima are the highest when the reflectivity of the front mirror is closer to unity, but then the full width at half maximum is also narrower.

Figure 17 shows the modulation due to multiple beam interference factor in Eq. (51), for three different structures. The figure in the middle is the same as described earlier ($R_- = 1$, $n_e = 3$, and $d_e = 500$ nm) with $R_+ = 0.8$. The diagram on the left shows the characteristics when the reflectivity of the front mirror is only $R_+ = 0.5$. In this case, the maxima are not as high, but the modes are significantly wider. The influence of the cavity thickness is illustrated by the figure to the right, which is calculated for a cavity with a thickness of only 100 nm, corresponding to one-half wavelength at $\lambda = 600$ nm. Compared with the 500-nm cavity, the mode is wider, but there is only one mode available.

5. Double Mirror Microcavities

When the wide-angle interference effect is combined with the multiple beam interference factor, then the actual emission from the dipole antennas in a microcavity can be estimated. In the following, it is again assumed that the dipoles are randomly oriented. The resulting emission (TM and TE) for the different structures of Fig. 18 are given in Fig. 19.

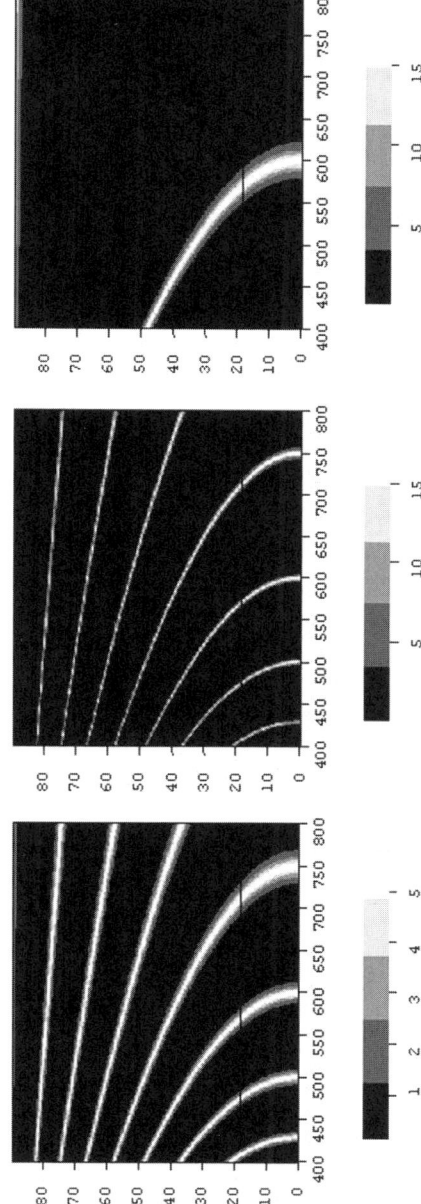

FIG. 17. Multiple-beam interference factor of Eq. (51) as a function of the wavelength λ and the emission angle α for three different structures. (Left) The front mirror $R_+ = 0.5$ at $d_e = 500\,\text{nm}$; (center) the front mirror $R_+ = 0.8$ at $d_e = 500\,\text{nm}$; and (right) the front mirror $R_+ = 0.8$ at $d_e = 100\,\text{nm}$.

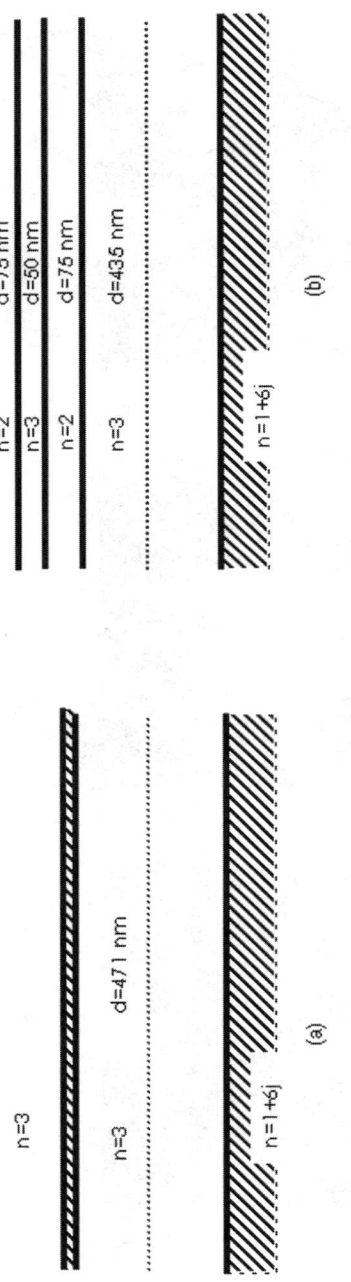

FIG. 18. Microcavity structures: (a) the thin layer between two aluminum mirrors and (b) the thin layer between aluminum and DBR mirror.

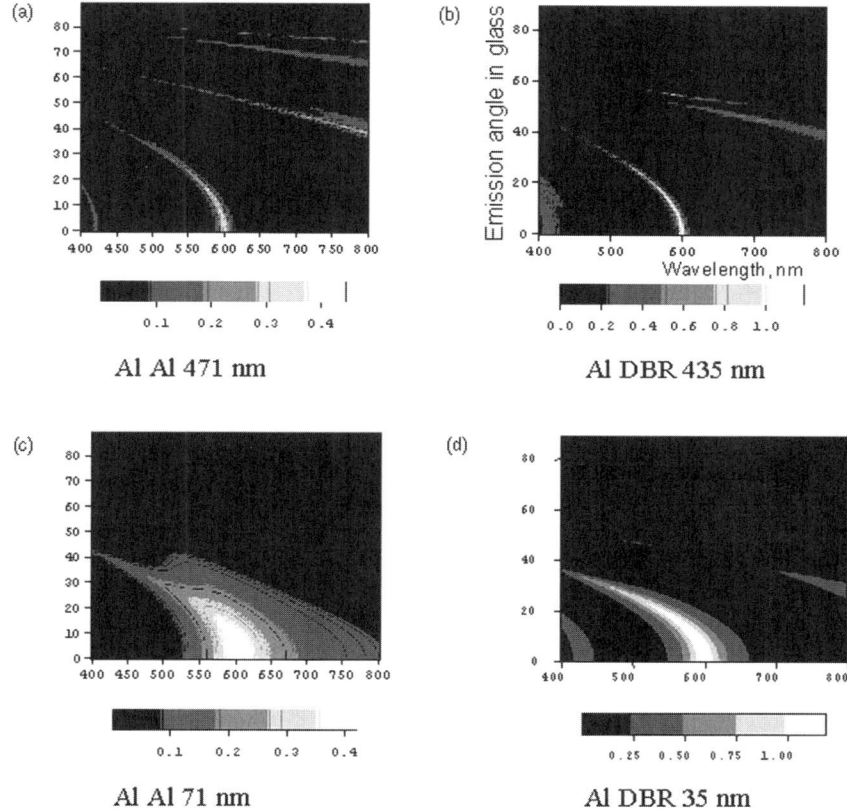

FIG. 19. Radiation P_e (TM + TE) for randomly oriented dipole antennas, as a function of the angle α and the wavelength of the antenna λ, (a) for the structure of Fig. 18a, (b) for the structure of Fig. 18b, (c) for the structure of Fig. 18a with $d_e = 71$ nm, and (d) for the structure of Fig. 18b with $d_e = 35$ nm.

Figure 18a shows a microcavity structure with a thick aluminum layer as back reflector and a 20-nm-thick semitransparent aluminum front electrode. The medium between the two mirrors has a thickness of 471 nm, which corresponds to 2.5 wavelengths when the phase shift at the mirrors is included in the optical thickness. The dipole antennas are located at a distance of 235 nm (see also Fig. 13), to enhance the wide-angle interference effect in the perpendicular direction. The resulting emission as a function of the wavelength and the angle of the emission is given in Fig. 19a.

The microcavity structure in Fig. 18b has a thick aluminum layer as back reflector and a five-layer DBR as a front mirror. The structure is also optimized for emission in the perpendicular direction, with the randomly oriented dipole antennas at 235 nm from the Al and 200 nm from the DBR, according to Figs. 13 and 14. Because there is no absorption in the DBR, the intensity in Fig. 19b is higher than for the structure with the semitransparent mirror. However, the maximum is somewhat narrower, because the effective thickness is larger for this structure.

The emission characteristics for the aluminum–aluminum and aluminum–DBR structures, for a much thinner microcavity, are given in Figs. 19c and 19d. Because the total effective thickness of the structure is now much smaller, there is a large emission intensity over a wider spectral and wavelength range.

IV. Applications

In this section, the use of microcavities is investigated for the three different types of electroluminescent devices treated in this book. The main advantage of microcavities for electroluminescent applications is the enhancement of the emission intensity. In the previous section, it became clear that large enhancements of the emission can only be obtained for a limited spectral range in a limited solid angle. The microcavity structure should thus be designed in such a way that the emission characteristics are adapted to the application. For some microcavity structures, it may be difficult to deposit a large stack of layers with accurate thicknesses, but in other applications the required accuracy is not high at all.

Because the index of refraction of many light-emitting materials is quite high, most of the light generated in a planar electroluminescent device is internally reflected and thus lost for the application. With a microcavity, it is possible to make the emission anisotropic, thereby substantially enhancing the emission in the small solid angle, which is useful for the application. In this case, the microcavity merely redistributes the emission toward the useful directions. If nonradiative decay is very important, it is also possible to enhance the total amount of photons which are generated, by increasing the radiative decay rate. But this may be more difficult to achieve in actual applications. In the following sections, the use of microcavities in inorganic EL devices, light-emitting diodes (LEDs), and organic LEDs are discussed separately. For each application there is an example to illustrate how microcavities can enhance the light emission properties.

1. Inorganic EL Devices

a. Introduction

The properties and behavior of thin film electroluminescent (TFEL) devices are treated in Chapter 7. The typical structure of a TFEL device is given in Fig. 20. The light is emitted in the phosphor layer, usually a II–VI compound such as ZnS or SrS, with a relatively high index of refraction. The phosphor layer is embedded between two insulator layers to prevent electrical breakdown. Finally, there are electrode materials on both sides, usually a metallic layer on one side and a transparent conductor, such as ITO, on the other side, but other electrodes, for example, thin metal layers are also possible. The entire phosphor layer is doped with activator atoms, and the light generation is thus distributed over the layer. The TFEL device is usually deposited on a glass substrate, but reflections on the glass–air interface are not considered for interference, because the thickness is beyond the coherence length of the emission.

The electrical operation limits the range of thicknesses for the phosphor and insulator. The dielectric layers must be thick enough to prevent breakdown of electrical charges between the two electrodes in the high electric fields. Also for the phosphor layer, there is a minimal thickness necessary to obtain efficient electroluminescence, probably because the crystalline properties of thin phosphor layers are relatively poor. The minimal optical thickness nd for appropriate TFEL light emission is of the order of 1 μm, which is equivalent to 2 wavelengths for the emission of 500 nm. When a DBR is used on the side of the ITO electrode, the effective optical thickness will be even larger. The light is generated in the entire phosphor layer, so wide-angle interference effects are reduced after averaging, and will be neglected in this section, as explained in Subsection 2(g).

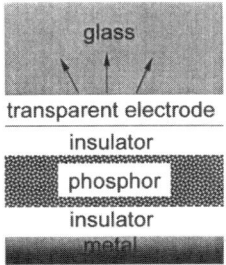

Fig. 20. Typical structure of a TFEL device.

The most important application of TFEL devices is in direct-view flat panel displays, with high-resolution and high-information content, which can be either full color or monochromatic. Some variation in the intensity as a function of the angle of observation can be tolerated, but as the eye is very sensitive for variations in color, the angular color variation should remain small. For monochromatic displays, the color variation of the emission can be reduced by choosing a phosphor with a narrow characteristic emission spectrum. For color displays, based on the red, green, and blue basic colors, the angular intensity variation of each primary color will have to be matched to reduce angular color variations. Because TFEL phosphor layers are thicker than a few wavelengths, a microcavity, which enhances the brightness in a given direction and for a given wavelength, will generate an unacceptable color variation over the angles of observation.

In the future, projection applications of TFEL devices such as head-mounted or virtual reality displays may become important. In these applications, only the light that is emitted into the acceptance cone of a lens or system of lenses will be projected on the eye of the observer. The acceptance cone of a projection system is relatively small compared with the angle of observation for direct view displays: the half angle is of the order of 10° in air. In this case, microcavities can be used to enhance the emission into this small solid angle by an important factor. The projection of light from TFEL devices by an optical system and the use of microcavities can also be useful for other nondisplay applications, where color variations are not important.

The use of microcavities for the enhancement of TFEL emission has been investigated in several papers. By using two metallic electrodes, a strong color modulation with reduced intensity was obtained by Vlasenko *et al.* (1970). Mauch *et al.* (1989) showed that only a moderate enhancement of the integrated output in air can be obtained by optimizing the thickness of the layers in the EL device. More recently, DBRs were used to obtain a strong enhancement of the emission in a narrow solid angle (Mueller *et al.*, 1996; Neyts *et al.*, 1997).

b. *An Example: 548-nm Emission Line from ZnS:Ho*

We consider here the application where the highest possible light flux from an electroluminescent device must be coupled into a lens system with given aperture. Let us assume that the acceptance cone of the optical system has a half angle of 10° in air (f-number = 2.84 or numerical aperture = 0.174). The angle of 10° in air corresponds with an angle of $\alpha_e = 4.2°$ in ZnS, and according to geometrical optics only a fraction of $(1 - \cos \alpha_e)/2$ or 0.14% of the emission will be emitted into this narrow cone. By using a good reflector, this value can be doubled to 0.27% of the total emission. It is possible to enhance the emission into this small cone by using the interference effects in a microcavity. The enhancement factor is, of course, wavelength-dependent, and the optimization of the microcavity must take into account the width of the phosphor spectrum. In this section, the optimization is carried out for the 548-nm emission line of holmium in ZnS with 8-nm full width at half maximum, as shown in Fig. 21. The holmium dopant is of special interest for this application, because it is not very sensitive to saturation and can be driven at high frequencies to obtain high brightness (Mueller-Mach et al., 1996).

Figure 22 shows the structure of the microcavity TFEL that was described in Mueller et al. (1996) and Neyts et al. (1997). It contains a ZnS:Ho phosphor layer between two SiON insulator layers, with an aluminum bottom electrode and a transparent ITO top electrode. The DBR consists of 10 layers with alternatingly SiO_2 and TiO_2. It was found that the emission characteristics of this device, and the color changes (blue, red, and green) as a function of angle can be simulated with the theory of Section II (Neyts et al., 1997).

In the following, the layer thicknesses of a similar microcavity TFEL device is optimized for a narrow emission cone around the perpendicular direction. To obtain higher enhancements, highly reflective silver is used as electrode instead of aluminum. The maximization of the green light flux into a 10° emission cone is practically obtained when the structure of the microcavity is optimized for the central wavelength ($\lambda_c = 548$ nm) and for the angle $\alpha_{air} = 7°$ (with the cosine in the middle between 1 and cos 10°). The optimized thicknesses are then given by

$$d_{optimal} = \left(\frac{k}{2} + \frac{l}{4}\right)\frac{\lambda_c}{\sqrt{n^2 - \sin^2\alpha_{air}}} \tag{52}$$

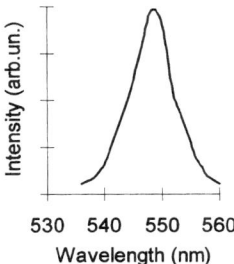

FIG. 21. Measured spectral dependence of the green emission line of holmium in ZnS.

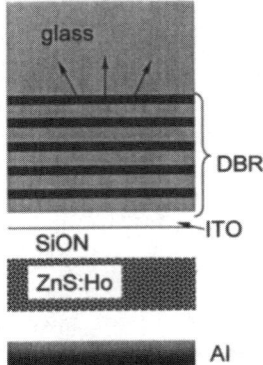

FIG. 22. Structure of a microcavity TFEL device with a metal and a DBR mirror.

with k an integer and l equal to 0 or 1, depending on the sequence of the refractive indices. Thus l is 0 for ZnS:Ho, SiON and ITO and 1 for the other layers. The value of k can be chosen arbitrarily, but in the structure proposed in the following, the thicknesses are compatible with TFEL devices. The resulting optimized thicknesses for the structure are given in Table I. In this structure, the SiO_2–TiO_2 couple represents the DBR, and has to be repeated n times, if it is an n-couple DBR (notation: n DBR). The first SiON layer is somewhat thinner than a quarter wavelength, because the silver mirror causes some additional phase shift.

For the given structure and different numbers of DBR couples, the reflectivities of the mirrors on both sides of the ZnS:Ho layer and the resulting integrated emission in the cone with half angle of 10° are given in Table II. The total emission reaches a maximum, which is 9.2 times larger than in the case without mirrors, when a DBR is used with three couples of layers. Further analysis of this structure yields

TABLE I

STRUCTURE OF A MICROCAVITY TFEL DEVICE OPTIMIZED FOR 548-nm EMISSION IN A 10° CONE

Ag	SiON	ZnS:Ho	SiON	ITO	SiO_2	TiO_2	glass
$0.055 + 3.32i$	2.	2.35	2.	1.95	1.455	2.38	1.5
	$<\lambda/4$	2.5λ	$\lambda/2$	$\lambda/2$	$\lambda/4$	$\lambda/4$	
	45 nm	584 nm	137 nm	141 nm	94 nm	58 nm	

4 MICROCAVITIES FOR ELECTROLUMINESCENT DEVICES 219

TABLE II

MIRROR CHARACTERISTICS, EMISSION INTO THE 10° CONE, AND ENHANCEMENT FOR THE MICROCAVITY STRUCTURE IN TABLE I

Mirror left	Mirror right	R_{left}	R_{right}	Total emission	Enhancement
None	None	0%	0%	0.14%	1.
Ag–SiON	Only EL (0 DBR)	97.5%	5%	0.39%	2.8
Ag–SiON	With 1 DBR	97.5%	38%	0.78%	5.7
Ag–SiON	With 2 DBR	97.5%	70%	1.13%	8.2
Ag–SiON	With 3 DBR	97.5%	87.6%	1.26%	9.2
Ag–SiON	With 4 DBR	97.5%	95.2%	1.12%	8.2
Ag–SiON	With 5 DBR	97.5%	98.2%	0.77%	5.6

- The maximum enhancement factor is 42
- For TM: $d_{\text{eff},\alpha} = 1050$ nm and FWHM $\Delta \cos \alpha_e = 0.0028$
- For TE: $d_{\text{eff},\alpha} = 1100$ nm and FWHM $\Delta \cos \alpha_e = 0.0027$
- For the wavelength: $d_{\text{eff},\lambda} = 900$ nm and FWHM $\Delta \lambda = 1.8$ nm

The discrepancy between the maximum enhancement factor 42 and the average enhancement of 9.2 is mainly due to the wide spectral width of the holmium emission line. Only a small fraction of the holmium spectral width of 8 nm is enhanced by the microcavity, which has an FWHM of only 1.8 nm. This example shows that it is important to use phosphor materials with very narrow emission characteristics. The angular dependence of the emission integrated over the holmium spectrum of Fig. 21, for the optimal structure, is given in Fig. 23. The actual shape of the emission fits very well

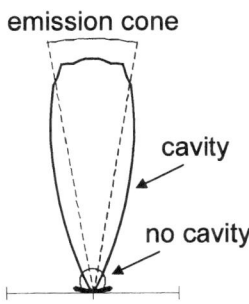

FIG. 23. Simulated angular dependence of the integrated emission of the structure of Fig. 22 compared with the emission without cavity.

into the 10° emission cone, for which it was optimized. For a phosphor layer without mirrors the emission in air is nearly Lambertian emitter (Fig. 23).

Neyts (1997) showed that for monochromatic light in the perpendicular direction, the optimal reflectivity for the DBR is equal to the reflectivity of the metal mirror. In practical cases, where a spectral width and nonzero solid angle are considered, the optimal DBR reflectivity is always lower than that of the metallic mirror. In the preceding example, the optical thickness of the phosphor layer is about 2.5 wavelengths. Reducing the thickness of the phosphor layer could increase the spectral range with strong enhancement. However, a thinner EL device will normally emit less light, due to a lower excitation efficiency, and the overall result may be not beneficial. When aluminum is used instead of silver as a back electrode, the best result for the total emission (0.79%) is obtained with only four layers in the DBR, because of the lower reflectivity.

2. LIGHT-EMITTING DIODES

a. Introduction

The properties and applications of light-emitting diodes are treated in detail in Chapters 1, 2, and 3 in this book. In this section some features relevant for microcavities are reviewed. In light-emitting diodes, the radiation is generated near the junction between n-type and p-type semiconductor materials. Under the forward bias condition, electrons can be injected across the junction into the p-type material and holes can reach the n-type material. The spontaneous emission of a photon is caused by the radiative recombination of electrons and holes across the bandgap of the semiconductor material. Often the recombination is confined to an intrinsic region with a slightly narrower bandgap located between the n- and p-type materials. To avoid nonradiative recombination and to obtain a high internal quantum efficiency, it is important to have a very low concentration of defects in the material, and therefore the different layers of the LED are grown epitaxially, lattice-matched to the substrate. This production method limits the possibilities for the optical structure. It is possible to deposit a metallic mirror on top of the structure, but the other mirror must be a DBR consisting of lattice-matched materials. Often the contrast in index of refraction for the DBR materials is relatively small and many layers are required to obtain a mirror with high reflectivity. Such DBRs have a large effective thickness, which limits the advantages of a microcavity structure. Instead of using a metallic mirror and a DBR, it is also possible to use DBRs on both sides of the active layer. The structure is then more complicated, but the mirrors are loss-free and high reflectivities can be obtained. For highly reflective DBRs, the structure is similar to a vertical-cavity surface-emitting laser, where the radiation of the LED in the forward direction is further enhanced by stimulated emission in the active region. In this section about microcavities, the discussion is limited to spontaneous emission; stimulated emission or optical gain are not considered.

In traditional homojunction LEDs and in devices with a double heterostructure, the thickness of the recombination region where light is generated is of the order of 1 μm. Microcavities based on these LEDs have an effective thickness of several wavelengths, because the index of refraction of III–V compounds is relatively large (order of 3). Because the thick DBR also contributes to the effective thickness, the intensity enhancement that can be realized in such microcavities depends strongly on the emission angle and the wavelength. The advantages of using a microcavity in this case are modest compared with the cost of the DBR.

In quantum-well LEDs, the recombination and light generation is confined to one or a few very thin layers (order of 5 nm). This means that it is possible to take full advantage of the wide-angle interference effect by placing the quantum well at the optimum distance from the mirrors. The distance between the mirrors can be reduced to values below one wavelength, yielding a minimal effective thickness, which is very suitable for microcavities. Because of this, most of the investigations on LED microcavities are devoted to quantum-well devices. The lattice constant of the material in the quantum-well layer is usually different from that of the substrate, so there is a lot of strain involved in the epitaxial growth of the quantum well. Due to this strain, the light generated by radiative recombination of electrons and holes is not isotropic. In a good approximation, the emission can be modeled as the radiation from an ensemble of dipole antennas, which are parallel to the plane of the quantum well, without any contribution from perpendicular dipole antennas (as in Fig. 10b).

The III–V (and II–VI) compounds used in LEDs have a large index of refraction (on the order of 3), and most of the light generated in the thin film system is internally reflected in the structure. For a thin film with $n = 3.6$, the critical angle for emission into air is 16°, which contains only a fraction of 4% of the solid angle. The light extraction from the LED into air can be improved by trying to avoid a planar structure [roughening of the surface (Yablanovitch *et al.*, 1998), mesa-etching of the layers, packaging the LED in a transparent spherical shape]. The other option is to avoid the consequences of geometrical optics and to use a microcavity to make the emission anisotropic and enhance the fraction emitted below the critical angle of the planar structure. Microcavities with a planar structure can be simulated with the methods of Section II. In three-dimensional microcavities, obtained after mesa-etching a planar microcavity, the light-emission region is confined between two DBRs in the perpendicular direction and air in the lateral direction. In this case, three-dimensional simulations are required (Baba et al., 1991).

In an LED, the active layer can also reabsorb the emission, which means that the energy of a photon is used to create a new electron–hole pair. This electron–hole pair can recombine nonradiatively, or generate a new photon, according to the mechanism of spontaneous emission. The photons that are trapped inside the structure have a relatively large chance to be reabsorbed by the active layer, and the mechanism of reabsorption will therefore enhance the total emission leaving the device. The importance of reabsorption for the overall emission characteristics is strongly determined by the internal quantum efficiency of the LED. If the quantum efficiency is small, then reabsorption can be considered as a loss process, similar to absorption.

The substrate material on which the LED is grown is not always transparent for the light generated in the quantum well. Microcavity devices with two DBRs can minimize the emission toward the bottom and enhance the emission toward the top electrode. An alternative is to reduce the thickness of the absorbing substrate, or to remove it completely, to enable emission toward the device bottom.

In the literature, simulations and experimental results for several types of III–V microcavity LEDs can be found. A list with studied LED microcavity structures and references is given in Table III. Typically large enhancements in the perpendicular direction and narrow spectra can be obtained for microcavity LEDs with highly reflective mirrors. De Neve et al. (1996) observed an enhancement of the integrated LED emission in air. In the next section, the behavior of a microcavity based on an InGaAs quantum well (QW) in GaAs is discussed, using a GaAs–AlAs DBR or a silver mirror.

TABLE III
OVERVIEW OF LED MICROCAVITY DEVICE REFERENCES

Substrate	QW active layer	Embedding layer	DBR materials	References
GaAs	InGaAs	GaAs	AlAs–GaAs	Bjork et al. (1991), Deppe and Lei (1991), Lin et al. (1994), and Rogers et al. (1990)
GaAs	InGaAs	AlGaAs	AlAs–GaAs	De Neve et al. (1996) and Hunt et al. (1992)
GaAs	GaAs	AlGaAs	AlAs–AlGaAs	Schubert et al. (1992)
InP	InGaAsP	InGaAsP	InGaAsP–InP	Hunt et al. (1992)

b. An Example: InGaAs Quantum Well in GaAs

In this section, the influence of the optical structure of an LED on the emission characteristics is investigated. The wavelength and angle dependence of the emission are calculated for several microcavity LED devices, based on an InGaAs quantum well in GaAs. This example is similar to the devices for which simulation results were presented in Bjork *et al.* (1991), Deppe and Lei (1991), Lin *et al.* (1994), and Rogers *et al.* (1990). Because the quantum well is much thinner than a wavelength, its index of refraction does not play a role in the calculation of the emission pattern. In fact, it can be assumed that the radiating dipoles are directly placed in the GaAs material ($n = 3.51$). The DBR consists of quarter-wavelength layers of the materials GaAs and AlAs ($n = 2.95$), which are assumed to be loss-free. Because the contrast in index of refraction is relatively small, many layers are needed to attain reflectivities of the order of 90%. Because of the strain in the quantum well, the simulation is based on the assumption that the electrical dipoles are parallel with the plane of the quantum well. Instead of using the particular spectral distribution from the quantum well, we assume that the quantum well emits a flat white spectrum. This will make it easier to evaluate the enhancement or reduction of the emission at different wavelengths. The angle dependence of the emission is determined in the GaAs substrate, which is also the medium with the largest index of refraction and makes it possible to represent all contributions of the emission as a function of this angle.

Figure 24a shows the structure and the emission of the quantum well in a homogeneous GaAs material, without the influence of coherent reflections. The spontaneous emission from the parallel dipoles is not isotropic, but directed slightly toward the perpendicular direction. As a reference for this example, the intensity (per unit solid angle) in the perpendicular direction is set to unity. For α close to $90°$ the intensity in GaAs is reduced to a value of 0.5. As illustrated in Fig. 10b, three-quarters of the emission is radiated isotropically with TE polarization, and the remaining quarter is radiated in TM waves (nonisotropically). The emission spectrum of the quantum well covers the 800–1200-nm range, and therefore the angular dependence of the emission is shown over this wavelength interval in Fig. 24b.

Figure 25a gives the structure of an LED based on the same quantum well, with one DBR consisting of 10 quarter-wavelength pairs of GaAs–AlAs. This DBR is designed to have a high reflectivity in the perpendicular direction for a wavelength of 1000 nm ($R_- = 88.4\%$). The reflectivity of the DBR depends on the angle, the wavelength, and the polarization of the incident light. It is close to unity when the angle is larger than $57.2°$, the critical angle for plane wave transmission in AlAs, and zero at the Brewster

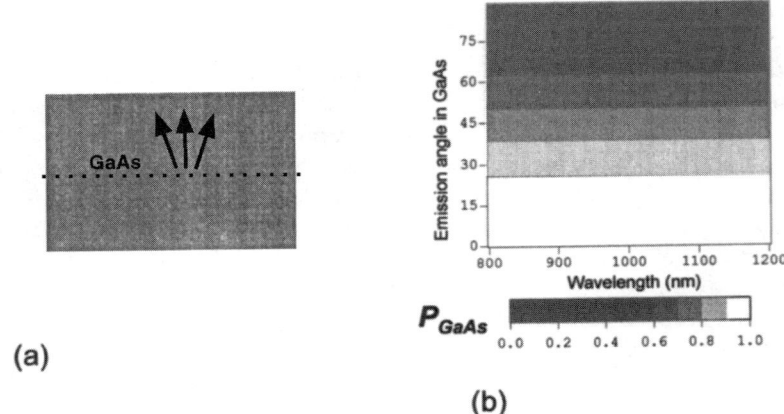

FIG. 24. (a) Homogeneous GaAs LED with an InGaAs quantum well and (b) emission characteristic P for parallel dipoles.

angle (40°) for TM polarization. The distance between the quantum well and the DBR is set to one half wavelength to enhance the wide-angle interference in the perpendicular direction. The emission intensity of the quantum well in is given in Fig. 25b. The intensity in the perpendicular direction is a factor 3.76 larger than in the reference case; this is almost twice

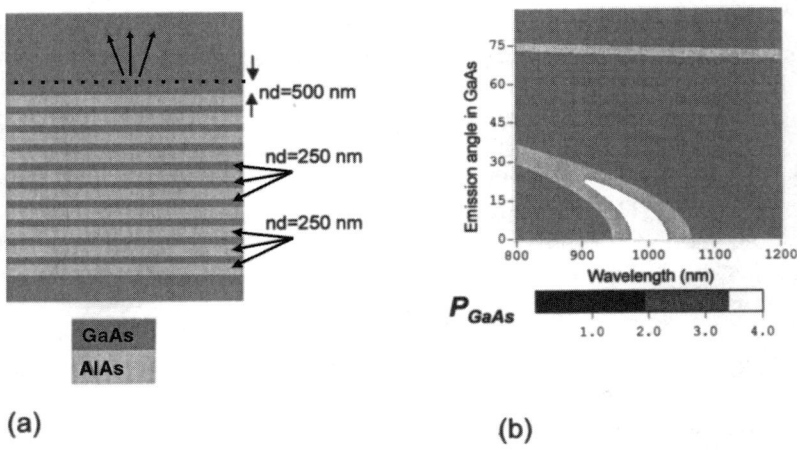

FIG. 25. (a) Quantum-well GaAs LED with one DBR and (b) emission characteristic P for parallel dipoles near the DBR.

the value $1 + R_-$ due to wide-angle interference. Below the critical angle, the variation of the intensity with angle and wavelength is roughly similar to the variation of the TM and TE reflectivities of the DBR. Above the critical angle, the reflectivity of the DBR is close to unity and the modulation in the emission (including the maximum around 73°) is due to wide-angle interference, mainly for TE polarization.

The structure of a symmetric quantum-well LED between two DBRs is given in Fig. 26, together with the emission intensity in the upward direction. With two highly reflective mirrors, the multiple-beam interference effect is very strong, and the emission can be enhanced by a very large factor (32.3 in the perpendicular direction at 1000 nm). The maximum has become much narrower, compared to the case with only one mirror, because the condition for constructive multiple-beam interference is more severe. There is also a very narrow, but intense TE mode around 73°, which is also present for the case with only one mirror. In the regions where the reflectivity of the DBR is small, the modulation of the emission is not so important.

When a silver mirror is used as bottom mirror instead of a DBR, the emission characteristic is very similar to that of Fig. 26b. Of course, the

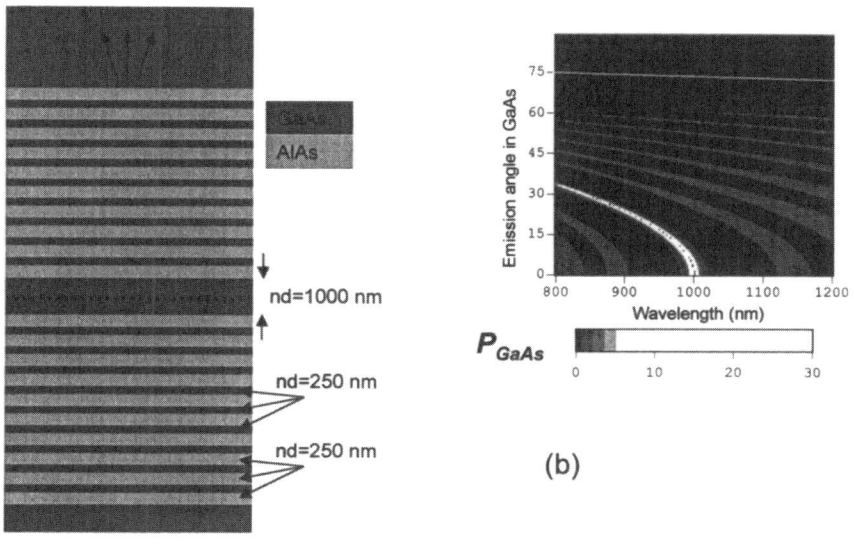

FIG. 26. (a) Quantum-well GaAs LED with two DBRs and (b) emission characteristic P for parallel dipoles in the microcavity with two DBRs.

optical distance to the quantum well must be adjusted to obtain the same phase shift ($nz_- = 177$ nm instead of 500 nm). For the wavelength of 1000 nm, the reflectivity of the mirror is 99%, leading to further enhancement in the perpendicular direction (factor 112 compared to the case without mirrors). Compared to the previous device with two DBRs, the narrow TE mode has shifted to a smaller angle (68°) and lost some intensity, because at these angles the reflectivity of the silver mirror is less than that of the DBR.

3. Organic Electroluminescent Devices

a. Introduction

Organic light-emitting diodes (OLEDs) are discussed in detail in Chapters 4, 5, and 6. They consist of one or more thin organic films, sandwiched between conductive layers. Electrons are injected in the structure at the cathodic electrode; the anodic electrode supplies the holes. The electrons and holes are transported through the structure and when they meet they can form an exciton. The recombination of the exciton can be a radiative process, and in that case, a photon is emitted. Usually the structure is deposited on a glass substrate, with ITO as a transparent anode. The other electrode is usually a metallic mirror.

In some organic thin films, hole transport is easier than electron transport. When such a layer is placed between two electrodes, the charge transfer will be dominated by transport of holes, and excitons (as well as photons) will mainly be generated near the cathode. Many organic LEDs use a combination of an electron transport layer and a hole transport layer to reduce the amount of charge carriers moving through the device without recombination. In such a structure, the creation of excitons and photons takes place near the boundary between the two layers. The fact that electrons or holes must be effectively injected from the electrodes imposes restrictions on the choice of electrode materials. The electronic properties of the interface are determined by a very thin layer of only a few nanometers, which is too thin to be of optical importance. It is thus in principle possible to use all kinds of mirror materials behind the transparent interface region.

Compared to inorganic EL and LEDs, organic EL devices are much thinner: the distance between the electrodes is only about 100 nm. Because the intrinsic thickness is so small, there is a lot of freedom to include it in an optical microcavity with an optimized structure. The application of organic EL devices in half-wavelength or full-wavelength microcavities is therefore relatively easy. The region where the light is generated is small compared to the wavelength of the emission. It is therefore possible to take full advantage of the wide-angle interference effect by placing the light generation region near the antinode of the cavity. Due to the small effective thickness of the cavity, the sensitivity of the emission for variations in angle and wavelength is relatively small. The blue shift in color with increasing angle of observation, typical for microcavity devices can be relatively small for the organic EL devices, which makes it possible to use microcavities in direct view display applications (Neyts et al., 1998).

In the literature, there are many reports of microcavity OLED devices with the OLED located between a metallic mirror and a DBR. By using highly reflective DBRs, the broad intrinsic spectrum of the OLED can be transformed into one or more high and narrow emission peaks (Dodabalapur et al., 1994, 1996; Fisher et al., 1995; Grüner et al., 1996; Jordan et al., 1996; Lemmer et al., 1995; Lidzey et al., 1996; Nakayama et al., 1993). In some papers, the measured spectral distributions are correlated with theoretical simulations (Dodabalapur et al., 1994; Grüner et al., 1996). Also OLED microcavities between two metallic mirrors have been investigated (Cimrová and Neher, 1996; Takada et al., 1993). By modifying the distance between the mirrors, different colors can be extracted from the same OLED device (Dodabalapur et al., 1996). For most structures presented the change of the emission spectrum with the angle of observation also is given (Tsutsui et al., 1994).

In the microcavity papers cited, multiple-beam interference is much more important than wide angle interference. There are a few reports that show that wide angle interference, determined by the distance between the light generation layer and the mirror, can also influence the emission considerably (Saito et al., 1993; So et al., 1999). Saito et al. (1993) also investigated the influence of the distance from the mirror on the decay time, and the measurements agree with the theoretical simulations. Lemmer et al. (1995) showed that the decay time can be influenced by the presence of a microcavity and also that the decay time can depend on the measurement wavelength. This latter effect is ascribed to inhomogeneous broadening: different kinds of excited states lead to different spectral regions. In this case, Eq. (47) does no longer hold because different values of F can apply to different states. Finally, stimulated emission or lasing has also been observed in microcavity OLEDs after intensive optical excitation (Tessler et al., 1996).

Interference effects are also important for the stacked OLEDs discussed in Chapter 5 (Subsection 5 of Section IV), because the stack contains two metallic mirrors, forming a microcavity (Burrows et al., 1997). The emission of the OLEDs between these mirrors will be modulated as a function of emission angle and wavelength.

b. An Example: Maximizing the Integrated Emission

In the TFEL and LED examples, the device structures were designed to maximize the emission near the perpendicular direction for a particular wavelength. For many applications where the light is collected in a narrow cone by an optical system, this may be appropriate. In the case of a display where the emission must be observable over a wide viewing angle, however, the integrated emission into air is a better quantity when evaluating the enhancement of a microcavity. Also it is important to estimate the enhancement over the entire emission spectrum of the device, not for one particular wavelength. Because OLEDs have a small effective thickness, and the display application is of particular importance, considering spectral and angle distribution is essential in OLED microcavity design.

In the following example an OLED is considered with the structure of Fig. 27a (Neyts et al., 1998):

$$Ag-Alq_3-\text{emitting } Alq_3 \text{ region } (10\,nm)-TPD-ITO-Ag-\text{glass}$$

The organic layers and ITO have an index of refraction of about 1.9, glass has 1.5, while silver has a wavelength-dependent complex index of refraction (Schulz and Tangherlini, 1954). Only the light emitted at angles below 41°

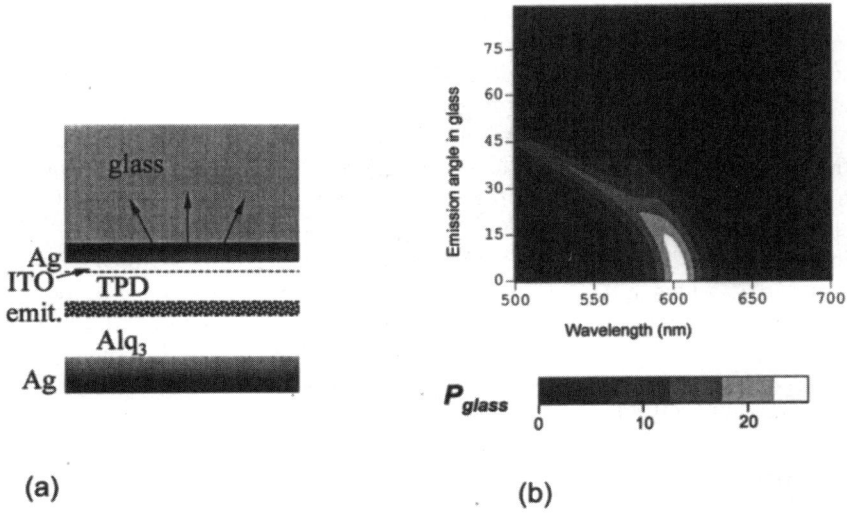

FIG. 27. (a) Half-wavelength microcavity OLED device between two silver mirrors and (b) emission characteristic P for the OLED microcavity with two silver mirrors.

in the glass substrate, or below 31.8° degrees in the Alq_3 material, will not suffer total internal reflection at the glass–air interface and is considered useful for a display. For an OLED structure without mirrors we expect that a fraction of $(1 - \cos 31.8°)/2$ or 0.075 of the emission will be transmitted into air. With the preceding microcavity structure this fraction can be enhanced considerably.

The probabilities for radiative and nonradiative decay in OLEDs are not well known, and for simplicity we first assume that nonradiative decay is dominant ($\Gamma_{nr} \gg \Gamma_r$). If the normalized spectral distribution is S_λ [as in Eq. (48), the total emission from the microcavity into air is given by [similar to Eqs. (46) and (47)]:

$$E_{air} = \int_{\lambda_{min}}^{\lambda_{max}} \int_{0°}^{41°} P_{glass}(\alpha_{glass}, \lambda) 2\pi \sin \alpha_{glass} S_\lambda(\lambda) \, d\alpha_{glass} \, d\lambda \quad (53)$$

with P_{glass} the power density defined in Eq. (12) including TM and TE contributions. The thickness of each layer can be optimized to find the maximum value for the intensity in air in Eq. (53). Using the spectrum of the widely used rubrene dopant, the maximum integrated intensity in air that can be obtained with the preceding structure is $E_{air} = 0.90$, or a factor of 12 larger than for the noncavity case (Neyts et al., 1998). The corresponding angle and wavelength dependence of the factor P_{glass} is given in Fig. 27b. For this example, the emission above the critical angle of 41° is larger than in the case without mirrors and the total emission F calculated from Eq. (47) is 3.0.

For a silver–DBR microcavity, based on the materials TiO_2 and SiO_2 for the DBR ($n = 2.3$ and 1.5), the optimum structure is obtained for a DBR with only three layers (Fig. 28a) (Neyts et al., 1998):

Ag–Alq_3–emitting Alq_3 (10 nm)–TPD–ITO–TiO_2–SiO_2–TiO_2–glass

The integrated emission in air for this case is $E_{air} = 0.57$, or 7.6 times larger than for the case without microcavity. This is smaller than for the silver–silver microcavity, which can be explained by the lower reflectivity, the larger effective thickness and the increased sensitivity for variations in angle and wavelength. The angle and wavelength dependence of the factor P_{glass} is given in Fig. 28b. The factor F for this device is 1.9. Of course, the optimum structure and the resulting intensity depend on the materials used for the DBR, so higher intensities could be obtained with a higher-contrast DBR.

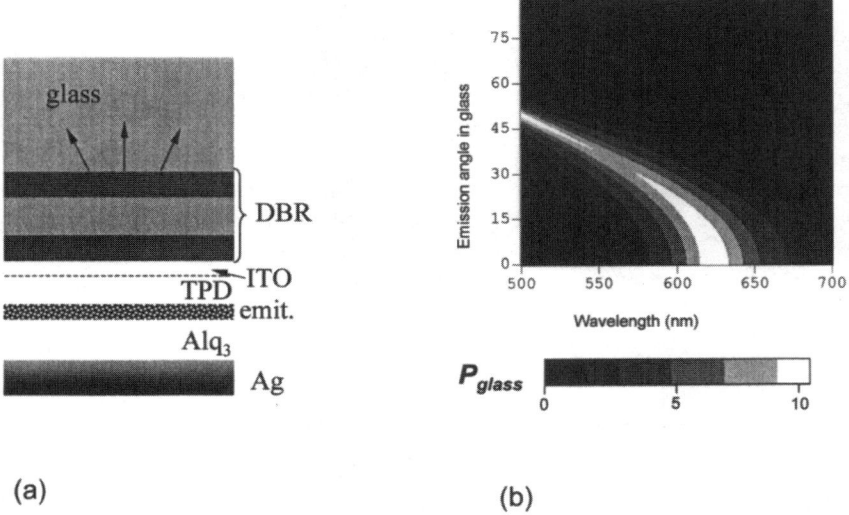

FIG. 28. (a) Half-wavelength microcavity OLED device with a silver mirror and a DBR and (b) emission characteristic P for the OLED microcavity with a silver mirror and a DBR.

For the two structures given here, the large values for the emission in air $E_{air} = 0.90$ and $E_{air} = 0.57$ are partly due to the higher directionality of the emission and partly due to the increased radiative decay rate (at the expense of less nonradiative decay). If we assume that the decay of the dopant is mainly radiative, the factor F has to be taken into account according to Eq. (46), and the emission in air for the two given structures is reduced to $E_{air} = 0.30$.

V. Conclusion

The fabrication of different kinds of electroluminescent devices is based on thin film processing, and as a result, the devices have a planar one-dimensional structure with layer thicknesses comparable to the wavelength of the light that is emitted. These properties match the requirements for interference or microcavity effects, and as a result the intensity of the emission is modulated as a function of the wavelength or the angle of the emission.

The different factors involved in the microcavity effects are analyzed in the theoretical description of Section II: multiple-beam and wide-angle interference of plane waves, power emitted into evanescent waves or discrete modes, dependence of the emission on angle, wavelength and polarization, and the change in the decay time of the excited state. The basic effects, which occur in a microcavity are illustrated in Section III. The source of the light emission is modeled as a group of electrical dipole antennas with randomly distributed orientation, emitting a mixture of TE and TM polarized light. The properties of metallic and distributed Bragg reflectors are analyzed: the metals have good reflectivity in a large angle and wavelength interval, but some of the radiation is absorbed; the DBRs are absorption-free, but are good reflectors only for specific types of waves. It is shown that a microcavity with only one mirror is able to enhance the emission by more than a factor of 2 (up to 4), because the radiation reflecting on the mirror must be added coherently to the initial radiation (wide-angle interference). When a second mirror is added, light waves can reflect back and forth between the mirrors, and all contributions have to be added coherently. This kind of multiple-beam interference leads to a very strong modulation of the emission, especially when the mirrors are highly reflective. The enhancement factor depends on the emission angle and the wavelength, and the sensitivity to variations of these parameters increases with the effective thickness of the microcavity.

Finally, the use of microcavities in electroluminescent devices of different kinds was investigated. Inorganic thin film electroluminescent devices have an optical thickness of several wavelengths, and therefore the enhancement of the emission is very sensitive to small variations in angle and wavelength. Thus, the use of microcavities is limited to applications where the phosphor material has a narrow spectral emission line, which is collected in a small solid angle. For the 548-nm emission line of the ZnS:Ho phosphor with FWHM 8 nm, an enhancement of a factor of 9 can be obtained for the total emission in a $10°$ cone in air. Light-emitting diodes based on quantum wells generate light in a very narrow layer and can profit from the wide-angle interference effect if the quantum well is placed in an antinode of the cavity. With this technology it is not easy to make distributed Bragg reflectors with high contrast of refractive indices, and this increases the effective thickness of the microcavity. Finally, for organic LEDs the effective thickness of the microcavity can be quite small: of the order of half a wavelength. If a thick and a semitransparent metallic mirror are used in the microcavity, then a very broad enhancement can be obtained, both as a function of wavelength and as a function of emission angle. For an Alq_3-based OLED, the total emission into air integrated over the entire spectrum can be enhanced by a factor of 12. Such OLEDs may prove very useful for display applications.

Acknowledgments

I would like to acknowledge the people who supported or contributed to the microcavity research: P. De Visschere, J. Van den Bossche, D. Corlatan, B. Soenen, G. Stuyven, and H. Pauwels of the ELIS Department at the University of Ghent, and further, E. Weber, G. Mueller, R. Mueller-Mach, D. Fork, and S. So. The University of Ghent and the Fund for Scientific Research, Flanders (Belgium), are acknowledged for financial support.

References

Baba, T., T. Hamano, F. Koyama, and K. Iga (1991). Spontaneous emission factor of a microcavity DBR surface-emitting laser. *IEEE J. Quantum Electron.* **27**, 1347–1358.

Bjork, G., and S. Machida, Y. Yamamoto, and K. Igeta. (1991). Modification of spontaneous emission rate in planar dielectric microcavity structures. *Phys. Rev. A* **44**, 669–681.

Brorson, S., and P. Skovgaard. (1996). Optical mode density and spontaneous emission in microcavities. Chapter 2 in *Optical Processes in Microcavities* (R. K. Chang and A. J. Campillo, eds.), World Scientific, Singapore.

Burrows, P. E., Z. Shen, and S. R. Forrest. (1997). Saturated full color stacked Organic light emitting devices. In *Conf. Record of the 17th IDRC* (J. Morreale, ed.), pp. 318–321. Society for Inf. Display, Santa Ana, CA.

Chance, R. R., A. Prock, and R. Silbey. (1974). Lifetime of an emitting molecule near a partially reflecting surface. *J. Chem. Phys.* **60**, 2744–2748.

Cimrová, V., and D. Neher. (1996). Microcavity effects in single-layer light emitting devices based on poly(p-phenylene vinylene). *J. Appl. Phys.* **79**, pp. 3299–3306.

De Neve, H., J. Blondelle, R. Baets, P. Demeester, P. Van Daele, and G. Borghs. (1996). Resonant cavity LED's. In *Microcavities and Photonic Bandgaps: Physics and Applications* (J. Rarity and C. Weisbuch, eds.), NATO ASI Series E: Applied Sciences, Vol. 324, pp. 333–342. Kluwer Academic Publishers, Dordrecht.

De Martini, F., M. Marrocco, P. Mataloni, L. Crescentini, and R. Loudon. (1991). Spontaneous emission in the optical microscopic cavity. *Phys. Rev. A* **43**, 2480–2497.

Deppe, D. G., and C. Lei. (1991). Spontaneous emission from a dipole in a semiconductor microcavity. *J. Appl. Phys.* **70**, 3443–3448.

Dodabalapur, A., L. J. Rothberg, T. M. Miller, and E. W. Kwock. (1994). Color variation with electroluminescent organic semiconductors in multimode resonant cavities. *Appl. Phys. Lett.* **64**, 2486–2488.

Dodabalapur, A., L. J. Rothberg, R. H. Jordan, T. M. Miller, R. E. Slusher, and J. M. Philips (1996). Physics and applications of organic microcavity light emitting diodes. *J. Appl. Phys.* **80**, 6954–6964

Drexhage, K. H., H. Kuhn, and F. P. Schäfer. (1968). Variation of the fluorescence decay time of a molecule in front of a mirror. *Berichte Bunsengesellschaft Phys. Chemie* **72**, 329.

Fisher, T., D. Lidzey, M. Pate, M. Weaver, D. Whittaker, M. Skolnick, and D. Bradley. (1995). Electroluminescence from a conjugated polymer microcavity structure. *Appl. Phys. Lett.* **67**, 1355.

Grüner, J., F. Cacialli, and R. H. Friend. (1996). Emission enhancement in single-layer conjugated polymer microcavities. *J. Appl Phys.* **80**, 207–215.

Haroche, S., and D. Kleppner. (1989). Cavity quantum electrodynamics. *Phys. Today*, pp. 24–30.
Ho, S.-T., D. Y. Chu, J.-P. Zhang, and M.-K. Chin. (1996). Dielectric photonic wells and wires and spontaneous emission coupling efficiency of microdisk and photonic wire semiconductor lasers. In *Optical Processes in Microcavities* (R. K. Chang, A. J. Campillo, eds.), Chapter 10. World Scientific, Singapore.
Hunt, N. E., E. F. Schubert, R. Logan, and G. Zydzik. (1992). Enhanced spectral power density and reduced linewidth at 1.3 μm in an InGaAsP quantum well resonant-cavity light-emitting diode. *Appl. Phys. Lett.* **61**, 2287–2289
Hunt, N. E. J., E. F. Schubert, R. F. Kopf, D. L. Sivco, A. Y. Cho, and G. Zydzik. (1993). Increased fiber communications bandwidth from a resonant cavity emitting diode emitting at 940 nm. *Appl. Phys. Lett.* **63**, 2600–2602
Jordan, R. H., L. J. Rothberg, A. Dodabalapur, and R. E. Slusher. (1996). Efficiency enhancement of microcavity organic light emitting diodes. *Appl. Phys. Lett.* **69**, 1997–1999
Lemmer, U., R. Hennig, W. Guss, A. Ochse, J. Pommerehne, R. Sander, A. Greiner, R. F. Mahrt, H. Bässler, J. Feldmann, and E. O. Göbel. (1995). Microcavity effects in a spin-coated polymer two-layer system. *Appl. Phys. Lett.* **66**, 1301.
Lidzey, D. G., T. A. Fisher, D. D. C. Bradley, M. A. Poate, M. S. Weaver, D. M. Whittaker, and M. S. Skolnick (1996). Probing the optical mode structure in conjugated polymer microcavities. In *Inorganic and Organic Electroluminescence/EL96* Berlin (R. H. Mauch, H.-E. Gumlich, eds.), pp. 219–222. Wissenschaft und Technik Verlag, Berlin.
Lin, C. C., D. G. Deppe, and C. Lei. (1994). Role of waveguide light emission in planar microcavities. *IEEE J. Quantum Electron.* **30**, pp. 2304–2313.
Lukosz W. (1980). Theory of optical-environment-dependent spontaneous-emission rates for emitters in thin layers. *Phys. Rev. B* **22**, 3030–3038.
Lukosz, W. (1981). Light emission by multipole sources in thin layers. I. Radiation patterns of electric and magnetic dipoles. *J. Opt. Soc. A* **71**, 744–754.
Lukosz, W., and R. E. Kunz. (1977). Light emission by magnetic and electric dipoles close to a plane interface. I. Total radiated power. *J. Opt. Soc. Am.* **67**, 1607–1619.
Lukosz, W., and R. E. Kunz. (1979). Changes in fluorescence lifetimes induced by variation of the radiating molecules' optical environment. *Opt. Commun.* **31**, 42–46.
Marcuse D. (1972). *Light Transmission Optics.* Van Nostrand Reinhold Publishers, New York.
Mauch, R. H., K. A. Neyts, and H.-W. Schock. (1989). Optical behaviour of electroluminescent devices. In *Electroluminescence, Proceedings of the 4th Workshop on Electroluminescence,* Springer Proceedings in Physics 38 (S. Shionoya and H. Kobayashi, eds.), pp. 291–295. Springer-Verlag, Berlin.
Mueller, G. O., R. Mach, E. Alinsog, H. Lee, and D. Harrison. (1996). Microcavity effects in thin film electroluminescence. In *Inorganic and Organic Electroluminescence/EL96* Berlin (R. H. Mauch, H.-E. Gumlich, eds.), pp. 399–402. Wissenschaft und Technik Verlag, Berlin.
Mueller-Mach, R., C. Fouassier, X. W. Fan, G. O. Mueller, G. Z. Zhong, A. Garcia, L. J. Zhao, and J. M. Sun. (1997). Holmium, a multi-color emitter in sulfides. *Extended Abstracts of the 3rd. Int. Conf. on Display Phosphors* in Huntington Beach, pp. 45–48.
Nakayama, Y., Y. Itoh, and A. Kakuta (1993). Organic photo- and electroluminescent devices with double mirrors. *Appl. Phys. Lett.* **63**, 594–595.
Neyts, K. (1996). Cavity effects in thin film phosphors based on ZnS. In *Microcavities and Photonic Bandgaps: Physics and Applications* (J. Rarity and C. Weisbuch, eds.), NATO ASI Series E: Applied Sciences, Vol. 324, pp. 397–406. Kluwer Academic Publishers, Dordrecht.
Neyts, K. (1997). Thin film microcavities for display applications. In *Conf. Record of the 17th IDRC* (J. Morreale, ed.), pp. 421–424. Society for Inf. Display, Santa Ana, CA.

Neyts, K. (1998). The simulation of light emission from thin film microcavities. *J. Opt. Soc. Am. A* **15**, 962–971.

Neyts, K., B. Soenen, and W. Ooghe (1995). Wide-angle interference and cavity effects in thin ZnS Layers doped with Mn. *J. Lumin.* **65**, 221–225.

Neyts, K., G. Mueller, R. Mueller-Mach, D. Corlatan, T. Yu, E. Weber. (1997). Simulation and measurement of microcavity effects in thin phosphor films. *Proc. 3rd Conf. on Display Phosphors*, Los Angeles, pp. 187–190.

Neyts, K., D. Fork, G. B. Anderson, P. De Visschere, and E. R. Weber. (1998). Half-wavelength microcavity with two metal mirrors for organic light emitting diodes. Spring Meeting of the MRS, San Francisco, CA, p. 141.

Rigneault, H., C. Amra, E. Pelletier, F. Flory, M. Cathelinaud, and L. Roux. (1996). Dielectric thin films for microcavity applications. In *Microcavities and Photonic Bandgaps: Physics and Applications* (J. Rarity and C. Weisbuch, eds.), NATO ASI Series E: Applied Sciences, Vol. 324, pp. 427–442. Kluwer Academic Publishers, Dordrecht.

Rogers, T. J., D. G. Deppe, and B. G. Streetman. (1990). Effect of an AlAs/GaAs mirror on the spontaneous emission of an InGaAs-GaAs quantum well. *Appl. Phys. Lett.* **57**, 1858–1860.

Saito, S., T. Tsutsui, M. Era, N. Takada, C. Adachi, Y. Hamada, and T. Wakimoto (1993). Progress in organic multilayer electroluminescent devices. In *Electroluminescent Materials, Devices, and Large-Screen Displays* (E. M. Conwell, M. Stolka, and M. R. Miller, eds.), *Proc. SPIE.* **1910**, 212–221.

Schubert, E. F., Y.-H. Wang, A. Y. Cho, L.-W. Tu, and G. J. Zydnik. (1992). Resonant cavity light-emitting diode. *Appl. Phys. Lett.* **60**, 921–923.

Schulz L. G., and F. R. Tangherlini. (1954). The optical constants of silver, copper, and aluminium. I and II. *J. Opt. Soc. Am.* **44**, 357–368.

So, S. K., W. K. Choi, L. M. Leung, and K. Neyts. (1999). Interference effects in organic bilayer light-emitting diodes. *Appl. Phys. Lett.* **74**, pp. 1939–1941.

Takada, N., T. Tsustui, and S. Saito. (1993). Control of emission characteristics in organic thin-film electroluminescent diodes using an optical-microcavity structure. *Appl. Phys. Lett.* **63**, 2032–2034.

Tessler, N., G. J. Denton, and R. H. Friend. (1996). Lasing from conjugated-polymer microcavities. *Nature* **382**, 695–679.

Tsutsui, T., N. Takada, and S. Saito. (1994). Sharply directed emission in organic electroluminescent diodes with an optical-microcavity structure. *Appl. Phys. Lett.* **65**, 1868–1870.

Vlasenko, N. A., S. A. Zynyo, and A. Pukhlii. (1970). Investigation of interference effects in thin electroluminescent ZnS-Mn films. *Optics Spectrosc.* **28**, 66–71.

Yablonovitch, E., T. J. Gmitter, and R. Bhat. (1988). Inhibited and enhanced spontaneous emission from optically thin AlGaAs/GaAs double heterostructure. *Phys. Rev. Lett.* **61**, 2546–2549.

Yokoyama, H. (1992). Physics and device applications of optical microcavities. *Science* **256**, 66–70.

Index

A

Active matrix TFEL devices, 89–91, 110
Applications
  microcavity effects, 215–230
  OLED devices, 22–23
  TFEL devices, 80–93

C

CaF_2 films, 151–152
Carrier confinement in PLEDs, 11
Carrier injection efficiency in PLEDs, 12–13
Carrier transport properties
  conjugated polymers, 3
  polymer thin films, 13–16, 22
  TFEL devices, 45–55, 73–77
CaS films, 41, 43, 145–147
Cathodoluminescence, 101
"Color-by-white" displays, 83–87, 152–153
Color emissions and doping, 39–41, 84–87, 113–116
Current-voltage characteristics
  PLEDs, 13–16
  TFEL devices, 46–50

D

Distributed Bragg reflectors, 200–202, 218–219, 220–221, 223–226
Doping
  concentrations, 121–122, 127–129
  and conduction properties, 4
  TFELs, 39–44, 62–65, 68–71, 84–87, 113–116, 121–122, 127–129
Dye diffusion process, 9

E

Edge-emitting TFEL, 56–58, 92–93
Electrode materials
  in PLEDs, 9–10, 12–13
  in TFELs, 154–157
Electron energy distribution, 75–77

F

Field quenching, 62–65
Film deposition methods, 6–7, 111–112, 117–118, 131
Flat panel displays
  PLEDs, 22–23
  TFELs, 30–41
Flexible PLED display panels, 11
Fowler-Nordheim theory, 12
Fractional sublimation, 5

H

High-field electronic transport, 73–77

I

Ink-jet printing, 8, 9–10
Inorganic *vs.* organic semiconductors, 17–22
Insulator materials in TFEL devices, 157–164

L

Light-emitting diodes (LEDs). *See also* Polymeric light-emitting diodes (PLEDs)

Light-emitting diodes (*Continued*)
  microcavity effects, 220–226
Light extraction in TFELs
  scattering to increase brightness, 37–39, 55–59
  using microcavity effects, 215–220
Light guiding in TFEL devices, 35–37, 55–59, 215–220
Luminous efficiency
  PLEDs, 16–17
  TFEL devices, 68–72, 88–89, 215–220

M

Masked dye diffusion process, 9
Materials in TFEL devices, 107–164
Measurement methods for TFELs
  brightness, 93–94
  cathodoluminescence, 101
  drive frequency, 95–96
  input power, 96–97
  photoluminescence, 99–100
  Rutherford backscattering, 100
  secondary ion mass spectroscopy (SIMS), 100–101
  space charge, 100
  spectral radiance, 94–95
Microcavity effects
  application to LEDs, 220–226
  application to OLEDs, 226–230
  application to TFELs, 215–220
  example structures, 200–214
  modes and effective thickness, 195–197
  theoretical description, 185–200
Monochrome displays, 81–82
Multicolor displays, 82–87, 109–110, 126–127, 152–153

O

Organic light-emitting diodes (OLEDs)
  microcavity effects, 226–230
Organic *vs.* inorganic semiconductors, 17–22

P

Patterning methods, 8–10

Planar light sources, 91
Polymeric light-emitting diodes (PLEDs), 1–23
  carrier confinement, 11
  carrier injection efficiency, 12–13
  carrier transport properties, 3, 13–16, 22
  chemical stability, 19
  color tuning with side chains, 3–4
  vs. crystalline inorganic LEDs, 18–19
  defects, 22
  device fabrication, 6–10
  doping, 4
  electrode materials, 9–10, 12–13
  film deposition methods, 6–7
  flexibility, 11, 17–18
  future trends, 22–23
  heterostructure, 11, 17
  historical overview, 1–2
  instability of organic materials, 19
  luminance efficiency, 16–17
  market potential, 22–23
  mechanical stability, 19, 22
  purity of polymers, 5
  single-polymer-layer, 10–11, 17
  stretched polymers, 4
  structures, 10–11
  thin film patterning, 8–10
  transparency, 18
Polymer-metal interfaces, 12–13
PPV (poly-(*para*-phenylenevinylene))
  carrier transport properties, 13–16
  film deposition, 6–7
  PLED structures, 10–11
Purification of organic compounds, 4–5

R

Reliability
  chemical stability of PLEDs, 19
  instability of organic materials, 19
  mechanical stability of PLEDs, 19, 22
  TFEL devices and insulator design, 162–163
Rutherford backscattering, 100

S

Scattering to increase brightness, 37–39

INDEX

Screen printing, 112
Secondary ion mass spectroscopy (SIMS), 100–101
Space charge, 100
Space-charge-limited conduction
　trap-free, in PLEDs, 13–15
　trap-limited, in PLEDs, 16
Spin-coating method, 6, 7
SrS:Ag,Cu films, 41, 138–139
SrS:Ce films, 41
　codoping, 131–132, 135
　in "color-by-white" applications, 136
　deposition methods, 131
　emission spectra, 131
　enhancing blue emissions, 132–135
SrS:Cu films, 136–138
SrS films, 43, 54, 55
SrS:Pr films, 146–147
Substrate materials for TFEL devices, 152–154

T

Thin film electroluminescence (TFEL), 27–101. *See also specific film materials*
　active matrix, 89–91, 110
　applications, 80–93
　brightness vs. voltage characteristics, 31–32
　color emissions and doping, 39–41, 84–87, 113–116
　device design, 77–80
　display panel design, 30–31, 35
　doping, 39–44, 62–65, 68–71, 84–87, 113–116, 121–122, 127–129
　edge-emitting, 56–58, 92–93
　efficiency factorization, 68–71
　electrical properties, 45–55
　electron energy distribution, 75–77
　emission mechanism, 44–45, 60–65
　field quenching, 62–65
　historical overview, 27–30
　insulator materials, 157–164
　interference effects, 58–59, 215–220
　light guiding, 35–37, 55–59
　materials, 41, 43–44, 107–164
　microcavity effects, 215–220
　monochrome displays, 81–82
　multicolor displays, 82–87, 109–110, 126–127, 152–153
　phosphor requirements, 112–117
　planar light sources, 91
　polarity "memory" effect, 33, 35
　power losses, 88–89
　scattering to increase brightness, 37–39, 56
　stability, 124–126
　structure, 30–31, 108–111
　substrate materials, 154–155
　trailing-edge current, 53–54, 72
　viewing angle, 39
　Zener diode model, 46–50
Thin film patterning, 8–10
Thiogallate films, 41, 43, 143–144
Trailing-edge current in TFEL devices, 53–54

V

Vapor deposition methods, 6–7, 118

Z

ZnF_2 films, 150–151
ZnS:Mn films
　aging behavior, 124–125, 126
　in color devices, 126–127
　deposition methods, 112, 117–118
　doping concentration, 121–122
　efficiency, 69–71, 121–122
　electroluminescence spectra, 39–40, 119–121
　emission mechanism, 62, 121
　historical overview, 117
　historical perspective, 29
　light guiding, 55
　stability, 125–126
　structure and properties, 113–116, 118–119
ZnS:RE films, 139–142
ZnS:Tb films
　doping concentration, 127–129
　emission mechanism, 127, 129
　historical overview, 127
　stability, 130
Zone melting, 4–5

Contents of Volumes in This Series

Volume 1 **Physics of III–V Compounds**

C. *Hilsum*, Some Key Features of III–V Compounds
F. *Bassani*, Methods of Band Calculations Applicable to III–V Compounds
E. O. *Kane*, The k-p Method
V. L. *Bonch-Bruevich*, Effect of Heavy Doping on the Semiconductor Band Structure
D. *Long*, Energy Band Structures of Mixed Crystals of III–V Compounds
L. M. *Roth and P. N. Argyres*, Magnetic Quantum Effects
S. M. *Puri and T. H. Geballe*, Thermomagnetic Effects in the Quantum Region
W. M. *Becker*, Band Characteristics near Principal Minima from Magnetoresistance
E. H. *Putley*, Freeze-Out Effects, Hot Electron Effects, and Submillimeter Photoconductivity in InSb
H. *Weiss*, Magnetoresistance
B. *Ancker-Johnson*, Plasma in Semiconductors and Semimetals

Volume 2 **Physics of III–V Compounds**

M. G. *Holland*, Thermal Conductivity
S. I. *Novkova*, Thermal Expansion
U. *Piesbergen*, Heat Capacity and Debye Temperatures
G. *Giesecke*, Lattice Constants
J. R. *Drabble*, Elastic Properties
A. U. *Mac Rae and G. W. Gobeli*, Low Energy Electron Diffraction Studies
R. *Lee Mieher*, Nuclear Magnetic Resonance
B. *Goldstein*, Electron Paramagnetic Resonance
T. S. *Moss*, Photoconduction in III–V Compounds
E. *Antoncik and J. Tauc*, Quantum Efficiency of the Internal Photoelectric Effect in InSb
G. W. *Gobeli and I. G. Allen*, Photoelectric Threshold and Work Function
P. S. *Pershan*, Nonlinear Optics in III–V Compounds
M. *Gershenzon*, Radiative Recombination in the III–V Compounds
F. *Stern*, Stimulated Emission in Semiconductors

Volume 3 Optical of Properties III–V Compounds

M. Hass, Lattice Reflection
W. G. Spitzer, Multiphonon Lattice Absorption
D. L. Stierwalt and R. F. Potter, Emittance Studies
H. R. Philipp and H. Ehrenveich, Ultraviolet Optical Properties
M. Cardona, Optical Absorption above the Fundamental Edge
E. J. Johnson, Absorption near the Fundamental Edge
J. O. Dimmock, Introduction to the Theory of Exciton States in Semiconductors
B. Lax and J. G. Mavroides, Interband Magnetooptical Effects
H. Y. Fan, Effects of Free Carries on Optical Properties
E. D. Palik and G. B. Wright, Free-Carrier Magnetooptical Effects
R. H. Bube, Photoelectronic Analysis
B. O. Seraphin and H. E. Bennett, Optical Constants

Volume 4 Physics of III–V Compounds

N. A. Goryunova, A. S. Borschevskii, and D. N. Tretiakov, Hardness
N. N. Sirota, Heats of Formation and Temperatures and Heats of Fusion of Compounds $A^{III}B^{V}$
D. L. Kendall, Diffusion
A. G. Chynoweth, Charge Multiplication Phenomena
R. W. Keyes, The Effects of Hydrostatic Pressure on the Properties of III–V Semiconductors
L. W. Aukerman, Radiation Effects
N. A. Goryunova, F. P. Kesamanly, and D. N. Nasledov, Phenomena in Solid Solutions
R. T. Bate, Electrical Properties of Nonuniform Crystals

Volume 5 Infrared Detectors

H. Levinstein, Characterization of Infrared Detectors
P. W. Kruse, Indium Antimonide Photoconductive and Photoelectromagnetic Detectors
M. B. Prince, Narrowband Self-Filtering Detectors
I. Melngalis and T. C. Harman, Single-Crystal Lead-Tin Chalcogenides
D. Long and J. L. Schmidt, Mercury-Cadmium Telluride and Closely Related Alloys
E. H. Putley, The Pyroelectric Detector
N. B. Stevens, Radiation Thermopiles
R. J. Keyes and T. M. Quist, Low Level Coherent and Incoherent Detection in the Infrared
M. C. Teich, Coherent Detection in the Infrared
F. R. Arams, E. W. Sard, B. J. Peyton, and F. P. Pace, Infrared Heterodyne Detection with Gigahertz IF Response
H. S. Sommers, Jr., Macrowave-Based Photoconductive Detector
R. Sehr and R. Zuleeg, Imaging and Display

Volume 6 Injection Phenomena

M. A. Lampert and R. B. Schilling, Current Injection in Solids: The Regional Approximation Method
R. Williams, Injection by Internal Photoemission
A. M. Barnett, Current Filament Formation

R. Baron and J. W. Mayer, Double Injection in Semiconductors
W. Ruppel, The Photoconductor-Metal Contact

Volume 7 Application and Devices
Part A

J. A. Copeland and S. Knight, Applications Utilizing Bulk Negative Resistance
F. A. Padovani, The Voltage-Current Characteristics of Metal-Semiconductor Contacts
P. L. Hower, W. W. Hooper, B. R. Cairns, R. D. Fairman, and D. A. Tremere, The GaAs Field-Effect Transistor
M. H. White, MOS Transistors
G. R. Antell, Gallium Arsenide Transistors
T. L. Tansley, Heterojunction Properties

Part B

T. Misawa, IMPATT Diodes
H. C. Okean, Tunnel Diodes
R. B. Campbell and Hung-Chi Chang, Silicon Junction Carbide Devices
R. E. Enstrom, H. Kressel, and L. Krassner, High-Temperature Power Rectifiers of $GaAs_{1-x}P_x$

Volume 8 Transport and Optical Phenomena

R. J. Stirn, Band Structure and Galvanomagnetic Effects in III–V Compounds with Indirect Band Gaps
R. W. Ure, Jr., Thermoelectric Effects in III–V Compounds
H. Piller, Faraday Rotation
H. Barry Bebb and E. W. Williams, Photoluminescence I: Theory
E. W. Williams and H. Barry Bebb, Photoluminescence II: Gallium Arsenide

Volume 9 Modulation Techniques

B. O. Seraphin, Electroreflectance
R. L. Aggarwal, Modulated Interband Magnetooptics
D. F. Blossey and Paul Handler, Electroabsorption
B. Batz, Thermal and Wavelength Modulation Spectroscopy
I. Balslev, Piezopptical Effects
D. E. Aspnes and N. Bottka, Electric-Field Effects on the Dielectric Function of Semiconductors and Insulators

Volume 10 Transport Phenomena

R. L. Rhode, Low-Field Electron Transport
J. D. Wiley, Mobility of Holes in III–V Compounds
C. M. Wolfe and G. E. Stillman, Apparent Mobility Enhancement in Inhomogeneous Crystals
R. L. Petersen, The Magnetophonon Effect

Volume 11 Solar Cells

H. J. *Hovel*, Introduction; Carrier Collection, Spectral Response, and Photocurrent; Solar Cell Electrical Characteristics; Efficiency; Thickness; Other Solar Cell Devices; Radiation Effects; Temperature and Intensity; Solar Cell Technology

Volume 12 Infrared Detectors (II)

W. L. *Eiseman, J. D. Merriam,* and *R. F. Potter,* Operational Characteristics of Infrared Photodetectors
P. R. *Bratt,* Impurity Germanium and Silicon Infrared Detectors
E. H. *Putley,* InSb Submillimeter Photoconductive Detectors
G. E. *Stillman, C. M. Wolfe,* and *J. O. Dimmock,* Far-Infrared Photoconductivity in High Purity GaAs
G. E. *Stillman* and *C. M. Wolfe,* Avalanche Photodiodes
P. L. *Richards,* The Josephson Junction as a Detector of Microwave and Far-Infrared Radiation
E. H. *Putley,* The Pyroelectric Detector — An Update

Volume 13 Cadmium Telluride

K. *Zanio,* Materials Preparations; Physics; Defects; Applications

Volume 14 Lasers, Junctions, Transport

N. *Holonyak, Jr.* and *M. H. Lee,* Photopumped III–V Semiconductor Lasers
H. *Kressel* and *J. K. Butler,* Heterojunction Laser Diodes
A *Van der Ziel,* Space-Charge-Limited Solid-State Diodes
P. J. *Price,* Monte Carlo Calculation of Electron Transport in Solids

Volume 15 Contacts, Junctions, Emitters

B. L. *Sharma,* Ohmic Contacts to III–V Compounds Semiconductors
A. *Nussbaum,* The Theory of Semiconducting Junctions
J. S. *Escher,* NEA Semiconductor Photoemitters

Volume 16 Defects, (HgCd)Se, (HgCd)Te

H. *Kressel,* The Effect of Crystal Defects on Optoelectronic Devices
C. R. *Whitsett, J. G. Broerman,* and *C. J. Summers,* Crystal Growth and Properties of $Hg_{1-x}Cd_xSe$ alloys
M. H. *Weiler,* Magnetooptical Properties of $Hg_{1-x}Cd_xTe$ Alloys
P. W. *Kruse* and *J. G. Ready,* Nonlinear Optical Effects in $Hg_{1-x}Cd_xTe$

Volume 17 CW Processing of Silicon and Other Semiconductors

J. F. *Gibbons,* Beam Processing of Silicon
A. *Lietoila, R. B. Gold, J. F. Gibbons,* and *L. A. Christel,* Temperature Distributions and Solid Phase Reaction Rates Produced by Scanning CW Beams

A. Leitoila and J. F. Gibbons, Applications of CW Beam Processing to Ion Implanted Crystalline Silicon
N. M. Johnson, Electronic Defects in CW Transient Thermal Processed Silicon
K. F. Lee, T. J. Stultz, and J. F. Gibbons, Beam Recrystallized Polycrystalline Silicon: Properties, Applications, and Techniques
T. Shibata, A. Wakita, T. W. Sigmon, and J. F. Gibbons, Metal-Silicon Reactions and Silicide
Y. I. Nissim and J. F. Gibbons, CW Beam Processing of Gallium Arsenide

Volume 18   Mercury Cadmium Telluride

P. W. Kruse, The Emergence of $(Hg_{1-x}Cd_x)Te$ as a Modern Infrared Sensitive Material
H. E. Hirsch, S. C. Liang, and A. G. White, Preparation of High-Purity Cadmium, Mercury, and Tellurium
W. F. H. Micklethwaite, The Crystal Growth of Cadmium Mercury Telluride
P. E. Petersen, Auger Recombination in Mercury Cadmium Telluride
R. M. Broudy and V. J. Mazurczyck, (HgCd)Te Photoconductive Detectors
M. B. Reine, A. K. Soad, and T. J. Tredwell, Photovoltaic Infrared Detectors
M. A. Kinch, Metal-Insulator-Semiconductor Infrared Detectors

Volume 19   Deep Levels, GaAs, Alloys, Photochemistry

G. F. Neumark and K. Kosai, Deep Levels in Wide Band-Gap III–V Semiconductors
D. C. Look, The Electrical and Photoelectronic Properties of Semi-Insulating GaAs
R. F. Brebrick, Ching-Hua Su, and Pok-Kai Liao, Associated Solution Model for Ga-In-Sb and Hg-Cd-Te
Y. Ya. Gurevich and Y. V. Pleskon, Photoelectrochemistry of Semiconductors

Volume 20   Semi-Insulating GaAs

R. N. Thomas, H. M. Hobgood, G. W. Eldridge, D. L. Barrett, T. T. Braggins, L. B. Ta, and S. K. Wang, High-Purity LEC Growth and Direct Implantation of GaAs for Monolithic Microwave Circuits
C. A. Stolte, Ion Implantation and Materials for GaAs Integrated Circuits
C. G. Kirkpatrick, R. T. Chen, D. E. Holmes, P. M. Asbeck, K. R. Elliott, R. D. Fairman, and J. R. Oliver, LEC GaAs for Integrated Circuit Applications
J. S. Blakemore and S. Rahimi, Models for Mid-Gap Centers in Gallium Arsenide

Volume 21   Hydrogenated Amorphous Silicon
Part A

J. I. Pankove, Introduction
M. Hirose, Glow Discharge; Chemical Vapor Deposition
Y. Uchida, di Glow Discharge
T. D. Moustakas, Sputtering
I. Yamada, Ionized-Cluster Beam Deposition
B. A. Scott, Homogeneous Chemical Vapor Deposition

F. J. Kampas, Chemical Reactions in Plasma Deposition
P. A. Longeway, Plasma Kinetics
H. A. Weakliem, Diagnostics of Silane Glow Discharges Using Probes and Mass Spectroscopy
L. Gluttman, Relation between the Atomic and the Electronic Structures
A. Chenevas-Paule, Experiment Determination of Structure
S. Minomura, Pressure Effects on the Local Atomic Structure
D. Adler, Defects and Density of Localized States

Part B

J. I. Pankove, Introduction
G. D. Cody, The Optical Absorption Edge of a-Si:H
N. M. Amer and W. B. Jackson, Optical Properties of Defect States in a-Si:H
P. J. Zanzucchi, The Vibrational Spectra of a-Si:H
Y. Hamakawa, Electroreflectance and Electroabsorption
J. S. Lannin, Raman Scattering of Amorphous Si, Ge, and Their Alloys
R. A. Street, Luminescence in a-Si:H
R. S. Crandall, Photoconductivity
J. Tauc, Time-Resolved Spectroscopy of Electronic Relaxation Processes
P. E. Vanier, IR-Induced Quenching and Enhancement of Photoconductivity and Photo luminescence
H. Schade, Irradiation-Induced Metastable Effects
L. Ley, Photoelectron Emission Studies

Part C

J. I. Pankove, Introduction
J. D. Cohen, Density of States from Junction Measurements in Hydrogenated Amorphous Silicon
P. C. Taylor, Magnetic Resonance Measurements in a-Si:H
K. Morigaki, Optically Detected Magnetic Resonance
J. Dresner, Carrier Mobility in a-Si:H
T. Tiedje, Information about band-Tail States from Time-of-Flight Experiments
A. R. Moore, Diffusion Length in Undoped a-Si:H
W. Beyer and J. Overhof, Doping Effects in a-Si:H
H. Fritzche, Electronic Properties of Surfaces in a-Si:H
C. R. Wronski, The Staebler-Wronski Effect
R. J. Nemanich, Schottky Barriers on a-Si:H
B. Abeles and T. Tiedje, Amorphous Semiconductor Superlattices

Part D

J. I. Pankove, Introduction
D. E. Carlson, Solar Cells
G. A. Swartz, Closed-Form Solution of I–V Characteristic for a a-Si:H Solar Cells
I. Shimizu, Electrophotography
S. Ishioka, Image Pickup Tubes

P. G. LeComber and W. E. Spear, The Development of the a-Si:H Field-Effect Transistor and Its Possible Applications
D. G. Ast, a-Si:H FET-Addressed LCD Panel
S. Kaneko, Solid-State Image Sensor
M. Matsumura, Charge-Coupled Devices
M. A. Bosch, Optical Recording
A. D'Amico and G. Fortunato, Ambient Sensors
H. Kukimoto, Amorphous Light-Emitting Devices
R. J. Phelan, Jr., Fast Detectors and Modulators
J. I. Pankove, Hybrid Structures
P. G. LeComber, A. E. Owen, W. E. Spear, J. Hajto, and W. K. Choi, Electronic Switching in Amorphous Silicon Junction Devices

Volume 22 Lightwave Communications Technology
Part A

K. Nakajima, The Liquid-Phase Epitaxial Growth of InGaAsP
W. T. Tsang, Molecular Beam Epitaxy for III–V Compound Semiconductors
G. B. Stringfellow, Organometallic Vapor-Phase Epitaxial Growth of III–V Semiconductors
G. Beuchet, Halide and Chloride Transport Vapor-Phase Deposition of InGaAsP and GaAs
M. Razeghi, Low-Pressure Metallo-Organic Chemical Vapor Deposition of $Ga_xIn_{1-x}As P_{1-y}$ Alloys
P. M. Petroff, Defects in III–V Compound Semiconductors

Part B

J. P. van der Ziel, Mode Locking of Semiconductor Lasers
K. Y. Lau and A. Yariv, High-Frequency Current Modulation of Semiconductor Injection Lasers
C. H. Henry, Special Properties of Semiconductor Lasers
Y. Suematsu, K. Kishino, S. Arai, and F. Koyama, Dynamic Single-Mode Semiconductor Lasers with a Distributed Reflector
W. T. Tsang, The Cleaved-Coupled-Cavity (C^3) Laser

Part C

R. J. Nelson and N. K. Dutta, Review of InGaAsP InP Laser Structures and Comparison of Their Performance
N. Chinone and M. Nakamura, Mode-Stabilized Semiconductor Lasers for 0.7–0.8- and 1.1–1.6-μm Regions
Y. Horikoshi, Semiconductor Lasers with Wavelengths Exceeding 2 μm
B. A. Dean and M. Dixon, The Functional Reliability of Semiconductor Lasers as Optical Transmitters
R. H. Saul, T. P. Lee, and C. A. Burus, Light-Emitting Device Design
C. L. Zipfel, Light-Emitting Diode-Reliability
T. P. Lee and T. Li, LED-Based Multimode Lightwave Systems
K. Ogawa, Semiconductor Noise-Mode Partition Noise

Part D

F. *Capasso*, The Physics of Avalanche Photodiodes
T. P. *Pearsall and M. A. Pollack*, Compound Semiconductor Photodiodes
T. *Kaneda*, Silicon and Germanium Avalanche Photodiodes
S. R. *Forrest*, Sensitivity of Avalanche Photodetector Receivers for High-Bit-Rate Long-Wavelength Optical Communication Systems
J. C. *Campbell*, Phototransistors for Lightwave Communications

Part E

S. *Wang*, Principles and Characteristics of Integrable Active and Passive Optical Devices
S. *Margalit and A. Yariv*, Integrated Electronic and Photonic Devices
T. *Mukai, Y. Yamamoto, and T. Kimura*, Optical Amplification by Semiconductor Lasers

Volume 23 Pulsed Laser Processing of Semiconductors

R. F. *Wood, C. W. White, and R. T. Young*, Laser Processing of Semiconductors: An Overview
C. W. *White*, Segregation, Solute Trapping, and Supersaturated Alloys
G. E. *Jellison, Jr.*, Optical and Electrical Properties of Pulsed Laser-Annealed Silicon
R. F. *Wood and G. E. Jellison, Jr.*, Melting Model of Pulsed Laser Processing
R. F. *Wood and F. W. Young, Jr.*, Nonequilibrium Solidification Following Pulsed Laser Melting
D. H. *Lowndes and G. E. Jellison, Jr.*, Time-Resolved Measurement During Pulsed Laser Irradiation of Silicon
D. M. *Zebner*, Surface Studies of Pulsed Laser Irradiated Semiconductors
D. H. *Lowndes*, Pulsed Beam Processing of Gallium Arsenide
R. B. *James*, Pulsed CO_2 Laser Annealing of Semiconductors
R. T. *Young and R. F. Wood*, Applications of Pulsed Laser Processing

Volume 24 Applications of Multiquantum Wells, Selective Doping, and Superlattices

C. *Weisbuch*, Fundamental Properties of III–V Semiconductor Two-Dimensional Quantized Structures: The Basis for Optical and Electronic Device Applications
H. *Morkoc and H. Unlu*, Factors Affecting the Performance of (Al,Ga)As/GaAs and (Al,Ga)As/InGaAs Modulation-Doped Field-Effect Transistors: Microwave and Digital Applications
N. T. *Linh*, Two-Dimensional Electron Gas FETs: Microwave Applications
M. *Abe et al.*, Ultra-High-Speed HEMT Integrated Circuits
D. S. *Chemla, D. A. B. Miller, and P. W. Smith*, Nonlinear Optical Properties of Multiple Quantum Well Structures for Optical Signal Processing
F. *Capasso*, Graded-Gap and Superlattice Devices by Band-Gap Engineering
W. T. *Tsang*, Quantum Confinement Heterostructure Semiconductor Lasers
G. C. *Osbourn et al.*, Principles and Applications of Semiconductor Strained-Layer Superlattices

Volume 25 Diluted Magnetic Semiconductors

W. Giriat and J. K. Furdyna, Crystal Structure, Composition, and Materials Preparation of Diluted Magnetic Semiconductors
W. M. Becker, Band Structure and Optical Properties of Wide-Gap $A_{I-x}^{II}Mn_xB_{IV}$ Alloys at Zero Magnetic Field
S. Oseroff and P. H. Keesom, Magnetic Properties: Macroscopic Studies
T. Giebultowicz and T. M. Holden, Neutron Scattering Studies of the Magnetic Structure and Dynamics of Diluted Magnetic Semiconductors
J. Kossut, Band Structure and Quantum Transport Phenomena in Narrow-Gap Diluted Magnetic Semiconductors
C. Riquaux, Magnetooptical Properties of Large-Gap Diluted Magnetic Semiconductors
J. A. Gaj, Magnetooptical Properties of Large-Gap Diluted Magnetic Semiconductors
J. Mycielski, Shallow Acceptors in Diluted Magnetic Semiconductors: Splitting, Boil-off, Giant Negative Magnetoresistance
A. K. Ramadas and R. Rodriquez, Raman Scattering in Diluted Magnetic Semiconductors
P. A. Wolff, Theory of Bound Magnetic Polarons in Semimagnetic Semiconductors

Volume 26 III–V Compound Semiconductors and Semiconductor Properties of Superionic Materials

Z. Yuanxi, III–V Compounds
H. V. Winston, A. T. Hunter, H. Kimura, and R. E. Lee, InAs-Alloyed GaAs Substrates for Direct Implantation
P. K. Bhattacharya and S. Dhar, Deep Levels in III–V Compound Semiconductors Grown by MBE
Y. Ya. Gurevich and A. K. Ivanov-Shits, Semiconductor Properties of Supersonic Materials

Volume 27 High Conducting Quasi-One-Dimensional Organic Crystals

E. M. Conwell, Introduction to Highly Conducting Quasi-One-Dimensional Organic Crystals
I. A. Howard, A Reference Guide to the Conducting Quasi-One-Dimensional Organic Molecular Crystals
J. P. Pouquet, Structural Instabilities
E. M. Conwell, Transport Properties
C. S. Jacobsen, Optical Properties
J. C. Scott, Magnetic Properties
L. Zuppiroli, Irradiation Effects: Perfect Crystals and Real Crystals

Volume 28 Measurement of High-Speed Signals in Solid State Devices

J. Frey and D. Ioannou, Materials and Devices for High-Speed and Optoelectronic Applications
H. Schumacher and E. Strid, Electronic Wafer Probing Techniques
D. H. Auston, Picosecond Photoconductivity: High-Speed Measurements of Devices and Materials
J. A. Valdmanis, Electro-Optic Measurement Techniques for Picosecond Materials, Devices, and Integrated Circuits.
J. M. Wiesenfeld and R. K. Jain, Direct Optical Probing of Integrated Circuits and High-Speed Devices
G. Plows, Electron-Beam Probing
A. M. Weiner and R. B. Marcus, Photoemissive Probing

Volume 29 Very High Speed Integrated Circuits: Gallium Arsenide LSI

M. Kuzuhara and T. Nazaki, Active Layer Formation by Ion Implantation
H. Hasimoto, Focused Ion Beam Implantation Technology
T. Nozaki and A. Higashisaka, Device Fabrication Process Technology
M. Ino and T. Takada, GaAs LSI Circuit Design
M. Hirayama, M. Ohmori, and K. Yamasaki, GaAs LSI Fabrication and Performance

Volume 30 Very High Speed Integrated Circuits: Heterostructure

H. Watanabe, T. Mizutani, and A. Usui, Fundamentals of Epitaxial Growth and Atomic Layer Epitaxy
S. Hiyamizu, Characteristics of Two-Dimensional Electron Gas in III–V Compound Heterostructures Grown by MBE
T. Nakanisi, Metalorganic Vapor Phase Epitaxy for High-Quality Active Layers
T. Nimura, High Electron Mobility Transistor and LSI Applications
T. Sugeta and T. Ishibashi, Hetero-Bipolar Transistor and LSI Application
H. Matsueda, T. Tanaka, and M. Nakamura, Optoelectronic Integrated Circuits

Volume 31 Indium Phosphide: Crystal Growth and Characterization

J. P. Farges, Growth of Discoloration-free InP
M. J. McCollum and G. E. Stillman, High Purity InP Grown by Hydride Vapor Phase Epitaxy
T. Inada and T. Fukuda, Direct Synthesis and Growth of Indium Phosphide by the Liquid Phosphorous Encapsulated Czochralski Method
O. Oda, K. Katagiri, K. Shinohara, S. Katsura, Y. Takahashi, K. Kainosho, K. Kohiro, and R. Hirano, InP Crystal Growth, Substrate Preparation and Evaluation
K. Tada, M. Tatsumi, M. Morioka, T. Araki, and T. Kawase, InP Substrates: Production and Quality Control
M. Razeghi, LP-MOCVD Growth, Characterization, and Application of InP Material
T. A. Kennedy and P. J. Lin-Chung, Stoichiometric Defects in InP

Volme 32 Strained-Layer Superlattices: Physics

T. P. Pearsall, Strained-Layer Superlattices
F. H. Pollack, Effects of Homogeneous Strain on the Electronic and Vibrational Levels in Semiconductors
J. Y. Marzin, J. M. Gerárd, P. Voisin, and J. A. Brum, Optical Studies of Strained III–V Heterolayers
R. People and S. A. Jackson, Structurally Induced States from Strain and Confinement
M. Jaros, Microscopic Phenomena in Ordered Superlattices

Volume 33 Strained-Layer Superlattices: Materials Science and Technology

R. Hull and J. C. Bean, Principles and Concepts of Strained-Layer Epitaxy
W. J. Schaff, P. J. Tasker, M. C. Foisy, and L. F. Eastman, Device Applications of Strained-Layer Epitaxy

S. T. Picraux, B. L. Doyle, and J. Y. Tsao, Structure and Characterization of Strained-Layer Superlattices
E. Kasper and F. Schaffer, Group IV Compounds
D. L. Martin, Molecular Beam Epitaxy of IV–VI Compounds Heterojunction
R. L. Gunshor, L. A. Kolodziejski, A. V. Nurmikko, and N. Otsuka, Molecular Beam Epitaxy of II–VI Semiconductor Microstructures

Volume 34 Hydrogen in Semiconductors

J. I. Pankove and N. M. Johnson, Introduction to Hydrogen in Semiconductors
C. H. Seager, Hydrogenation Methods
J. I. Pankove, Hydrogenation of Defects in Crystalline Silicon
J. W. Corbett, P. Deák, U. V. Desnica, and S. J. Pearton, Hydrogen Passivation of Damage Centers in Semiconductors
S. J. Pearton, Neutralization of Deep Levels in Silicon
J. I. Pankove, Neutralization of Shallow Acceptors in Silicon
N. M. Johnson, Neutralization of Donor Dopants and Formation of Hydrogen-Induced Defects in n-Type Silicon
M. Stavola and S. J. Pearton, Vibrational Spectroscopy of Hydrogen-Related Defects in Silicon
A. D. Marwick, Hydrogen in Semiconductors: Ion Beam Techniques
C. Herring and N. M. Johnson, Hydrogen Migration and Solubility in Silicon
E. E. Haller, Hydrogen-Related Phenomena in Crystalline Germanium
J. Kakalios, Hydrogen Diffusion in Amorphous Silicon
J. Chevalier, B. Clerjaud, and B. Pajot, Neutralization of Defects and Dopants in III–V Semiconductors
G. G. DeLeo and W. B. Fowler, Computational Studies of Hydrogen-Containing Complexes in Semiconductors
R. F. Kiefl and T. L. Estle, Muonium in Semiconductors
C. G. Van de Walle, Theory of Isolated Interstitial Hydrogen and Muonium in Crystalline Semiconductors

Volume 35 Nanostructured Systems

M. Reed, Introduction
H. van Houten, C. W. J. Beenakker, and B. J. van Wees, Quantum Point Contacts
G. Timp, When Does a Wire Become an Electron Waveguide?
M. Büttiker, The Quantum Hall Effects in Open Conductors
W. Hansen, J. P. Kotthaus, and U. Merkt, Electrons in Laterally Periodic Nanostructures

Volume 36 The Spectroscopy of Semiconductors

D. Heiman, Spectroscopy of Semiconductors at Low Temperatures and High Magnetic Fields
A. V. Nurmikko, Transient Spectroscopy by Ultrashort Laser Pulse Techniques
A. K. Ramdas and S. Rodriguez, Piezospectroscopy of Semiconductors
O. J. Glembocki and B. V. Shanabrook, Photoreflectance Spectroscopy of Microstructures
D. G. Seiler, C. L. Littler, and M. H. Wiler, One- and Two-Photon Magneto-Optical Spectroscopy of InSb and $Hg_{1-x}Cd_xTe$

Volume 37 The Mechanical Properties of Semiconductors

A.-B. Chen, A. Sher and W. T. Yost, Elastic Constants and Related Properties of Semiconductor Compounds and Their Alloys
D. R. Clarke, Fracture of Silicon and Other Semiconductors
H. Siethoff, The Plasticity of Elemental and Compound Semiconductors
S. Guruswamy, K. T. Faber and J. P. Hirth, Mechanical Behavior of Compound Semiconductors
S. Mahajan, Deformation Behavior of Compound Semiconductors
J. P. Hirth, Injection of Dislocations into Strained Multilayer Structures
D. Kendall, C. B. Fleddermann, and K. J. Malloy, Critical Technologies for the Micromachining of Silicon
I. Matsuba and K. Mokuya, Processing and Semiconductor Thermoelastic Behavior

Volume 38 Imperfections in III/V Materials

U. Scherz and M. Scheffler, Density-Functional Theory of sp-Bonded Defects in III/V Semiconductors
M. Kaminska and E. R. Weber, El2 Defect in GaAs
D. C. Look, Defects Relevant for Compensation in Semi-Insulating GaAs
R. C. Newman, Local Vibrational Mode Spectroscopy of Defects in III/V Compounds
A. M. Hennel, Transition Metals in III/V Compounds
K. J. Malloy and K. Khachaturyan, DX and Related Defects in Semiconductors
V. Swaminathan and A. S. Jordan, Dislocations in III/V Compounds
K. W. Nauka, Deep Level Defects in the Epitaxial III/V Materials

Volume 39 Minority Carriers in III–V Semiconductors: Physics and Applications

N. K. Dutta, Radiative Transitions in GaAs and Other III–V Compounds
R. K. Ahrenkiel, Minority-Carrier Lifetime in III–V Semiconductors
T. Furuta, High Field Minority Electron Transport in p-GaAs
M. S. Lundstrom, Minority-Carrier Transport in III–V Semiconductors
R. A. Abram, Effects of Heavy Doping and High Excitation on the Band Structure of GaAs
D. Yevick and W. Bardyszewski, An Introduction to Non-Equilibrium Many-Body Analyses of Optical Processes in III–V Semiconductors

Volume 40 Epitaxial Microstructures

E. F. Schubert, Delta-Doping of Semiconductors: Electronic, Optical, and Structural Properties of Materials and Devices
A. Gossard, M. Sundaram, and P. Hopkins, Wide Graded Potential Wells
P. Petroff, Direct Growth of Nanometer-Size Quantum Wire Superlattices
E. Kapon, Lateral Patterning of Quantum Well Heterostructures by Growth of Nonplanar Substrates
H. Temkin, D. Gershoni, and M. Panish, Optical Properties of $Ga_{1-x}In_xAs/InP$ Quantum Wells

Volume 41 High Speed Heterostructure Devices

F. Capasso, F. Beltram, S. Sen, A. Pahlevi, and A. Y. Cho, Quantum Electron Devices: Physics and Applications
P. Solomon, D. J. Frank, S. L. Wright, and F. Canora, GaAs-Gate Semiconductor–Insulator–Semiconductor FET
M. H. Hashemi and U. K. Mishra, Unipolar InP-Based Transistors
R. Kiehl, Complementary Heterostructure FET Integrated Circuits
T. Ishibashi, GaAs-Based and InP-Based Heterostructure Bipolar Transistors
H. C. Liu and T. C. L. G. Sollner, High-Frequency-Tunneling Devices
H. Ohnishi, T. More, M. Takatsu, K. Imamura, and N. Yokoyama, Resonant-Tunneling Hot-Electron Transistors and Circuits

Volume 42 Oxygen in Silicon

F. Shimura, Introduction to Oxygen in Silicon
W. Lin, The Incorporation of Oxygen into Silicon Crystals
T. J. Schaffner and D. K. Schroder, Characterization Techniques for Oxygen in Silicon
W. M. Bullis, Oxygen Concentration Measurement
S. M. Hu, Intrinsic Point Defects in Silicon
B. Pajot, Some Atomic Configurations of Oxygen
J. Michel and L. C. Kimerling, Electical Properties of Oxygen in Silicon
R. C. Newman and R. Jones, Diffusion of Oxygen in Silicon
T. Y. Tan and W. J. Taylor, Mechanisms of Oxygen Precipitation: Some Quantitative Aspects
M. Schrems, Simulation of Oxygen Precipitation
K. Simino and I. Yonenaga, Oxygen Effect on Mechanical Properties
W. Bergholz, Grown-in and Process-Induced Effects
F. Shimura, Intrinsic/Internal Gettering
H. Tsuya, Oxygen Effect on Electronic Device Performance

Volume 43 Semiconductors for Room Temperature Nuclear Detector Applications

R. B. James and T. E. Schlesinger, Introduction and Overview
L. S. Darken and C. E. Cox, High-Purity Germanium Detectors
A. Burger, D. Nason, L. Van den Berg, and M. Schieber, Growth of Mercuric Iodide
X. J. Bao, T. E. Schlesinger, and R. B. James, Electrical Properties of Mercuric Iodide
X. J. Bao, R. B. James, and T. E. Schlesinger, Optical Properties of Red Mercuric Iodide
M. Hage-Ali and P. Siffert, Growth Methods of CdTe Nuclear Detector Materials
M. Hage-Ali and P Siffert, Characterization of CdTe Nuclear Detector Materials
M. Hage-Ali and P. Siffert, CdTe Nuclear Detectors and Applications
R. B. James, T. E. Schlesinger, J. Lund, and M. Schieber, $Cd_{1-x}Zn_xTe$ Spectrometers for Gamma and X-Ray Applications
D. S. McGregor, J. E. Kammeraad, Gallium Arsenide Radiation Detectors and Spectrometers
J. C. Lund, F. Olschner, and A. Burger, Lead Iodide
M. R. Squillante, and K. S. Shah, Other Materials: Status and Prospects
V. M. Gerrish, Characterization and Quantification of Detector Performance
J. S. Iwanczyk and B. E. Patt, Electronics for X-ray and Gamma Ray Spectrometers
M. Schieber, R. B. James, and T. E. Schlesinger, Summary and Remaining Issues for Room Temperature Radiation Spectrometers

Volume 44   II–IV Blue/Green Light Emitters: Device Physics and Epitaxial Growth

J. Han and R. L. Gunshor, MBE Growth and Electrical Properties of Wide Bandgap ZnSe-based II–VI Semiconductors

S. Fujita and S. Fujita, Growth and Characterization of ZnSe-based II–VI Semiconductors by MOVPE

E. Ho and L. A. Kolodziejski, Gaseous Source UHV Epitaxy Technologies for Wide Bandgap II–VI Semiconductors

C. G. Van de Walle, Doping of Wide-Band-Gap II–VI Compounds — Theory

R. Cingolani, Optical Properties of Excitons in ZnSe-Based Quantum Well Heterostructures

A. Ishibashi and A. V. Nurmikko, II–VI Diode Lasers: A Current View of Device Performance and Issues

S. Guha and J. Petruzello, Defects and Degradation in Wide-Gap II–VI-based Structures and Light Emitting Devices

Volume 45   Effect of Disorder and Defects in Ion-Implanted Semiconductors: Electrical and Physiochemical Characterization

H. Ryssel, Ion Implantation into Semiconductors: Historical Perspectives

You-Nian Wang and Teng-Cai Ma, Electronic Stopping Power for Energetic Ions in Solids

S. T. Nakagawa, Solid Effect on the Electronic Stopping of Crystalline Target and Application to Range Estimation

G. Müller, S. Kalbitzer and G. N. Greaves, Ion Beams in Amorphous Semiconductor Research

J. Boussey-Said, Sheet and Spreading Resistance Analysis of Ion Implanted and Annealed Semiconductors

M. L. Polignano and G. Queirolo, Studies of the Stripping Hall Effect in Ion-Implanted Silicon

J. Stoemenos, Transmission Electron Microscopy Analyses

R. Nipoti and M. Servidori, Rutherford Backscattering Studies of Ion Implanted Semiconductors

P. Zaumseil, X-ray Diffraction Techniques

Volume 46   Effect of Disorder and Defects in Ion-Implanted Semiconductors: Optical and Photothermal Characterization

M. Fried, T. Lohner and J. Gyulai, Ellipsometric Analysis

A. Seas and C. Christofides, Transmission and Reflection Spectroscopy on Ion Implanted Semiconductors

A. Othonos and C. Christofides, Photoluminescence and Raman Scattering of Ion Implanted Semiconductors. Influence of Annealing

C. Christofides, Photomodulated Thermoreflectance Investigation of Implanted Wafers. Annealing Kinetics of Defects

U. Zammit, Photothermal Deflection Spectroscopy Characterization of Ion-Implanted and Annealed Silicon Films

A. Mandelis, A. Budiman and M. Vargas, Photothermal Deep-Level Transient Spectroscopy of Impurities and Defects in Semiconductors

R. Kalish and S. Charbonneau, Ion Implantation into Quantum-Well Structures

A. M. Myasnikov and N. N. Gerasimenko, Ion Implantation and Thermal Annealing of III-V Compound Semiconducting Systems: Some Problems of III-V Narrow Gap Semiconductors

Volume 47 Uncooled Infrared Imaging Arrays and Systems

- R. G. Buser and M. P. Tompsett, Historical Overview
- P. W. Kruse, Principles of Uncooled Infrared Focal Plane Arrays
- R. A. Wood, Monolithic Silicon Microbolometer Arrays
- C. M. Hanson, Hybrid Pyroelectric-Ferroelectric Bolometer Arrays
- D. L. Polla and J. R. Choi, Monolithic Pyroelectric Bolometer Arrays
- N. Teranishi, Thermoelectric Uncooled Infrared Focal Plane Arrays
- M. F. Tompsett, Pyroelectric Vidicon
- T. W. Kenny, Tunneling Infrared Sensors
- J. R. Vig, R. L. Filler and Y. Kim, Application of Quartz Microresonators to Uncooled Infrared Imaging Arrays
- P. W. Kruse, Application of Uncooled Monolithic Thermoelectric Linear Arrays to Imaging Radiometers

Volume 48 High Brightness Light Emitting Diodes

- G. B. Stringfellow, Materials Issues in High-Brightness Light-Emitting Diodes
- M. G. Craford, Overview of Device issues in High-Brightness Light-Emitting Diodes
- F. M. Steranka, AlGaAs Red Light Emitting Diodes
- C. H. Chen, S. A. Stockman, M. J. Peanasky, and C. P. Kuo, OMVPE Growth of AlGaInP for High Efficiency Visible Light-Emitting Diodes
- F. A. Kish and R. M. Fletcher, AlGaInP Light-Emitting Diodes
- M. W. Hodapp, Applications for High Brightness Light-Emitting Diodes
- I. Akasaki and H. Amano, Organometallic Vapor Epitaxy of GaN for High Brightness Blue Light Emitting Diodes
- S. Nakamura, Group III-V Nitride Based Ultraviolet-Blue-Green-Yellow Light-Emitting Diodes and Laser Diodes

Volume 49 Light Emission in Silicon: from Physics to Devices

- D. J. Lockwood, Light Emission in Silicon
- G. Abstreiter, Band Gaps and Light Emission in Si/SiGe Atomic Layer Structures
- T. G. Brown and D. G. Hall, Radiative Isoelectronic Impurities in Silicon and Silicon-Germanium Alloys and Superlattices
- J. Michel, L. V. C. Assali, M. T. Morse, and L. C. Kimerling, Erbium in Silicon
- Y. Kanemitsu, Silicon and Germanium Nanoparticles
- P. M. Fauchet, Porous Silicon: Photoluminescence and Electroluminescent Devices
- C. Delerue, G. Allan, and M. Lannoo, Theory of Radiative and Nonradiative Processes in Silicon Nanocrystallites
- L. Brus, Silicon Polymers and Nanocrystals

Volume 50 Gallium Nitride (GaN)

- J. I. Pankove and T. D. Moustakas, Introduction
- S. P. DenBaars and S. Keller, Metalorganic Chemical Vapor Deposition (MOCVD) of Group III Nitrides
- W. A. Bryden and T. J. Kistenmacher, Growth of Group III-A Nitrides by Reactive Sputtering
- N. Newman, Thermochemistry of III–N Semiconductors
- S. J. Pearton and R. J. Shul, Etching of III Nitrides

S. M. Bedair, Indium-based Nitride Compounds
A. Trampert, O. Brandt, and K. H. Ploog, Crystal Structure of Group III Nitrides
H. Morkoc, F. Hamdani, and A. Salvador, Electronic and Optical Properties of III–V Nitride based Quantum Wells and Superlattices
K. Doverspike and J. I. Pankove, Doping in the III-Nitrides
T. Suski and P. Perlin, High Pressure Studies of Defects and Impurities in Gallium Nitride
B. Monemar, Optical Properties of GaN
W. R. L. Lambrecht, Band Structure of the Group III Nitrides
N. E. Christensen and P. Perlin, Phonons and Phase Transitions in GaN
S. Nakamura, Applications of LEDs and LDs
I. Akasaki and H. Amano, Lasers
J. A. Cooper, Jr., Nonvolatile Random Access Memories in Wide Bandgap Semiconductors

Volume 51A Identification of Defects in Semiconductors

G. D. Watkins, EPR and ENDOR Studies of Defects in Semiconductors
J.-M. Spaeth, Magneto-Optical and Electrical Detection of Paramagnetic Resonance in Semiconductors
T. A. Kennedy and E. R. Glaser, Magnetic Resonance of Epitaxial Layers Detected by Photoluminescence
K. H. Chow, B. Hitti, and R. F. Kiefl, μSR on Muonium in Semiconductors and Its Relation to Hydrogen
K. Saarinen, P. Hautojärvi, and C. Corbel, Positron Annihilation Spectroscopy of Defects in Semiconductors
R. Jones and P. R. Briddon, The Ab Initio Cluster Method and the Dynamics of Defects in Semiconductors

Volume 51B Identification of Defects in Semiconductors

G. Davies, Optical Measurements of Point Defects
P. M. Mooney, Defect Identification Using Capacitance Spectroscopy
M. Stavola, Vibrational Spectroscopy of Light Element Impurities in Semiconductors
P. Schwander, W. D. Rau, C. Kisielowski, M. Gribelyuk, and A. Ourmazd, Defect Processes in Semiconductors Studied at the Atomic Level by Transmission Electron Microscopy
N. D. Jager and E. R. Weber, Scanning Tunneling Microscopy of Defects in Semiconductors

Volume 52 SiC Materials and Devices

K. Järrendahl and R. F. Davis, Materials Properties and Characterization of SiC
V. A. Dmitriev and M. G. Spencer, SiC Fabrication Technology: Growth and Doping
V. Saxena and A. J. Steckl, Building Blocks for SiC Devices: Ohmic Contacts, Schottky Contacts, and p-n Junctions
M. S. Shur, SiC Transistors
C. D. Brandt, R. C. Clarke, R. R. Siergiej, J. B. Casady, A. W. Morse, S. Sriram, and A. K. Agarwal, SiC for Applications in High-Power Electronics
R. J. Trew, SiC Microwave Devices

J. Edmond, H. Kong, G. Negley, M. Leonard, K. Doverspike, W. Weeks, A. Suvorov, D. Waltz, and C. Carter, Jr., SiC-Based UV Photodiodes and Light-Emitting Diodes
H. Morkoç, Beyond Silicon Carbide! III–V Nitride-Based Heterostructures and Devices

Volume 53 Cumulative Subject and Author Index Including Tables of Contents for Volume 1–50

Volume 54 High Pressure in Semiconductor Physics I

W. Paul, High Pressure in Semiconductor Physics: A Historical Overview
N. E. Christensen, Electronic Structure Calculations for Semiconductors under Pressure
R. J. Neimes and M. I. McMahon, Structural Transitions in the Group IV, III-V and II-VI Semiconductors Under Pressure
A. R. Goni and K. Syassen, Optical Properties of Semiconductors Under Pressure
P. Trautman, M. Baj, and J. M. Baranowski, Hydrostatic Pressure and Uniaxial Stress in Investigations of the EL2 Defect in GaAs
M. Li and P. Y. Yu, High-Pressure Study of DX Centers Using Capacitance Techniques
T. Suski, Spatial Correlations of Impurity Charges in Doped Semiconductors
N. Kuroda, Pressure Effects on the Electronic Properties of Diluted Magnetic Semiconductors

Volume 55 High Pressure in Semiconductor Physics II

D. K. Maude and J. C. Portal, Parallel Transport in Low-Dimensional Semiconductor Structures
P. C. Klipstein, Tunneling Under Pressure: High-Pressure Studies of Vertical Transport in Semiconductor Heterostructures
E. Anastassakis and M. Cardona, Phonons, Strains, and Pressure in Semiconductors
F. H. Pollak, Effects of External Uniaxial Stress on the Optical Properties of Semiconductors and Semiconductor Microstructures
A. R. Adams, M. Silver, and J. Allam, Semiconductor Optoelectronic Devices
S. Porowski and I. Grzegory, The Application of High Nitrogen Pressure in the Physics and Technology of III-N Compounds
M. Yousuf, Diamond Anvil Cells in High Pressure Studies of Semiconductors

Volume 56 Germanium Silicon: Physics and Materials

J. C. Bean, Growth Techniques and Procedures
D. E. Savage, F. Liu, V. Zielasek, and M. G. Lagally, Fundamental Crystal Growth Mechanisms
R. Hull, Misfit Strain Accommodation in SiGe Heterostructures
M. J. Shaw and M. Jaros, Fundamental Physics of Strained Layer GeSi: Quo Vadis?
F. Cerdeira, Optical Properties
S. A. Ringel and P. N. Grillot, Electronic Properties and Deep Levels in Germanium-Silicon
J. C. Campbell, Optoelectronics in Silicon and Germanium Silicon
K. Eberl, K. Brunner, and O. G. Schmidt, $Si_{1-y}C_y$ and $Si_{1-x-y}Ge_xC_y$ Alloy Layers

Volume 57 Gallium Nitride (GaN) II

R. J. Molnar, Hydride Vapor Phase Epitaxial Growth of III-V Nitrides
T. D. Moustakas, Growth of III-V Nitrides by Molecular Beam Epitaxy
Z. Liliental-Weber, Defects in Bulk GaN and Homoepitaxial Layers
C. G. Van de Walle and N. M. Johnson, Hydrogen in III-V Nitrides
W. Götz and N. M. Johnson, Characterization of Dopants and Deep Level Defects in Gallium Nitride
B. Gil, Stress Effects on Optical Properties
C. Kisielowski, Strain in GaN Thin Films and Heterostructures
J. A. Miragliotta and D. K. Wickenden, Nonlinear Optical Properties of Gallium Nitride
B. K. Meyer, Magnetic Resonance Investigations on Group III-Nitrides
M. S. Shur and M. Asif Khan, GaN and AlGaN Ultraviolet Detectors
C. H. Qiu, J. I. Pankove, and C. Rossington, III-V Nitride-Based X-ray Detectors

Volume 58 Nonlinear Optics in Semiconductors I

A. Kost, Resonant Optical Nonlinearities in Semiconductors
E. Garmire, Optical Nonlinearities in Semiconductors Enhanced by Carrier Transport
D. S. Chemla, Ultrafast Transient Nonlinear Optical Processes in Semiconductors
M. Sheik-Bahae and E. W. Van Stryland, Optical Nonlinearities in the Transparency Region of Bulk Semiconductors
J. E. Millerd, M. Ziari, and A. Partovi, Photorefractivity in Semiconductors

Volume 59 Nonlinear Optics in Semiconductors II

J. B. Khurgin, Second Order Nonlinearities and Optical Rectification
K. L. Hall, E. R. Thoen, and E. P. Ippen, Nonlinearities in Active Media
E. Hanamura, Optical Responses of Quantum Wires/Dots and Microcavities
U. Keller, Semiconductor Nonlinearities for Solid-State Laser Modelocking and Q-Switching
A. Miller, Transient Grating Studies of Carrier Diffusion and Mobility in Semiconductors

Volume 60 Self-Assembled InGaAs/GaAs Quantum Dots

Mitsuru Sugawara, Theoretical Bases of the Optical Properties of Semiconductor Quantum Nano-Structures
Yoshiaki Nakata, Yoshihiro Sugiyama, and Mitsuru Sugawara, Molecular Beam Epitaxial Growth of Self-Assembled InAs/GaAs Quantum Dots
Kohki Mukai, Mitsuru Sugawara, Mitsuru Egawa, and Nobuyuki Ohtsuka, Metalorganic Vapor Phase Epitaxial Growth of Self-Assembled InGaAs/GaAs Quantum Dots Emitting at 1.3 μm
Kohki Mukai and Mitsuru Sugawara, Optical Characterization of Quantum Dots
Kohki Mukai and Mitsuru Sugawara, The Photon Bottleneck Effect in Quantum Dots
Hajime Shoji, Self-Assembled Quantum Dot Lasers
Hiroshi Ishikawa, Applications of Quantum Dot to Optical Devices
Mitsuru Sugawara, Kohki Mukai, Hiroshi Ishikawa, Koji Otsubo, and Yoshiaki Nakata, The Latest News

Volume 61 Hydrogen in Semiconductors II

Norbert H. Nickel, Introduction to Hydrogen in Semiconductors II
Noble M. Johnson and Chris G. Van de Walle, Isolated Monatomic Hydrogen in Silicon
Yurij V. Gorelkinskii, Electron Paramagnetic Resonance Studies of Hydrogen and Hydrogen-Related Defects in Crystalline Silicon
Norbert H. Nickel, Hydrogen in Polycrystalline Silicon
Wolfhard Beyer, Hydrogen Phenomena in Hydrogenated Amorphous Silicon
Chris G. Van de Walle, Hydrogen Interactions with Polycrystalline and Amorphous Silicon — Theory
Karen M. McNamara Rutledge, Hydrogen in Polycrystalline CVD Diamond
Roger L. Lichti, Dynamics of Muonium Diffusion, Site Changes and Charge-State Transitions
Matthew D. McCluskey and Eugene E. Haller, Hydrogen in III-V and II-VI Semiconductors
S. J. Pearton and J. W. Lee, The Properties of Hydrogen in GaN and Related Alloys
Jörg Neugebauer and Chris G. Van de Walle, Theory of Hydrogen in GaN

Volume 62 Intersubband Transitions in Quantum Wells: Physics and Device Applications I

Manfred Helm, The Basic Physics of Intersubband Transitions
Jerome Faist, Carlo Sirtori, Federico Capasso, Loren N. Pfeiffer, Ken W. West, Deborah L. Sivco, and Alfred Y. Cho, Quantum Interference Effects in Intersubband Transitions
H. C. Liu, Quantum Well Infrared Photodetector Physics and Novel Devices
S. D. Gunapala and S. V. Bandara, Quantum Well Infrared Photodetector (QWIP) Focal Plane Arrays

Volume 63 Chemical Mechanical Polishing in Si Processing

Frank B. Kaufman, Introduction
Thomas Bibby and Karey Holland, Equipment
John P. Bare, Facilitization
Duane S. Boning and Okumu Ouma, Modeling and Simulation
Shin Hwa Li, Bruce Tredinnick, and Mel Hoffman, Consumables I: Slurry
Lee M. Cook, CMP Consumables II: Pad
François Tardif, Post-CMP Clean
Shin Hwa Li, Tara Chhatpar, and Frederic Robert, CMP Metrology
Shin Hwa Li, Visun Bucha, and Kyle Wooldridge, Applications and CMP-Related Process Problems

Volume 64 Electroluminescence I

M. G. Craford, S. A. Stockman, M. J. Peanasky, and F. A. Kish, Visible Light-Emitting Diodes
H. Chui, N. F. Gardner, P. N. Grillot, J. W. Huang, M. R. Krames, and S. A. Maranowski, High-Efficiency AlGaInP Light-Emitting Diodes
R. S. Kern, W. Götz, C. H. Chen, H. Liu, R. M. Fletcher, and C. P. Kuo, High-Brightness Nitride-Based Visible-Light-Emitting Diodes
Yoshiharu Sato, Organic LED System Considerations
V. Bulović, P. E. Burrows, and S. R. Forrest, Molecular Organic Light-Emitting Devices

ISBN 0-12-752174-7